Building Lean, Building BIM

Building Lean, Building BIM is the essential guide for any construction company that wants to implement Lean Construction and Building Information Modelling (BIM) to gain a strategic edge over its competition.

The first of its kind, this book outlines the principles of Lean, the functionality of BIM, and the interactions between the two, illustrating them through the story of how Tidhar Construction has implemented Lean Construction and BIM in a concerted effort over four years. Tidhar is a small-to-medium-sized construction company that pioneered a new way of working that significantly improved its operations and its margins. The company's story serves as a case study that explains the various facets of Lean Construction and BIM. Each of the core chapters defines a principle of Lean and/or BIM, describes the achievements and failures with that principle in Tidhar's implementation based on the experiences of the key people involved, and reviews the relevant background and theory.

The implementation at Tidhar was not a pure success, but by examining their motives alongside their achievements and failures, readers who seek to implement Lean and BIM will learn about what pitfalls and pinnacles to expect. A number of chapters also present the experiences of other companies who are leaders in their fields: Skanska Finland, Lease Crutcher Lewis and Fira.

This book is highly relevant and useful to a wide range of readers from the construction industry and beyond, especially those who are frustrated with the inefficiencies in their companies and major projects. It is also essential reading for Lean and BIM enthusiasts, researchers, and students from a variety of industries and backgrounds.

Rafael Sacks is a Professor in the Faculty of Civil and Environmental Engineering, Technion – Israel Institute of Technology.

Samuel Korb is a PhD candidate in the Faculty of Civil and Environmental Engineering, Technion – Israel Institute of Technology.

Ronen Barak is a Civil Engineer and researcher of BIM and computer applications for construction. He is responsible for Lean and BIM activities at Tidhar Construction, and he has been directly involved with the entire Lean and BIM journey at Tidhar.

"This is a real tour-de-force for understanding BIM, lean and the human issues that are so important in its implementation."

– Paul Teicholz, Professor of Civil and Environmental Engineering, Stanford University, USA

"I highly recommend this book to both industry professionals and academics interested in successfully aligning construction processes, methods and tools. The book offers valuable insights and honest in-depth descriptions of the cultural challenges and opportunities of companies striving for higher performance through modernizing their current practices. I especially enjoyed the great balance between conceptually explaining the factors at play and providing practical examples and case studies."

– Roar Fosse, Regional Manager – Lean Construction, Skanska, Oslo, Norway

Building Lean, Building BIM

Improving Construction the Tidhar Way

Rafael Sacks, Samuel Korb and Ronen Barak

LONDON AND NEW YORK

First published 2018
by Routledge
2 Park Square, Milton Park, Abingdon, Oxon OX14 4RN

and by Routledge
711 Third Avenue, New York, NY 10017

Routledge is an imprint of the Taylor & Francis Group, an informa business

British Library Cataloguing-in-Publication Data
A catalogue record for this book is available from the British Library

Library of Congress Cataloging-in-Publication Data
A catalog record for this book has been requested

ISBN: 978-1-138-23722-3 (hbk)
ISBN: 978-1-138-23723-0 (pbk)
ISBN: 978-1-315-30051-1 (ebk)

Typeset in Frutiger
by Fish Books Ltd.

Printed in Great Britain by Ashford Colour Press Ltd

Visit the companion website: www.routledge.com/cw/sacks

In memory of Cheni Kerem: our teacher, our guide, and our friend.
Thank you for helping us learn to see.

R.S., S.K., and R.B.

Contents

Acknowledgments

Building Lean, Building BIM is, first and foremost, the story of the people of Tidhar. If not for their overwhelmingly positive attitude to the idea of the book and their willingness to be badgered for interviews, figures and data, their story could not have been written. None of them ever asked how they would be portrayed, whether as heroes or as villains; they simply smiled when approached and made the time to work with us. We are grateful to all of Tidhar's people, those who appear in the book by name and those who will recognize themselves in the stories and anecdotes. We are especially grateful to those who provided interviews and who contributed texts and images: thank you (in strictly alphabetical order) Arye Bachar, Eli Renzer, Gil Geva, Guy Frumer, Maia Davidson, Ronen Ben Dor, Tal Hershkovitz, Tanya Pankratov and Zohar Raz.

Cheni Kerem, of Lean Israel, became a permanent fixture at the Rosh Ha'ayin project site offices. Though he didn't come from a background in construction, he learned a great deal about the industry in his time working with Tidhar. But far more than that, he taught us a great deal about Lean production, Lean tools and Lean thinking. Tragically, Cheni passed away before this book went to print. We sorely miss his wisdom, his good-natured approach and his wry smile. Thank you Cheni for all that you contributed – may your memory be for a blessing.

As the Helsinki winter is cold, so are the people of Skanska Finland warm. They are old hands in Lean and BIM. Whether in the halls of UC Berkeley or the Ateljee Bar on Hotel Torni's roof, on the beach in Fortaleza or in Manskun Rasti, they have always been willing to devote their time to teach us all, and to help others learn from their experience. Thank you Jan Elfving, Ilkka Romo and Sara Troberg.

We are indebted to Lease Crutcher Lewis's employee owners for allowing us to present their company's procedures as a superb example of a successful Lean VDC process. For sharing their story, for their professional and enthusiastic explanations of Lewis's approach to VDC and of Lewis Lean, we especially thank Jeff Cleator, Lana Gochenauer and Lowrey Pugh.

Where would we be without the inspirational challenge to everything sacred in traditional construction without the brilliantly novel thinking that Fira's inspiring people apply to the industry? Not only are they innovative, they are also enthusiastic to share their passion. Thank you Otto Alhava, Sakari Pesonen, Antti Kauppila and Anneli Holmberg (of Wetrain).

The Technion in Haifa, Israel, is a magnet and a meeting place for all sorts of intelligent and talented people. Every one of the graduate students who has studied or pursued research in the Virtual Construction Laboratory through the years has contributed in one way or another to this book, by challenging our thinking, by inventing and developing Lean Construction and BIM theory and applications, and generally by being good humoured. Vitaliy Priven stands out among them in the context of the book thanks to his fundamentally significant role in the research with Tidhar. Thank you Vitaliy!

Two students of the Technion International Engineering School, both research intern volunteers in the Virtual Construction Lab, made direct contributions to the book, for which we are tremendously grateful. Meir Katz did all of the groundwork for Chapter 5, "The Skanska Finland way", and Sooraj Kumar calmly collated all of the figures, negotiated all of the permissions and helped compile the bibliography. Thank you Meir, thank you Sooraj.

To our incredibly generous volunteer editor, Lynne Federman, and to our supremely patient and highly professional publishers at Routledge, Ed Needle and Catherine Holdsworth, thanks to you too.

We have written this book for all the construction workers, supervisors, managers, engineers, architects and owners; all those who strive to improve their industry, sometimes against seemingly insurmountable obstacles, by introducing innovations like Lean Construction and BIM.

Although you may often face conservative attitudes and prejudice, you are on the right path. **Have faith, and have fun!**

Preface

Why should construction companies adopt Lean Construction and Building Information Modelling (or "Lean and BIM", for short)? Why should they invest in these two major innovations in an integrated way? And how can they be implemented in the best possible way?

In this book, we examine and try to answer these questions through the story of Tidhar Ltd, a small-to-medium-sized construction company that developed new ways of building that enabled it to grow its turnover and its income at rates unheard of in its market. In 2011, Tidhar embarked on a journey to improve and grow through collaboration, technology and innovation. It grew larger but also become leaner. At the time of writing the process was still ongoing, but Tidhar's story, with its successes and failures, was already one of leadership, professionalism, persistence and learning in the effort to apply Lean thinking and BIM to the complex world of building construction.

Building Lean, Building BIM is a set of lessons learned and an explanation of the theory and practice of Lean and BIM. Using examples from the history of Tidhar's journey of improvement and those of three other pioneering companies – Skanska Finland, Lease Crutcher Lewis and Fira – we explore what Lean Construction is, how BIM can be applied and exploited, and how the innovations of Lean Construction and BIM interact. How can they be applied successfully and what are the pitfalls? How might readers and their companies create their own journeys to greater profitability?

Tidhar's story is particularly special given the context of the construction industry in Israel. The local market is small (averaging US$19.9 billion from 2000 to 2016, compared with US$636 billion in the USA), yet demand is stable over time as a result of the controlled release of land for construction. As in most developed economies, local construction labour is in short supply and the industry employs many foreign workers with a variety of languages. As is the case in most countries, the industry is highly fragmented, with many small companies and very few large ones. Given the small geographical area, all companies compete with one another and most have some international activity. Yet the industry is slow to adopt innovation, because it is relatively isolated and protected from foreign competition, and thus insular and inward looking. Unless and until one company disrupts the status quo, no one feels any pressure to change. Once they do, however, the others scramble to catch up.

In this context, both Lean Construction and BIM remained essentially unknown to industry practitioners in Israel until Tidhar's initiative began. In this sense, Tidhar was not just an early adopter, it was *the* early adopter, with no local role models to emulate or spur progress. This meant opportunity, but it also meant that the impetus for change was driven solely by its senior management. In 2012, Tidhar was the 13th largest construction company in Israel;[1] by 2016 it was ranked 6th overall by annual turnover, was the country's largest privately held construction company and had more apartments under construction at any one time than any other construction company.

Each of the core chapters in the book begins with a statement of the guiding principle of a Lean, BIM or integrated process improvement. The text that follows presents a story from Tidhar's own experience to illustrate the principle and interpret the successes and failures in light of the principles of Lean and BIM. Three of the chapters (5, 12 and 19) tell the stories of pioneering construction companies from other countries: one large international company, one medium-sized regional company and a small (but growing) construction start-up company. We reflect on the motivations of the people involved, their skills, their knowledge and their learning, and how they arrived at the solutions that worked for them. Four chapters (4, 7, 10 and 13) profile four construction projects built by Tidhar, from which many of the examples in the other chapters are drawn. Finally, two chapters present some of the theoretical and academic aspects of Tidhar's Lean and BIM journey: Chapter 15 discusses the nature of flow in building construction and Chapter 21 examines the interplay of research and practice.

The book is annotated with reference papers from construction research journals, books and online sources. Where these are specific to a chapter, they appear as endnotes within each chapter. Where they are generic and referred to in multiple chapters, they are listed in the Bibliography at the end of the book.

One may read *Building Lean, Building BIM* as a narrative, focusing on Tidhar's Lean and BIM journey. The thread of the story is followed from case to case, with many of the characters reappearing in different chapters. Alternatively, one may read the book as a textbook, using the principles and the lessons learned to focus on the particular aspects of Lean Construction and BIM that are of specific interest.

We believe that the principles, the story and the lessons learned are universal in that they are relevant for managers and engineers in industries well beyond the construction industry.

Finally, two brief notes about our writing style. First, Israel has a low power distance culture, in which workers readily challenge the authority of managers, and leaders encourage independent thinking. People commonly use first names, even in formal work situations. Some readers may be surprised by the way in which we refer to characters by their first names, but the characters themselves would find this quite natural. We have provided an index of characters on page 398, which will help keep track of the people and their roles. Second, since the majority of Tidhar's front-line workers are male, we have used masculine prepositions for most of the examples in the book. This should not be construed as a prescription or the opinion of the authors.

Note

1 Dun's 100 Israel, Construction and Development. From www.duns100.co.il/en/rating/Construction_&_Real_Estate_/Construction_&_Development.

Introduction

Growing the Margin

"I read in a journal recently that 'Lean construction and building information modelling (BIM) are quite different initiatives, but both are having profound impacts on the construction industry. A rigorous analysis of the myriad specific interactions between them indicates that a synergy exists which, if properly understood in theoretical terms, can be exploited to improve construction processes beyond the degree to which they might be improved by application of either of these paradigms independently'. It's quite a mouthful, but I think it means that we should adopt both Lean Construction and BIM. But can that really work?"

Why do Lean and BIM offer great opportunity for construction companies?

The profit margins of most construction companies, measured as the ratio of profit to contract value for construction projects, are considered to be very low when compared with other industrial companies. This reflects the fact that companies compete in markets where the competitive edge is rooted in differences in local knowledge and negotiation skills rather than intellectual property or uniquely developed methods. The barriers to entry are relatively low and most contracts are bid and won on the basis of price alone.

In theory, in business environments like this, there is a great potential benefit to be gained from early adoption of new technologies or production methods that directly reduce production costs. This is because early adoption can allow a company to develop and maintain a fundamental competitive edge until the rest of the industry catches up. This can take some years, and these are years of growth opportunity for the early adopters. Henry Ford made a fortune and grew his company by implementing a radically new approach to mass production. Taiichi Ohno was instrumental in developing the Toyota Production System (TPS), which led to Toyota's growth from a small sectorial company to an international leader.[1] The leaders of DPR Construction in the USA grew their company from a start-up in the West Coast construction industry in 1990 to an international construction contractor with $2.9 billion in turnover by 2014, largely as a result of their democratic management approach but also in no small part due to their integrated adoption of Lean Construction and BIM.[2] Tidhar's story is a similar one, as we shall see. The leaders of all of these companies realized that "business as usual" was not enough for growth or even profit. An ethos of excellence led them to look for additional ways to excel.

There are risks, of course. Change requires investment and success is not guaranteed. Many companies have failed to make changes in appropriate ways and have gone into bankruptcy or have been taken over as a result. If it were easy, everyone would do it, and in fact, very few even make the attempt until they are left with no choice. This is particularly the case for both Lean Construction and for BIM adoption in the construction industry. Lean requires a change of thinking about production in any environment, and perhaps even more strongly so in construction because of the differences between manufacturing plants and construction sites. BIM requires new skills in technology and in systems processes. Changes in mindset, skills and technology are not easy for any individual and are even more difficult to implement across a company that cannot simply stop producing while changing. It has to be done "on the fly".

Lean thinking offers a great opportunity to construction companies because the ways in which they work, in almost every case, fail to deliver the full value possible and are rife with waste. Numerous studies of the poor performance of construction industries in many countries, published in books, industrial reports and academic research,[3] have illustrated the pervasiveness of the types of waste defined by Taiichi Ohno and others.[4] The common phenomenon of projects delivered later than promised and/or at greater expense than estimated reflects the prevalence of waste. The studies also point out the shortcomings of the products delivered, showing how buildings often fail to deliver the expected levels of performance their occupants expect. Thus, wherever Lean Construction can increase productivity, it reduces the resource consumption required to build a building while simultaneously enhancing the building's value to its customers.

BIM offers construction companies great opportunity too, because it solves many of the design quality, communication and control problems that plague the industry as a result of the use of 2D drawings. Here too, numerous studies have documented the potential to improve value to customers (primarily through simulation and analysis tools that enable much better quality control at the design stage) and to reduce costs (by reducing miscommunication and misunderstanding that lead to waiting, rework and other wastes directly, and by enabling the move to pre-assembly and off-site production which entirely removes many of the non-value-adding activities inherent in production on site).[5]

The initial notions of "Lean Construction" were crystallized circa 1992.[6] Although the term "BIM" was coined in 2002, BIM software has been available since the 1990s in different forms.[7] Why then did full-fledged adoption of these innovations only begin in the mid 2000s, and then only in pioneering companies? We hope that the answers to this question will become clearer through the remainder of the book, as we describe the significant changes that are necessary.

Why Lean and BIM together?

Lean Construction is a management paradigm that tends to disrupt traditional patterns of work. It can be implemented without any technology, but technological tools can support its implementation. BIM is first and foremost a technology, but it is not only a technology. Its successful implementation is entirely dependent on introducing BIM-appropriate workflows that are different from traditional workflows.

Introducing two novel systems simultaneously could cause disruption and failure of both efforts, posing a serious threat to the health of any construction company, if they are not compatible. Are the process changes required for BIM and Lean Construction compatible? This is a necessary (although not sufficient) condition for simultaneous implementation. The question has been researched extensively, and the answer is that for the most part they are compatible, although some points of friction have been identified.

In a 2010 article entitled *The Interaction of Lean and Building Information Modelling in Construction*, the authors analysed the possible interactions between 24 Lean Construction principles and 18 BIM functionalities.[8] By examining many examples from construction case studies, they identified 54 points of direct interaction, of which 50 were positive (mutually reinforcing) interactions and just 4 were negative. For example, when designers collaborate closely using BIM (functionalities 9 and 10), the result is a significant reduction in the cycle time needed for each design iteration (principle C; these are interactions 23 and 24 in Figure 1.1). This will be seen clearly in Chapter 14, "BIM in the big room", later in the book. However, the principle of reducing inventory (D) can be violated if planners generate large numbers of construction plan alternatives just because it is easy to do so using the power of the computer with BIM technology (interaction 29 in Figure 1.1).

The authors concluded that implementing BIM and Lean Construction simultaneously was not only possible, but indeed highly recommended, because many BIM functionalities improved the flow of design, planning, supply chain and construction processes, a key part of Lean. Thus, to successfully integrate the work processes that will replace traditional ones in any given construction company, those guiding the effort should be aware of these interactions, and consciously plan to amplify the positive ones while taking precautions not to fall into the traps set by the negative ones.

Tidhar Group, Ltd

Tidhar's founders, Arye Bachar and Gil Geva, established a small construction contracting company in 1990. Arye had worked as a salaried construction engineer and a freelance project manager for some 11 years, mostly in residential construction. Gil was a landscaping contractor. They had met when Arye hired Gil to do the landscaping and gardening development work on a housing project he was building. The two quickly realized that they shared a similar work ethic and business values. They looked with disdain on the behaviours of many of the construction companies they worked with in the small-to-medium residential sector. They saw that the construction companies were unable to plan or to work to a schedule, payments to suppliers were commonly made late and sometimes not at all, the quality of work was low, and they treated employees and even homebuyers poorly. They wanted to be different – to meet schedules, to pay on time and to nurture their employees to relate to their subcontractors as long-term partners rather than as opportunistically engaged service providers.

In 1993, the company was awarded a large residential project and wanted to begin construction quickly, but the company did not have the necessary government certification. To solve the problem quickly, they acquired a 50 per cent stake in a small, almost bankrupt, but appropriately certified company: Tidhar Construction Ltd. Two years later, they acquired the remaining 50 per cent.

Tidhar's relaunch under its two new owners coincided with the start of a decade of rapid growth in the local construction industry, which stemmed directly from the mass

Table 1.1 Lean Construction principles

Area	*Principle and means*
Flow process	Reduce variability
	Get quality right the first time (reduce product variability) (A)
	Focus on improving upstream flow variability (reduce production variability) (B)
	Reduce cycle times
	Reduce production cycle durations (C)
	Reduce inventory (D)
	Reduce batch sizes (strive for single piece flow) (E)
	Increase flexibility
	Reduce changeover times (F)
	Use multi-skilled teams (G)
	Select an appropriate production control approach
	Use pull systems (H)
	Level the production (I)
	Standardize (J)
	Institute continuous improvement (K)
	Use visual management
	Visualize production methods (L)
	Visualize production process (M)
	Design the production system for flow and value
	Simplify (N)
	Use parallel processing (O)
	Use only reliable technology (P)
	Ensure the capability of the production system (Q)
Value generation process	Ensure comprehensive requirements capture (R)
	Focus on concept selection (S)
	Ensure requirement flow down (T)
	Verify and validate (U)
Problem solving	Go and see for yourself (V)
	Decide by consensus, consider all options (W)
Developing partners	Cultivate an extended network of partners (X)

Source: Sacks et al. (2010). Reproduced with permission from ASCE.

immigration to Israel over that time of over one million people. Most of the new immigrants came from the Jewish communities of the various ex-Soviet Union countries who had gained the freedom to emigrate after the fall of the Berlin Wall and the collapse of the Iron Curtain. Providing housing and community buildings for this influx was a significant challenge; the immigrants represented approximately a 21 per cent increase in the total population of the country. Among other steps it took to provide housing, the

Table 1.2 BIM functionality

Stage	*Functionality*
Design	Visualization of form
	Aesthetic and functional evaluation (1)
	Rapid generation of multiple design alternatives (2)
	Reuse of model data for predictive analyses
	Predictive analysis of performance (3)
	Automated cost estimation (4)
	Evaluation of conformance to programme/client value (5)
	Maintenance of information and design model integrity
	Single information source (6)
	Automated clash checking (7)
	Automated generation of drawings and documents (8)
	Design and fabrication detailing
	Collaboration in design and construction
	Multiuser editing of a single discipline model (9)
	Multiuser viewing of merged or separate multidiscipline models (10)
	Preconstruction and construction
	Rapid generation and evaluation of construction plan alternatives
	Automated generation of construction tasks (11)
	Construction process simulation (12)
	4D visualization of construction schedules (13)
	Online/electronic object-based communication
	Visualizations of process status (14)
	Online communication of product and process information (15)
	Computer-controlled fabrication (16)
	Integration with project partner supply chain databases (17)
	Provision of context for status data collection on site/off site (18)

Source: Sacks et al. (2010). Reproduced with permission from ASCE.

government embarked on a crash construction programme to build small (one- to three-bedroom) apartments by issuing contracts for construction of apartment buildings with large bonuses for early completion. The industry rose to the challenge, increasing housing completions from an average of 20,000 per annum in 1990 to a peak of 70,000 in 1992. Unfortunately, construction quality was not a high priority at the time. The market returned to stable levels of construction at an average rate of about 35,000 housing units per annum by the end of the decade.

From many points of view, Tidhar was a typical building construction contracting company during its first decade. It began with contract housing construction for real estate developers. Yet its experience in one particular large project, coupled with its leaders' values, skills and ambition, created an environment in which the company's DNA changed from that of its construction contracting peers. The most significant

Lean Principles / BIM Functionality		Reduce Variability		Reduce cycle times		Reduce batch sizes	Increase flexibility		Select an appropriate production control approach		Standardise	Institute continuous improvement	Use visual management		Design the production system for flow and value				Ensure comprehensive requirements capture	Focus on concept selection	Ensure requirements flowdown	Verify and Validate	Go and see for yourself	Decide by consensus consider all options	Cultivate an extended network of partners
		A	**B**	**C**	**D**	**E**	**F**	**G**	**H**	**I**	**J**	**K**	**L**	**M**	**N**	**O**	**P**	**Q**	**R**	**S**	**T**	**U**	**V**	**W**	**X**
Visualization of form	**1**	1,2													3				4		11	5	6	4	
Rapid generation and evaluation of multiple design alternatives	**2**	1		22									7	7		8									
	3	9	9	22			51												1	16		5			
	4		10	12												8				16		5			
	5	1,2	1	12															1	1	1	5			
Maintenance of information and design model integrity	**6**	11	11																		11				
	7	12	12	22																		12			
Automated generation of drawings and documents	**8**	11		22	(52)	53											54	54							
Collaboration in design and construction	**9**			23						36						36									
	10	2,13		24				33											43			46		49	
Rapid generation and evaluation of multiple construction plan alternatives	**11**	14		25	(29)		31								(41)					44					
	12		15	25	(29)					37					(41)					44		47			
	13	2	40	25	(29)						17		40	40		40				44		47		49	
Online/electronic object-based communication	**14**		29	26	30	30			34					34			(42)					47	48		
	15	18		26	30	30			34		38		38	34			(42)				45			49	
	16	19		27			32										(42)								
	17		20	28					35								(42)								50
	18		21		30	30			34			39					(42)					47	48		

Figure 1.1 Positive and negative interactions between 24 Lean Construction principles and 18 BIM functionalities

Source: Sacks et al. (2010). Reproduced with permission from ASCE.

project from 1994 to 1999 was a series of building clusters in a neighbourhood called Aviv Gardens, near Tel Aviv, which had four contract stages. By the time it was finished, Tidhar had built a total of 2,516 apartments and a commercial centre. This was a formative project for Tidhar in many ways, leading to rapid growth. But perhaps the most significant effect was that it forced the partners and their engineering team to focus on the notion of flow in construction. That learning persisted as part of the ethos of the company, as a majority of the members of that team continued to work for the company for many years. Many of them still hold senior positions at the time of writing these lines.

At the peak of the Aviv Gardens project, with a contract to deliver 1,008 apartments within 14 months,[9] Arye realized that they would have to begin work on eight apartments every day. The implication was that every trade would have to complete the equivalent of eight apartments a day. This was the Takt time for the project, and although they had never heard the term nor learned about production management, the concept was crystal clear. The number of workers in each trade crew would be driven directly by the need to meet the required flow. They also quickly identified the bottleneck process – the buildings had load-bearing precast concrete façade panels that were cast on site in a field factory. The factory had to produce 20 panels every 24 hours to meet the required pace of eight apartments per day. Fortunately, there were no client changes or customizations at all. The concepts flow, bottleneck, cycle time and batching became deeply ingrained in Tidhar's outlook on construction at that time.

However, Aviv Gardens did not prepare them for two aspects that would become critical much later and lead to their readiness to adopt Lean Construction and BIM. The first was that there was little or no variation in the product or in the production. Until the end of the 1990s most of their work – and that of much of the industry at the time – was in projects in which the clients were not given any opportunity to make design changes or adaptations to their apartments. A majority of the workers were still employed directly, making them less reliant on the vagaries of subcontractors, and materials could be buffered because they had negotiated excellent credit terms and there was ample space on site. Thus, both product and production were relatively stable. The second aspect was that customer service was not a priority, largely due to the imbalance of very high demand vs limited supply of apartments on the market. Real estate developers did not need to offer customization to attract clients.

In 2004, Tidhar Group CEO Gil Geva launched a strategic plan designed to stimulate Tidhar's growth from a small to a medium-sized company. The plan's goals were defined in a new, ambitious vision and mission statement (see Box 1.1). Tidhar was to become a real estate developer in its own right, as well as a construction contractor. Capital was raised in part through the investment of a new partner from the US, Mark Weissman, and Tidhar indeed began to grow.

As the company grew, it took on projects in the higher-end housing market in which most homebuyers expected, and demanded, significant degrees of customization. They were forced to pay attention to the softer issues of management. Gil led Tidhar to adopt a clear declaration of principles by which its people agreed to work (see Box 1.2). Very few other companies in its market had developed guidelines of this kind. Tidhar became known for its professionalism and its skilled management. Its employees took great pride in being viewed as providing better service to customers than their competitors.

Figure 1.2 (above) Arye Bachar and Gil Geva, the founders of Tidhar Construction, in 1995; (below) Mark Weissman, Gil Geva and Arye Bachar

Box 1.1 Tidhar's vision and mission statement[10]

- To be the leading construction company in Israel, to strive for global growth, to cultivate values of excellence and innovation, and to be a role model for service, performance and quality.
- To be a growing, expansive and solid company that nurtures and rewards its employees, and constitutes a home as well as a source of empowerment and pride.
- To contribute to Israel's society and economy through community outreach, the creation of jobs, and the constant improvement of construction and management methods.

Box 1.2 Tidhar's staff commitment compact

1 Integrity

Reliability is fundamental to all Tidhar's activities. Reliability creates trust, which leads to loyalty. Tidhar is convinced that the relationships within and outside the organization, based on credibility, integrity, reliability and accuracy, ensure loyalty and build cooperation over time.

2 Service

Satisfied, cared-for customers are the foundation for the growth of Tidhar's reputation. Tidhar service is based on quality construction of the building, accompanied by professional, reliable and rapid customer response. Company employees are committed to providing service at the highest level.

3 Quality and Excellence

Tidhar is committed to quality and excellence in all areas of its activity. Adherence to this guiding principle is essential for every employee in the group. Quality and excellence mean performing every activity as well as possible. Forward planning, attention to detail, simplicity, consistent monitoring and continuous learning will ensure maintenance of quality and excellence and thus Tidhar's place at the top of Israel's construction industry.

4 Human Capital

The main resource underlying Tidhar's strength is a high-quality and dedicated staff. All Tidhar employees work in the spirit of this compact and excel in their fields. Employees will conduct their work with personal responsibility, discipline and teamwork. The culture of discussion is a basic component of the way we work. Tidhar is committed to maintain, promote and develop its employees.

5 Schedule

The schedule is the heart of the project management system. Completing work on schedule and fulfilling other commitments that arise from it, allows maximum profitability, and is the basis of Tidhar's reputation as a superior construction company. Keeping to schedule fulfils all of the company's basic values – reliability, quality and efficiency. All Tidhar employees and managers are committed to meeting the schedule.

6 Safety

Tidhar is responsible for the safety of all its employees. Tidhar is committed to developing and enforcing safety procedures, and will conduct routine safety management on sites to ensure a safe working environment. Tidhar will serve as an inspiration and example of compliance with safety requirements.

7 Relationships

Tidhar believes that correct and harmonious working relations between employees and with people outside the company are the basis for achieving targets of quality, schedule, cost and service. Company employees are committed to do everything in their power so that the company can fulfil its contracts, meet payment dates and all other obligations. Open and honest communication is the key to establishing these relationships.

8 Order and Cleanliness

Tidhar believes that order and cleanliness are integral to proper management of all activities. Construction sites, offices and employees of Tidhar are always to be neat and orderly, and construction equipment will be well organized. An orderly working environment enables people to work according to the guiding values of quality, excellence and good service.

9 Courtesy and Fairness

The relationships between human beings underlie every action we take. Tidhar believes that courtesy, decency and modesty are the cornerstones of all activity. Gentle and honest behaviour even when clarifying disputes will ensure the ability to maintain long-term cooperation within the business environment.

10 Continuous Improvement

Tidhar is an organization that strives toward excellence and continuous improvement – a learning organization, drawing conclusions and managing knowledge. Continuous improvement will preserve Tidhar's competitiveness in all market conditions, and deepen its comparative advantage.

11 Connection with the Community

In its building projects the company conserves the environment, uses eco-friendly materials, avoids damage to the landscape and considers the needs of the community. Tidhar considers contribution to the community and the personal involvement of its people in what is happening in the country and society at large to be an integral part of its activity.

Tidhar's commitment compact is a statement of ideals. As such, it expresses the core beliefs of its founders. In practice, it is essentially a tool that aims to inculcate these values in its employees. In reality, as with many such statements of values in contemporary society, it would be naïve to expect that all of the values would be maintained all of the time. As a company grows, it brings in new people who have learned to behave in the milieu of a highly competitive and less than perfect construction industry. Lofty ideals are difficult to maintain when the competition is profiting from less than perfect behaviour.

As Tidhar grew, maintaining the qualitative differences between its construction sites and those of its competitors became more challenging. In areas such as construction quality, recycling of waste materials, order and cleanliness, construction safety and continuous improvement, exerting control over workers employed by subcontractors became increasingly difficult.

Tidhar, 2011

In March 2010, Gil Geva began preparing a second strategic plan. Within three years of the 2004 effort, the company's turnover had grown to some $90 million, but then it stagnated. He made changes to the senior management, decided to focus solely on office and residential projects, and began exploring operational excellence, with BIM and Lean as the leading components. By the end of 2011, when he formally launched Tidhar on its Lean and BIM journey, it had grown into a small-to-medium-sized enterprise, with some 250 employees, of whom some 60 were engineers and the others were construction superintendents, site and head office administrative staff. Beyond the core of direct company staff, Tidhar also indirectly employed around 1,000 construction craft workers through its subcontractors. The company's activities were mostly focused in Israel, although it had a small number of projects in Bulgaria, Germany, Serbia and the US. In addition to building construction, Tidhar had two smaller divisions, one for real estate development and one for property management and income-yielding assets. A new division that performed interior fit-out of Tidhar's office projects was established at the start of 2012. To put the scale of activity of the building construction division in proportion, at the end of 2011 it had 22 projects with 2,000 apartments under construction, another 18 projects at various stages of planning, and another 6 under negotiation.

The core of Tidhar's business in 2011 was apartment buildings for the middle-class market in the Tel Aviv metropolitan area, accounting for 85 per cent of its income. This was a very generic product because the town planning codes of most of the municipalities in the area were similar, dictating a fairly standard design of apartments and of buildings. Almost all of the buildings had four apartments on each floor, arranged around a common core with elevators, stairwells, floor lobbies and service shafts, and almost all were eight or more storeys tall. This was a highly competitive market with very low profit margins, because the construction contractors, subcontractors and suppliers had become specialized in this product type. Common wisdom held that there was little room for competition through innovation or engineering expertise.

In this climate, Tidhar was facing a dead-end. Given its higher-than-average-industry-standard overhead costs, which were necessary to maintain the level of customer service and construction quality upon which its reputation was based, it found it increasingly difficult to compete. The minimum margin it considered viable for real estate projects

Figure 1.3 A typical floor layout of an apartment building, with four apartments arranged around a central core

Note: Every apartment has a security room with thicker than average concrete walls.

was a 15 per cent return on capital, if it was to avoid unpalatable risks. Yet most of the projects that it evaluated had single-figure margins. For every project Tidhar actually built, on average 20 bids were rejected, representing a huge loss of opportunity for growth. But with gross profit margins for construction work at the time in the range of 9–11 per cent, Tidhar Group's CEO was not prepared to assume the associated risk. Equally, he was not willing to take the easy short-term alternative, as many other companies did – to compromise on construction quality or on customer service.

The more challenging and more courageous option was to find a way to grow the margin itself. If that could be done, the company could outperform its competitors and easily grow its market without any compromise. For group CEO Gil Geva, remaining stuck as a medium-sized construction company was simply not an option.

Growing the Margin

"Growing the Margin" was the name and the goal of the strategic effort Gil Geva launched the company on at the start of 2011. It was a three-pronged strategy:

1 Grow turnover: Significantly increase the scale of operations, both in construction contracting and in real estate development.
2 Increase value: Increase the perceived value of the company's products and services by generating added value for customers; provide a higher quality apartment, better service and customer experience; brand the company such that its customers would be willing to pay a premium price for its products and services.
3 Decrease costs: Improve product design to achieve the most efficient design in every project; improve process design to improve productivity; increase operational efficiency and increase the operating profit.

The underlying goal was to leverage the improved profit margin to grow the company's turnover by enabling it to safely take on projects that would previously have been considered too financially risky.

The strategy was expressed as a set of five specific targets:

1 Triple the company's annual turnover to more than NIS1 billion (approx. $300 million).
2 Achieve a price premium of 5 per cent in the sale prices of Tidhar apartments.
3 Achieve a price premium of 2 per cent in the construction division's contract work.
4 Increase the construction division's gross margin to 15 per cent within four years.
5 Reduce the construction division's overhead to 4 per cent in four years by increasing volume.

These in turn were given substance in a list of action items:

1 Initiate and brand an effort to "create added value for customers".
2 Develop and apply a thorough value management process at the project programming stage to achieve accurate budget planning.
3 Standardize the design elements and details of Tidhar buildings.
4 Implement Building Information Modelling in the design and construction process.
5 Establish and manage a performance improvement programme with an emphasis on creating a culture of continuous improvement.
6 Upgrade information systems and processes in order to support enterprise knowledge management.
7 Develop a "Tidhar index" to measure development and construction performance.

Examining this list reveals that action items 1, 2, 3 and 5 are Lean Construction initiatives, and action item 4 deals directly with BIM. Item 6 deals with IT systems and 7 with performance measurement. The notion of Lean Construction was not mentioned directly, and nor was the subsidiary concept of removing waste from business and production processes. That level of specificity came later, as the company began to look for ways to implement the action items.

Box 1.3 "Growing the Margin" compared

It is interesting to compare Tidhar's "Growing the Margin" strategic effort with those of other construction companies. NCC, a major northern European construction, property development and infrastructure company with sales of €6.3 billion and 18,000 employees in 2014, published its "New Strategy for Global Growth" in November 2015.[11] The goals of the 2016–2020 strategy were:

- an operating margin of at least 4 per cent during the strategy period;
- average annual sales growth of 5 per cent during the strategy period;
- annual return on equity after tax of at least 20 per cent;
- net indebtedness of less than 2.5 times EBITDA (earnings before interest, tax, depreciation and amortization);
- equity/assets ratio of at least 20 per cent;
- reduction of the accident frequency rate by half by 2020 (compared with the 2015 outcome);
- reduction of NCC's CO_2 emissions by half by 2020 (compared with the 2015 outcome).

Apart from the emphasis on the need to improve gross operating margin, what is strikingly similar about NCC's strategy to that of Tidhar is the first of NCC's "must win battles" policy:

> Operational excellence: NCC will increase its focus on strengthening existing expertise by strengthening the company's position close to the customer, improving the efficiency of working methods and processes using Lean principles, and improving support for digitized information flows.

Just as they formed the foundation of Tidhar's policy, Lean Construction and BIM are the operational innovations that form the foundation of the company's competitive strategy.

Gil first presented the comprehensive "Growing the Margin" initiative to senior management at a meeting in October 2011. The meeting was held in the rooftop conference room of a Tel Aviv beachfront hotel. All division and department heads attended, as well as Zohar Benor and other Lean coaches from Lean Israel Ltd, a Lean consultancy company, Professor Rafael Sacks and two of his graduate students from the Technion – Israel Institute of Technology. The response was mixed: enthusiasm and excitement, disbelief and doubt, appreciation and apprehension. Many of the department heads and senior project managers felt that they knew how to build. As a company, they had a reputation for being the best, so why – and how – did Gil think they could make such substantial improvements? However, they were well aware of the endemic problems. They knew that construction was wasteful, they knew that things on site were far from ideal, but to date none of them had found solutions to these problems. Two senior managers had the courage to suggest that the turnover growth target, NIS1 billion within five years, was no less than "megalomaniacal"! Gil's leadership was critical at this point. If he had not exuded the confidence, commitment and determination that he did, very little would have been accomplished.

Gil next laid out the "Growing the Margin" initiative to the whole staff gathered at the company's annual staff meeting in December 2011. The staff, all 257 of them, were all direct Tidhar employees. In the presentation on the ways to reduce costs, Gil emphasized the ideas of "continuous improvement to improve operational efficiency". Operational efficiency was to be improved by rigorous design of the product and planning of the construction process to remove the wastes of defective products, overproduction, excess inventory, unneeded processing, unnecessary movement of people, wasted transport of materials and equipment, and waiting. It was no coincidence that these are the seven wastes identified by Taiichi Ohno and repeated by most Lean authors. During the preceding months, Gil had come to see the "Growing the Margin" goal of improvement in operational efficiency as being achieved through the paradigm change of Lean Construction. The first slide he used to begin to explain Lean Construction to all of the gathered employees is shown in Figure 1.4.

Gil wanted to convey two ideas: first, not everything Tidhar's employees did during their day brought value to their clients; and second, that when one carefully considered the work processes, they were in reality doing a lot of waiting and moving around that they perhaps did not think of as activities but were nonetheless consuming time without creating anything of value. This is precisely the approach taken in Lean value-stream mapping, in which non-value-adding activities are identified with a view to redesigning processes to remove them as much as possible. The ideas of waste and of non-value-adding work are covered thoroughly in Chapter 8, "The waste of non-value-adding work", later in the book.

At the same staff meeting, Gil also introduced Building Information Modelling. BIM was presented as the primary mechanism through which Tidhar would improve value and remove waste from operations. To introduce BIM, he showed a series of images of the model of a typical floor of a residential building that the company had recently built. He showed axonometric views, perspectives, a full set of plans and façades, room schedules and quantity take-off sheets, all produced directly from the model. He listed seven benefits that Tidhar could expect from the use of BIM:

1. Productivity: reduce the amount of work needed per unit product in construction and in design.
2. Quality: prevent design errors and inconsistencies between drawings and between design disciplines.
3. Project duration: reduce the time needed for design.
4. Accuracy: improved accuracy in all calculations, including in quantity take-offs and cost estimates.

- How we see our process:

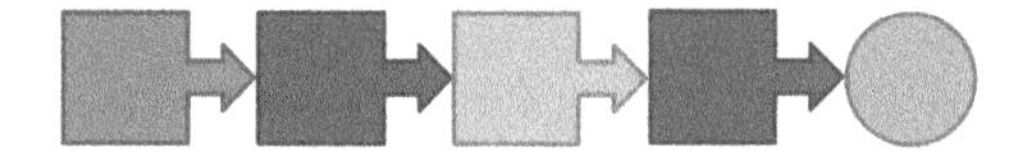

- What the client is willing to pay for:

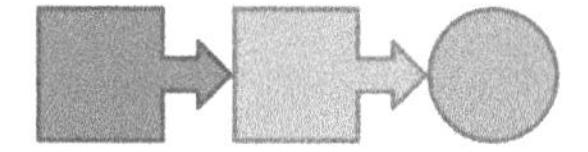

- What our process really looks like:

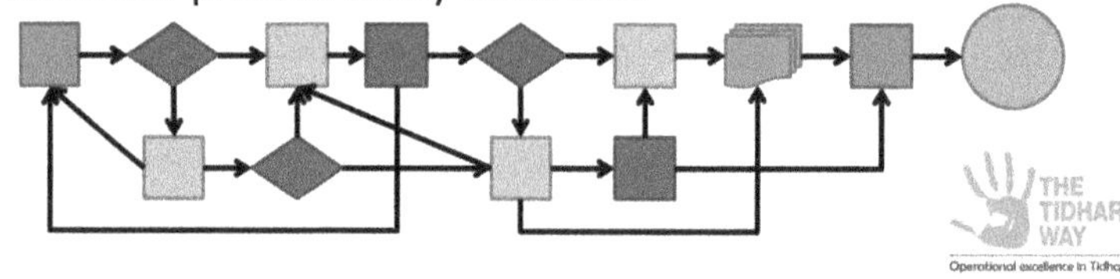

Figure 1.4 The client test – Tidhar's processes

5 Industrialization: produce detailed fabrication data, increased use of prefabrication and of pre-assembly off site.

6 Improved work flow in design and in production: improved flow of information to all who need it, better visualization of product and process.

7 Avoid the waste of over-design through sharper definition of customer needs and better visualization and analysis of proposed designs.

Gil was also realistic; he told the group that BIM would first be used for design and project development, and only later for the construction activities: budget control, schedule control, quality assurance, purchasing, client change coordination and cost accounting. As we shall see through the following chapters of the book, Tidhar managed to leverage BIM for use in many of these areas through the first four years of its implementation, but in other areas implementation has proved to be beyond the company's capacity or capability thus far.

In concluding his rallying cry at the staff evening, Gil told them why he thought they could improve both design and construction simultaneously:

> Our ability to succeed in this, adopting both BIM and Lean Construction, rests on the fact that we are both a real estate development company and a construction company. We are a united, purposeful company! If we can work together as a team, and be prepared to move out of our individual and collective comfort zones, we can succeed.

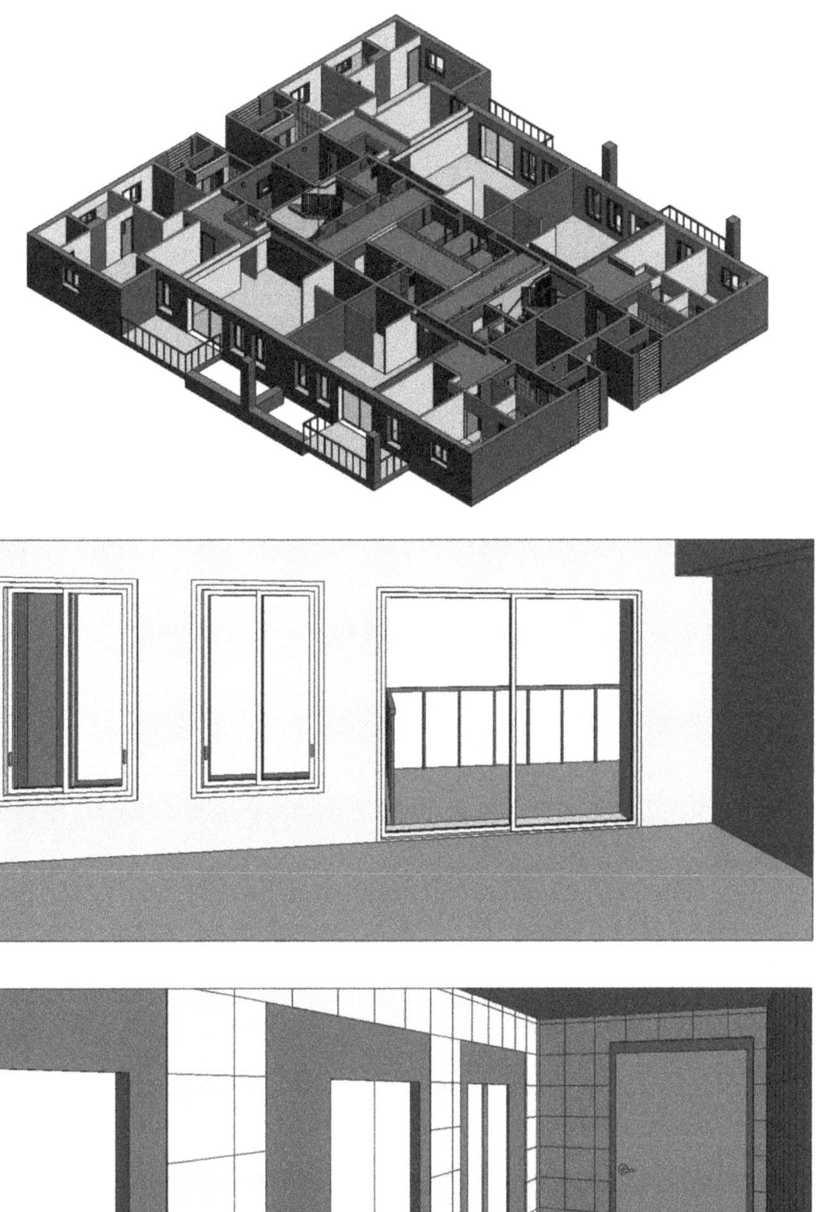

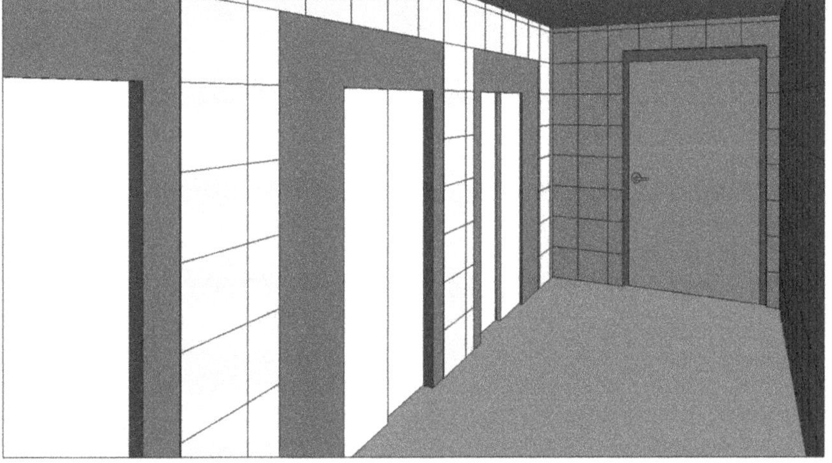

Figure 1.5 The first BIM models of a typical Tidhar apartment building, December 2011

Why did Tidhar believe it could be a pioneer in its market?

As we will see in Chapter 3, "Education and motivation", Tal Hershkovitz, the CEO of the Construction Division, returned from a study tour to Japan in late 2011 with a clear message: the construction industry was ripe for change. He returned with strong personal motivation – he was convinced that Tidhar could make that change and that it would be profitable. Yet even with strong personal commitment and guidance from the company's leaders, an important question remains. What gave Tidhar's leaders the confidence that the company was capable of being the first in Israel to take the enormous risk of initiating deep change, adopting both Lean and BIM? The industry environment had not developed any of the support systems, such as BIM standards; none of their suppliers or subcontractors had any experience with either innovation. With the exception of a few architectural firms who had begun using BIM for the sake of improved productivity in preparing drawings, none of their design consultants had BIM software, nor did their staff have the necessary skills. Why then did Tidhar want to be first, to be a pioneer?

The main explanation can be found in the company's leadership. With very few exceptions, Tidhar's senior managers had all held leadership positions of one kind or another in the Israel Defence Forces. The willingness to move forward, at times with calculated risks, was part of their DNA. They had always improved and always been at the forefront. The current situation or tools were never good enough to meet their requirements. The company's information systems are a good example: Tidhar built purchasing and project control systems in-house, because the available software was not good enough. Steel formwork shutters for their exterior wall construction system are another: when it became clear that they could not procure the equipment they needed from the major formwork companies, Tidhar set up a dedicated steel workshop to produce and maintain the custom formwork needed for its projects. When subcontracted suppliers of aluminium windows proved to be unreliable, causing delays in projects, Tidhar acquired a workshop and began supplying its own windows. They had built a strong customer service department and won national awards even when competing against companies from industries whose core business is customer service, like cellular phone providers.

Another part of the explanation can be found in the intersection of ambition with opportunity. Tidhar had a better reputation than its competitors, which guaranteed a steady and growing stream of clients and a desire to be bigger and better. Opportunity presented itself in the stories of the successes achieved by Lean Construction adopters in the US and Europe and in the form of maturing BIM technology.

Tidhar also had the capacity to make the changes. Unlike most of its competitors, Tidhar's senior engineers had an appreciation for the notions of flow and waste in production. They had always applied more engineering resources than other companies, and were known to have significantly higher site overheads, but also much closer control over their projects. They welcomed technology where it could be applied effectively, and they had access to knowledge and to the research capabilities of the Technion; Rafael Sacks was actively researching both Lean Construction and BIM and had introduced both subjects into the curriculum at the Technion, Israel's premier technological university (and until the early 2000s, its only school of civil engineering). He and a graduate student, Ronen Barak, had made BIM a central part of first year studies for all civil engineering students since 2006; Ronen was now working as R&D manager at Tidhar. Lastly, Tidhar had one more small advantage – Gil's lack of a formal civil engineering or construction management education left him free of the fixed conceptions that many construction managers hold about contracting, scheduling and planning, allowing him to lead Tidhar to "do things differently".[12]

As we will see through the remainder of the book, despite many conditions in its favour, Tidhar also faced serious obstacles in its path to adoption of BIM and Lean Construction. While the company made great strides forward in the five years on which we report and is more profitable and professional than it was, its design and construction standards are still far from perfect, far from the ideal vision of how things could be done in an idealized "Tidhar Way". The book is the story therefore not of the success, but of the struggle. Just as construction management should be, it is as much about the process as it is about the results.

Back to the theory

Tidhar's story, like those of Skanska, Lease Crutcher Lewis and Fira (told in Chapters 5, 12 and 19 respectively), is interesting in its own right, but the four stories are also valuable as a medium to explain some of the theory behind Lean Construction, Building Information Modelling, the processes to adopt them and some of their interactions. Underlying the different stories there is a unifying theme that presents production in construction from the interdependent points of view of information and of product flow.

A fundamental question concerns the nature of flow in construction. What, after all, is the construction equivalent of a consumer product, like a car? Is it a whole building, an apartment, a shop or a room? Or is it a set of physical products, such as walls, doors, floors? More specifically, what are the interim products that flow through the production system? How should we measure their cycle times and the amounts of work in progress (WIP)? The main idea of location-based management is that production should be planned, monitored and controlled around the physical spaces (the locations), through which trade crews flow.[13] This is a good starting point, but it does not capture the full view of production in construction, nor does it go beyond the boundaries of a project to consider the constraints of the surrounding environment. We will explore this idea throughout, culminating in Chapter 15, which asks (and attempts to answer) the question "What Flows in Construction?"

Among the many things that flow in construction, information is perhaps the most difficult to coordinate, measure and control. Information comprises not only product information (design information, as we may think of being contained in a "Building Information Model"), but also process information (which describes the status of the various flows – flow of work, flow of materials, flow of crews and equipment, flow of product information and flow of money). Research has shown that communicating process information clearly and fully, using visual aids such as production control and Andon boards, pull kanbans and Building Information Models to collect, carry and deliver process information as well as product information, can all enhance the flow of work itself.[14] Chapters 11, 14, 16, 17 and 18 all deal with the challenge of getting the right information to the right people at the right time.

And lastly, a question that readers may ask themselves throughout the book, is "Why are Lean Construction and BIM so difficult to adopt?" The question is not simply one of organizational consulting or structural change; it has to do with peoples' behaviour and willingness to change long-held habits. Embedding Lean Construction processes requires sustained attention and energy to make an effort on the input end, in order to get net value out of the process. BIM is similar; its adoption requires close monitoring and great effort. Both require people with new skills to fill new roles in construction projects, such as project information officer or production control engineer. We will meet a number of these people through the four stories.

Notes

1 The story of Taiichi Ohno's leading role in developing the Toyota Production System is told in Ohno (1988).
2 The adoption of Lean and BIM through integrated project delivery is detailed thoroughly in Fischer et al. (2017).
3 Some examples: LePatner's book, *Broken Buildings, Busted Budgets*, originally published in 2007, is a manifesto of the sorry state of the construction industry in the US at the time (see www.brokenbuildings.com/press.html). The Egan and the Latham reports, commissioned by the UK government in 1994 and 1998 respectively, presented a strong critique of the construction industry and set explicit and far-reaching targets for improvement. Teicholz's discussion of construction productivity (Teicholz, 2001) revealed stagnation in construction productivity. Josephson and Hammarlund (1999), Love and Edwards (2005) and many others have documented the waste in construction and reported statistical analyses of project delivery data that document widespread failure to deliver on time, on budget or with the required quality.
4 In Chapter 8, "The waste of non-value-adding work", we explain the eight forms of waste and how they manifest in many construction work methods.
5 Chapter 6 of the *BIM Handbook* (Eastman et al., 2011a and 2011b) provides a thorough discussion of the potential benefits of BIM for construction companies.
6 Koskela's CIFE report, *Application of the New Production Philosophy to Construction*, published in 1992, is arguably the first to make a significant contribution to laying the foundations for Lean Construction.
7 The original version of ArchiCAD software, called Radar CH, offered 3D modelling with object-oriented building entities as early as 1984. 3D CAD software tools from the manufacturing industry such as CATIA, CADAM and others were used for special buildings since the late 1980s, although they were too expensive for everyday use. The "Digital Project" BIM software is a direct offshoot of CATIA.
8 See Sacks et al. (2010).
9 The construction schedule was much shorter than usual because the real estate developers, who were Tidhar's clients, wanted to secure the government bonuses that were still available to companies who provided housing within stipulated deadlines.
10 See Tidhar's "Vision and Values" document at www.tidhar.co.il/en/vision.aspx.
11 New strategy for profitable growth at NCC; see www.ncc.group/media/pressrelease/a148296cac91f7a2/new-strategy-for-profitable-growth-at-ncc/.
12 This "freedom of thought" is reminiscent of the introduction of Lean thinking into the manufacturing plants of the Danaher Corporation, a group of companies assembled by two real estate entrepreneurs who had no experience or knowledge of manufacturing, but had the insight to bring disciples of Taiichi Ono to the US in 1987 to support adoption of Lean production practices in their firms. See Womack and Jones (2003, p. 111).
13 The definitive work on location-based management in construction is Kenley and Seppanen, (2010).
14 For discussion of the flow of work, see Gurevich and Sacks (2013) and Sacks et al. (2009).

2 False starts

A "critical mass of intent" is essential to initiate Lean and BIM change in a construction company. The commercial conditions and the intellectual context must also be right, but without sufficient intent on the part of those with the power to effect change, initiatives for change tend to have little or no lasting effect.

Tidhar's "Growing the Margin" initiative, started late in 2011, was not the company's first attempt to adopt either Lean Construction or BIM. The CEO of the construction division had tried to introduce the ideas as early as 2007, but his efforts failed to gain traction within the company. Why did this earlier attempt to bring change fail? Some of the reasons included lack of support from the company's board of directors, lack of knowledge on the part of most of Tidhar's executives and engineers, and a preoccupation with business as usual. To fully understand why the company failed to adopt the programme until 2011, in this chapter we explore the conditions under which the earlier attempt was made. Our goal is to define the "critical mass of intent" necessary for any construction company to succeed in implementing its own adoption programme.

Awareness of Lean Construction and BIM

The first author of this book brought Lean Construction thinking to Israel in 2000, when he began introducing Lean Construction ideas into research and classes at the Technion – Israel Institute of Technology. However, for many years, this exposure was limited to a handful of graduate students of construction management. The language barrier limited access for most practising engineers to new knowledge in the field , as most of the information in those years was in English. Throughout most of the decade from 2000–2010, very few people in the Israeli construction industry were aware of the growing momentum of Lean Construction thinking in the US and Europe that followed publication of Lauri Koskela's and Glenn Ballard's theses (on Transformation-Flow-Value theory and on the Last Planner® System, respectively). The expanding influence of the Lean Construction Institute and the International Group for Lean Construction (IGLC) had not reached the shores of the eastern Mediterranean. Awareness of Lean Construction within Israeli construction companies grew as more Technion civil engineering students graduated and joined the workforce, but it was only brought to the attention of local construction executives when Israel hosted the annual IGLC conference, including the associated Industry Day workshop, in 2010.

During this period, BIM, by contrast, was known but misunderstood. 3D modelling of buildings had been available in Israel as elsewhere since the advent of Graphisoft's ArchiCAD software in the late 1980s. While the majority of architects continued to use 2D CAD exclusively, a small but devoted minority of architects worked with ArchiCAD and later with Revit to produce 2D construction drawings. Structural engineers performed analyses in 3D of course, but when the level of detail reached down to the level of construction details (shop drawings), the information was provided to builders exclusively as 2D drawings. The mass immigration from the ex-Soviet Union in the early 1990s had provided a ready supply of highly trained engineers skilled in 2D CAD. As long as the cost of that resource remained low, 3D modelling and BIM were considered a luxury, used only on high-end projects.

In this context, construction company executives and engineers perceived 3D modelling (and later BIM, once the term itself was coined in the early 2000s), as entirely within the preserve of designers. They were largely unaware of the potential benefits for construction. The very idea of using apparently "high-tech" modelling hardware and software at construction site offices, let alone in the field, seemed unnecessary, an unjustifiable expense and a waste of time and effort.

A Lean approach to housing construction

The experience of one of Tidhar's direct competitors, when it attempted to improve its housing construction operations using both Lean Construction and BIM, provides insight into the background of the construction industry at the time. It also underlines the importance of the "critical mass of intent" needed for successful adoption.

In 2004, when this story began, Alpha Construction Ltd ("Alpha") was a much larger company than Tidhar. Its operations extended beyond the apartment building market (in which it competed with Tidhar), and included commercial and public buildings, industrial construction and infrastructure projects.

Alpha's apartment construction problem

At the time, Alpha was one of the largest builders of apartment block projects in the country. It had completed many projects with hundreds of apartments each. The company had decades of experience, and its engineers and managers believed they knew how to build apartments. But the market was changing, with apartment buyers becoming increasingly sophisticated and discerning. Buyers had begun demanding that the contractor customize the interiors of the apartments: kitchens, floor tiles, bathroom fixtures, partitions, HVAC (heating, ventilation and air conditioning) units, electrical systems and many other components needed to be tailored for each customer. The ability to customize became a key competitive advantage in a market that had, at the time, an excess of supply over demand. One construction company had billboards advertising its housing projects that showed nothing but a construction worker "genie" coming out of a magic lamp and promising "With us, your wishes are unlimited …"

To serve the market need, Alpha had established a centralized "customer changes" department at their head office, where company architects and engineers met with customers, coordinated the design changes with them, documented everything and prepared files for the project engineers. They were proud of the department – it offered a level of service their competitors could not match. Yet Alpha found itself facing a serious problem – it was losing money in its high-end residential projects. The apparent advantage they had in catering to customer changes had won them big projects, because real estate developers liked the idea of leaving the changes to the general contractor. In the past, limited customer changes on lower-cost projects had always worked to Alpha's advantage, because the unit prices they charged individual buyers for changed items were much higher than the prices for the same line items in their construction contracts with the real estate developers. As their clientele became increasingly upmarket, the scope of the changes increased, and so did the line item prices.

But something else had also changed. In the lower-end market, the company found it could dictate deadlines to the homeowners for making design change decisions. The deadlines were set according to the standard construction schedules they used for the buildings, which called for each trade to progress up the building, floor by floor. Customers could be pressured into choosing floor tiles by a certain date, for example, with ample lead time to allow purchase and delivery of floor tiles to the site well in time before the tiling crews reached their apartments. And if some of the customers' choices were unavailable, they could be persuaded to choose something that was available. Senior management measured the success of the customer change department by its ability to bring customers in line and provide each site with the information and supplies it needed to match the traditional construction methods and schedules they had always used.

Yet in the high-end market, homeowners employed interior designers. Design changes became extensive and included, for example, false ceilings with cornices and drops, completely redesigned bathrooms and kitchens, under-floor heating, intelligent home control systems, central vacuum systems and changed room layouts. Buyers who paid high purchase prices to the developers and agreed to pay high unit prices for their changes were not about to toe the line for clerks of the contracting company who demanded they make decisions according to a timetable that required the information months in advance of completion dates.

The result was chaos and waste. Typically, a subcontractor crew would arrive to work in an apartment, only to discover that the designs had been put on hold or that the materials had been changed. To avoid falling behind schedule, the project managers' solutions were all too often to simply complete the work at hand using the "standard" design and materials, on the assumption that returning to change the work later would result in minimum delay to the progress of the project as a whole.

Project managers on high-end projects reported spending up to 60 per cent of their time managing customer changes and the resulting complexity of instructions to specialty contractors. As the delays grew due to late change orders, management teams lost control of the flow of work; contractors, always searching for high rates of productivity, continually redirected their efforts to those apartments in which larger quantities of work were available, leaving many minor details unfinished in incomplete apartments. This appeared to be a major source of high levels of work in progress (WIP),[1] as apartments with little work left were simply not completed. Push scheduling was another important cause of high WIP, since work was begun in apartments that either had not been sold or where customer changes had not been confirmed, with the result that stoppages in their execution were inevitable.

To make matters worse, in the latter stages of projects, trade subcontractors tended to reduce the sizes of their crews. Toward a project's end, many small pieces of work remained – the "loose ends" that had accumulated when parts of larger work packages were left incomplete. Reduced crew sizes avoided the loss of productivity that would result from crews with too many workers for the small work packages. The instability in their work and the unpredictability of their work durations increased as each trade delayed the others, and projects could not be completed on time.

Box 2.1 A built-in incentive for waste

The structure of project financing for residential projects created an insidious cash flow incentive that made rework the default choice for many developers and general contractors. Developers purchased land and funded building construction using bank financing packages. Banks provided a line of credit against which the developer could draw money to finance construction. To reduce the banks' exposure to risk, cash was released at specific milestones as the construction progressed. Monies were only released once the banks' own inspectors determined that the contractors had reached the milestones.

For example, one common milestone was the completion of floor tiling. There was thus intense pressure on project managers and site supervisors to complete the flooring as soon as possible, because contractors depended on cash flow for survival. But what should a manager do if a customer had not chosen their floor tile when the flooring crews arrived? In the majority of cases, they elected to keep to schedule by laying the floors with the project's default standard floor tiles. This was done with the clear understanding that there was a high probability that the floors would need to be ripped up and redone once the customer reached a decision. This was a glaring waste that was directly built into the system, and one which some companies found hard to avoid.

For low-end projects, Alpha and other companies had largely solved the problem by offering significant discounts for timely design choices. But for high-end apartments, the discount was not a sufficient incentive to hurry a vacillating customer.

Analysis and a theoretical solution

When one of the company's flagship projects began to slide well beyond its construction budget, despite charging what were perceived to be very high prices for the customer changes, the vice president (VP) of Project Planning and Control turned to academia for advice. Rafael Sacks and a graduate student, studied the problem together with a core team from the company, including the VP, the manager of the customer changes department, a senior construction engineer, the project manager and three site supervisors.

Using historical data from multiple projects, they identified the high variability in the time when customers provided their design change decisions to be the main source of uncertainty that disrupted the flow of work. The statistics showed that the average cycle times for the interior finishing works on individual apartments were very long when compared with the net time actually required to perform the work. On average, a gross 49 weeks were needed to perform work that, were it done continuously with no interruptions, would have taken just 12 weeks. In almost every project, the quantity of WIP reached 100 per cent of the apartments for a significant period of the project's life, i.e. all of the apartments were simultaneously in some stage of production.

The root cause was instability in the customer design change process. The average duration for this process, measured from the time of the first to the last meeting between the customers and the company's customer change representative, was 25.5 weeks, and the standard deviation was 17.5 weeks! When subcontractor trade crews arrived to work in apartments according to the standard construction schedule, they often found that the prerequisite design information was unavailable. This lack of reliability forced the subcontractors to withdraw crews from the site, preferring to allow buffers of ready apartments to accumulate before returning to continue work.

This behaviour was rooted in the way they were paid: like many general contractors throughout the world, Alpha preferred to buy products rather than services, which meant that subcontractors were paid according to fixed unit prices for the quantity of work done, not for their time. Thus, any contractor whose workers were idle was losing money directly. The erratic flow of design change information was disrupting the flow of work for most of the trade crews performing interior finishing and building system works. It resulted in discontinuities in trade crews' presence on site, unreliable commitments to return when crews did leave, and excessive re-entrant flow for rework in all cases where changes were requested after work was completed. This of course snowballed and had a feedback effect, where unreliable progress on the part of any one trade directly affected the reliability of the following trades. The results were low productivity, long cycle times for completing work in apartments and increased load on the project and production management team. These translated into losses for subcontractors, frustration for customers when apartment delivery deadlines were missed and ballooning overhead costs for the general contractor.

The focal point of the problem was that the planned work flow for the interior finishing works – trade crews moving up through the building in the natural sequence dictated by construction of the structure – was quite different to the flow of information needed for completion of the works. Customer design change information was provided at a relatively steady rate of apartment units per month by Alpha's customer change department, but the sequence in which design information for apartments was provided was quite different to the sequence in which the construction plan called for trade crews to progress through the buildings.

The solution the team developed called for disconnecting construction of the building from construction of the apartments with regard to any and all works that were subject to customer changes. This would allow the trade crews to follow a stable path of work through the structural works and the interior finishing of the common spaces (basements, floor lobbies and corridors, etc.). It would also allow the interior apartment finishing works to be performed in the same sequence as the provision of information, independently of the progression from ground floor to top floor. The guiding Lean Construction principle was *pull*: the design information file for each apartment, with complete plans and material delivery information, was to be used as a *kanban* to pull the trade crews to perform the interior finishing work in the apartments, regardless of which floor they were on. The solution extended to two more Lean principles:

1 *Cell production* – ideally, multi-skilled crews could each perform more than one trade's work in an apartment, thus reducing the number of interfaces between crews and the complexity of the process.
2 *Reduced batch size* – by performing work on single apartments rather than on a whole floor of the building as a batch, the buffers of locations would be reduced, and each crew could progress from apartment to apartment without the need to wait for complete information for all of the apartments on a floor.

In theory, these interventions – pull, cell production and reduced batch size – would reduce cycle times, reduce the amounts of work in progress and all but eliminate rework. To test the ideas, the research team designed and implemented a simulation game to simulate the processes before and after the proposed interventions, called the LEAPCON™ (Lean Apartment Construction)[2] game. They subsequently implemented a discrete-event computer simulation to test and validate the outcomes of the game. The simulations showed that the Lean approach could reduce the quantities of resources consumed and project durations, and perhaps most importantly, improve cash flow for the general contractor. Reducing the batch size reduced the cycle times for each apartment, directly reducing the cash flow. Pull all but eliminated rework, because receipt of final design change information was a prerequisite for working in an apartment. Cell production reduced the complexity of the system and eliminated waiting at the interfaces between trades.

Practical Implementation

The theoretical solution was apparently a good one, but implementing it in its entirety would require fundamental changes in design, engineering, management and commercial relations, not to mention complete restructuring of the subcontractor organizations. Screening and sequencing the apartment files according to the relative maturity of their design decisions was straightforward, and Alpha's planning and control department compiled a software application for it. Yet constraints on the way work was structured in the industry at large prevented full implementation.

The major limitation was that multi-skilled crews were very difficult to find, and the few that might have been employed asked for prices much higher than the piecework prices the company obtained from trade-specific crews. Without the ability to reduce the cycle time for individual apartments with multi-skilled crews, the degree of coordination that would be needed to correctly route trade crews through a building in the desired sequence was considered prohibitive when compared with the simplicity of the natural progress from bottom to top of a building. The planning and control department suggested a software application for this too, and built a prototype (see Figure 2.1), but the resistance from the construction management teams proved too difficult to overcome.

Nevertheless, the direction for improvement was apparent from the proposed solution, and the company tested a number of partial steps on a pilot project site. To reduce batch size, crews were directed to work on fewer apartments at a time, releasing work for other crews and shortening the cycle times of apartments. Some of the interior work sequences were changed to reduce the number of interfaces and handovers from trade to trade. Where design changes were expected but information was not yet provided, customer change representatives flagged the information as incomplete.

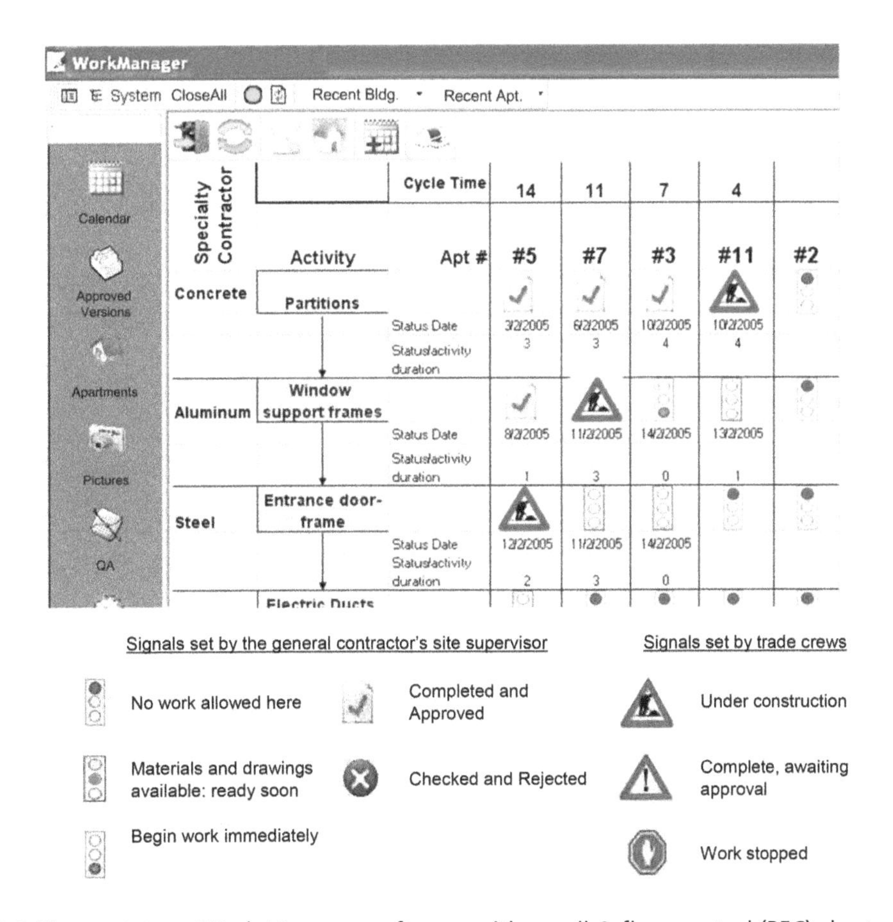

Figure 2.1 The prototype Work Manager software with a pull & flow control (PFC) chart and work package status signals

The results were good enough to capture the attention of the company's CEO. He made the decision to begin rolling out Lean Construction practices across the company, and held a workshop retreat with all of the senior managers to plan the process. It looked as if the VP's initiative was taking root and that major changes would follow.

Alpha's false start

Unfortunately, Alpha's Lean Construction initiative faltered and died when the CEO resigned and left to lead a different construction company. He took with him some of the people he valued most, including the VP of the Planning and Control Department. A senior construction engineer, who had opposed the Lean initiative on the grounds that subcontractor crews could not be trained to work or think any differently than they always had, was appointed to the CEO position. He took the view that the construction industry could not be changed and that Alpha's survival therefore depended on the time-honoured practice of tough price negotiation with subcontractors and coercing them to carry most of the risks of instability in production. With the key Lean leaders gone and a contrary philosophy promoted, Lean Construction thinking disappeared rapidly at all levels.

Tidhar's first encounter with Lean Construction

By 2006, both BIM software and Lean Construction practices had become sufficiently robust (and commercially available, in the case of BIM) for many construction companies to begin using them. Autodesk Inc. had acquired and begun vigorously promoting Revit software, and the Last Planner® System guide had been developed and refined through extensive use within industry, to name just two available tools.

Tidhar learned of Alpha's problems and the failed Lean initiative and began to consider its own situation. It faced many of the same issues, not only with the increasing need to customize apartments, but with the interrupted flow of interior finishing works. The CEO of Tidhar's construction division invited Professor Sacks to speak to a select group of Tidhar project managers and site engineers. The group also played the LEAPCON™ game and discussed its relevance to their everyday experience.

But the company was not ready for the kind of wholesale change needed. Its leadership was open to new technology, and numerous information system initiatives had been initiated and implemented successfully, such as a state-of-the-art in-house purchasing system. Tidhar was generally recognized as a company that had higher overheads than most other construction companies, evidence of a strong perception of the value of management and engineering in construction. Yet the company's business and construction engineering leadership at the time was still stuck in the traditional ways of thinking about the construction industry. Local optimization, ruthless negotiation with subcontractors to secure lowest cost and avoid risk, and a general ignorance of the waste in the processes, were all commonplace.

When introduced to the concept of waste as something more than wasted materials and time spent simply waiting, the initial reaction of Tidhar's project managers and site superintendents was similar to that of many construction professionals around the world. Some typical responses were:

- *The subcontractor's productivity is not our problem – we are paying them per square metre, not per hour. Let them work it out.*
- *Subcontractors are generally unreliable, they tend to promise mountains and deliver much less. It's a behavioural problem and we cannot change that.*
- *We offer them incentives to waste less material and to work faster, but it doesn't work for more than few days at a time. The behaviour is too deeply ingrained.*

These are typical attitudes which obstruct learning and change.

Two cases of BIM adoption in precast concrete construction

Robust and sophisticated BIM tools for precast concrete construction became available around 2005 with the maturing of Tekla's precast concrete design and fabrication software. This tool grew out of the initiative of the Precast Concrete Software Consortium in North America, which funded research and development to ensure that BIM would be available for its member companies and for the industry at large. Tekla's software built on its solid foundation for structural steel design and fabrication, adding a host of functions for precast concrete (e.g. modelling of reinforcement and pre-stressed strands, the ability to model camber and warping of slender sections such as double-tees, and features for production process monitoring and control).

Adoption was slow everywhere because of the deep change from 2D CAD practice, but the experience of two particular precast concrete manufacturing companies, working and competing with one another in the same market, provides some insight into the distinction between success and failure in adopting BIM. Cebus-Rimon Ltd was a precast concrete company producing components for residential housing projects, highway bridge girders, box sections and other products. Ackerstein Ltd was a competitor, with an orientation toward less-sophisticated precast concrete products, such as paving, water and sewerage systems. Like many precast concrete companies, both outsourced their design and shop-drawing services to independent engineering consultants.

In the period from 2005 to 2007, both companies participated in research projects in which they were exposed to the potential benefits of BIM for design and fabrication. They had equivalent opportunity to examine the benefits and the challenges. Yet the results were diametrically opposite: Ackerstein applied BIM to complex construction projects and expanded its market for precast buildings significantly, while Cebus-Rimon continued to work with 2D CAD with its precast building market remaining static. The difference in fortunes can be ascribed to many factors, of course, but in the case of Ackerstein the types of precast buildings in which it has grown its expertise and expanded its market are heavily, if not entirely, dependent on BIM technology. Three examples include:

- A precast parking garage for a large shopping mall required 120 precast beams. Each beam had a different geometry at its ends because the garage was laid out on the side of a hill, with a curved shape in plan and different slopes for ramps. When the general contractor failed to find a precast fabricator who could mobilize the resources for preparing shop drawings for fabrication within the scheduled time (they needed 120 distinct shop drawings), Ackerstein was able to work with its engineering consultant to deliver the shop drawings and the beams themselves, on schedule and without error (see Figure 2.2).
- The company added precast product lines and automated much of the associated BIM fabrication modelling to cater to the need for modular buildings for industrial construction (see Figure 2.3). A variety of buildings for army bases require complex geometry and short construction schedules, making them ideal candidates for modular precast concrete construction.
- Construction of 94 battery-replacement stations for electric vehicles had to be completed within two years to coincide with the launch of the "Better Place" network. Ackerstein built many of them using a unique modular precast concrete design for the structures, allowing the company to assemble two-storey, 500 m^2 battery-replacement stations within 72 hours.[3]

The major difference between the two companies' experience is that Ackerstein chose to invest, together with their engineering consultant, in BIM technology and training, whereas Cebus-Rimon did not. Investment in BIM required not just capital, but the clear vision of what the "future state" of precast concrete with BIM could look like and the

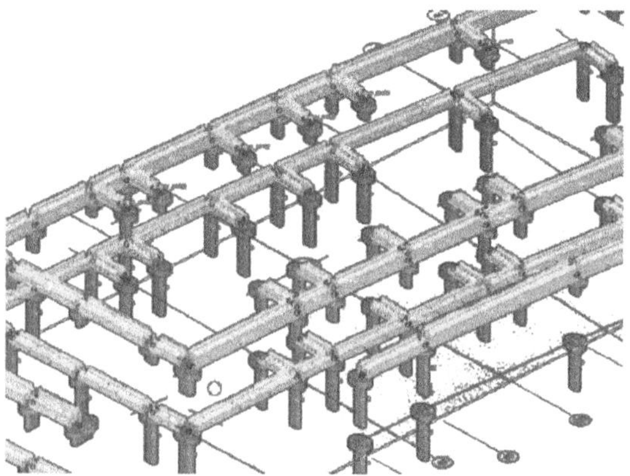

Figure 2.2 A precast parking garage for a large shopping mall laid out on a hillside required some 120 precast beams, each with different geometry at its ends

Source: Images courtesy of Israel Kaner, Star Engineers.

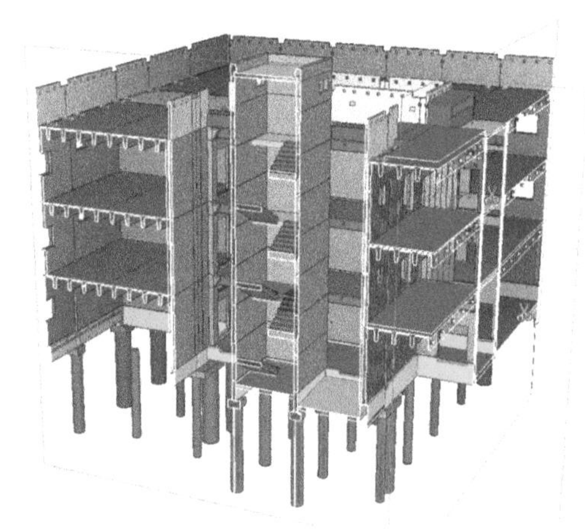

Figure 2.3 Modular precast buildings with complex geometry and short construction schedules

Source: Images courtesy of Israel Kaner, Star Engineers.

leadership to drive the collaboration between the precast fabricator and the engineering consultant despite the inherent risk. The commercial independence of the engineering consultant from the precast concrete company, which is typical wherever such services are outsourced, meant that the leaders of both companies had to make long-term commitments both to one another and to the symbiotic investment in BIM and BIM-dependent fabrication technology. These commitments are risky, and thus require strong intent and mutual trust. It is not difficult to see why the odds were against BIM adoption, and thus why Cebus-Rimon missed the opportunity.

Lesson learned – The critical mass of intent

These three stories – Alpha's false start, Tidhar's first encounter with Lean Construction and Cebus-Rimon's missed opportunity for BIM adoption in precast concrete construction – all have one common thread: *a lack of focused and informed intent among the individuals who have the power to make change*.

In all three cases, the information about Lean and BIM was available. Fairly robust and effective tools were available for both Lean and for BIM. Skills were lacking but opportunities for learning were there. The economic incentive was understood but not internalized, and the cost of change was perceived to be an investment too difficult to make. In each case there were internal pioneers, but the "rank and file" needed a change of culture that only those wielding real power within the companies could bring about. As the Alpha story shows quite clearly by the collapse of the Lean initiative when leadership was withdrawn, leadership of change is key. Alpha lost the leadership; Tidhar's construction division's CEO had neither the power or the conviction needed; and Cebus-Rimon did not have the business foresight or focus needed to change its own practice or that of its engineering consultants. None of them had the critical mass of intent needed for change.

Notes

1 Work in progress, or WIP for short, is the quantity of unfinished products in a production system. In manufacturing, WIP includes all of the partially machined components and partially assembled products that have not reached the end of the line. In construction, WIP includes all of the spaces where work has begun but that are not complete. Construction spaces are formally defined in Chapter 15, "What flows in construction?" In the context of Alpha's residential buildings, WIP was defined as the count of all apartments whose interior finishing works had been begun but not completed.

2 The Technion Lean Apartment Construction simulation game. The game is explained in Chapter 3, "Education and motivation". More information can be found at http://sacks.net.technion.ac.il/research/lean-construction/technion-leapcon-management-simulation-game/.

3 A time-lapse video of the assembly of a complete battery-replacement station within 72 hours can be seen at https://youtu.be/zQrw-junP2s.

3 Education and motivation

Lean and BIM represent new ways of thinking about construction. Adoption requires not only learning new ways of designing and building, but also new ways of thinking about designing and building.

Seeing is believing

When people are fixed in their ways of thinking and doing, finding reasons why change is not possible in their context is often easier than making the leap of imagination needed to envisage a different way of working. Seeing other people working differently reduces the perception of risk and threat. Successful examples provide a starting point for talking about how things could be, an essential precursor to action.

This is true for both BIM and Lean. Statements expressing hesitation, disbelief, fear or other negative attitudes abound:

> "Lean requires collaboration, but our subcontractors are never going to consider the interests of their peers. They're only interested in short-term gain."
>
> "BIM is great for making slick presentations, but it requires far too many resources to be economical in day-to-day production work."
>
> "As contractors, we need to protect ourselves against the client. Building a file of evidence for claims is an essential part of our site management. Open books for IPD?[1] Never."
>
> "We need to protect our investment in compiling Revit families. Make sure we only send PDF or DWG files out of the office."

These sentiments and others like them were commonplace among Tidhar's design consultants, construction managers and subcontractors prior to starting their Lean and BIM learning. They are no different from those echoed in the industry worldwide. Lean and BIM education must go beyond teaching techniques and skills: it needs to reframe people's mindset.

Changing minds, ways of thinking, ways of seeing

Organizations begin Lean journeys for different reasons: the desire to reap the benefits that a successful Lean implementation offers, a fear of losing out to the competition or simply a desire to be the best they can be. Whatever the impetus, the next question is practical: what should we do next? What is the best way to begin implementing Lean? And more importantly, how can we maximize the chances of long-term success by starting off on the right foot? Answering these questions requires some understanding of schools of thought beyond Lean, because the shift from traditional approaches to Lean thinking represents nothing less than a paradigm change.

Paradigms

Paradigms are the frameworks through which we see the world, and they colour the information we receive. Paradigms explain why two people on opposite ends of the political spectrum can read a news article and interpret it in radically different ways – usually reinforcing their a priori beliefs (this is known as confirmation bias, when people tend to construe facts in such a way that they support what they already believe). Paradigms also explain why most of the innovations in technology and science tend to come from upstarts as opposed to established stalwarts; the latter are so invested in the existing paradigm that they are hard-pressed to break out of established thought patterns, a prerequisite for the next leap.

Paradigms help explain the incredulity seasoned professional managers feel when they are first exposed to Lean. Many of the tenets of Lean and many of the achievements of successful Lean implementations fly in the face of what the traditional paradigm considers within the realm of the possible. Consider shrinking batch size: in a construction project with multiple buildings, rather than start building them all at once in parallel, Lean thinking says a more effective way to build would be to build them one at a time in sequence. Instead of pushing to start all the buildings simultaneously, wait on the start of the second and third. In a traditional view, intentionally waiting seems unthinkable since it is construed as a delay. But what Lean thinking realizes is that working in smaller batches allows stability and better flow. If problems arise, they can be discovered early, when the effort to fix them and take corrective measures is still small. By rushing ahead, the traditional approach inadvertently sets itself up for inevitable future delays. For someone mired in this worldview, the suggestion that "slowing down" might enable them to go faster in the long run seems impossible, and any proffered counter-explanation would be rejected as so much wishful thinking.

As the batch size example shows, Lean can sound absurd to someone steeped in traditional thinking. Thus the Lean claims themselves are suspect and often rejected outright. The first step for change leaders is not more telling, but showing, as the goal is not only to educate people, but to change the way they see the world – to change their paradigm.

Paradigm shifts for Lean Construction and for BIM

Within the context of construction, the traditional paradigm for understanding production is "Transformation", which expresses the notion that construction production is performed through individual activities that transform inputs into outputs (raw materials into building components). In 1992, Professor Lauri Koskela published his seminal report *Application of the New Production Philosophy to Construction*, in which he described this paradigm formally and suggested two additional ways of viewing production in construction: the "Flow" paradigm, which emphasizes the flow of work, workers, materials and information, and the "Value" paradigm, which focuses attention on creation of value for the customer.[2]

BIM also represents a paradigm change for construction. BIM replaces symbolic representation of the building in construction drawings, whether drawn by hand or by using CAD software, with an object-oriented parametric model that can be used to drive analytical simulations and fabrication directly. BIM enables builders to create digital prototypes of buildings, allowing testing and refinement of product and process before production, a process referred to as virtual design and construction (VDC). Wherever BIM information can be carried through from design to fabrication directly, as is already the case for a variety of building components, drawings become superfluous.

Change management

The field that deals directly with the question of how to change the way organizations work is called change management. Organizational behaviour scholars have long sought to understand the dynamics at work when companies manage tectonic shifts of direction or culture, and they have developed theories and models to explain typical mechanisms at work during the periods of change.

One well-known model was developed by Harvard Business School's Professor John Kotter. Kotter postulated eight prototypical phases of change (Kotter, 1996)[3]:

1 establishing a sense of urgency;
2 creating the guiding coalition;
3 developing a vision and strategy;
4 communicating the change vision;
5 empowering broad-based action;
6 generating short-term wins;
7 consolidating gains and producing more change;
8 anchoring new approaches in the culture.

Only in step six does implementation begin. All of the preceding steps prepare the organization to be sufficiently receptive to the change when it actually occurs. Before processes can be changed, hearts and minds must be opened; otherwise the new way of doing things is almost guaranteed to fail. The tricky part is that for hearts and minds to be opened, individuals want to see tangible proof. A "catch 22" is created where people are unwilling to change until they see proof, but without them changing there is no proof.

This is why proper education is crucial during the fragile first steps. Lean has its own set of concepts, vocabulary and vantage point into organizational processes, and all of these must be explained and illustrated for people to grasp and learn.

A robust course of instruction is multi-faceted and multi-sensory. Lectures have their place, particularly when it comes to learning jargon. But infrequently is a lecture sufficient. Field trips to visit other organizations that are implementing Lean can be very powerful, since they give the participants a chance to get away from their day-to-day work and see what the system looks like when the parts begin fitting together and creating synergy. These treks can be local or global (for many years a trip to Japan was a badge of honour among Lean enthusiasts, but now with so many stellar examples outside Japan, it is possible to find organizations to visit in almost every part of the world). These visits can be to other companies in the same industry or to vastly different industries. Two of the most famous implementations of Lean in the healthcare industry began with visits to a snow-blower manufacturer and an aircraft manufacturer, respectively.[4]

Beyond the passive receipt of information, active hands-on learning is even more powerful. By being involved, learners get to translate the spoken word into practical results, and by trying their own hand, they learn more deeply than otherwise. Here too there are options. Interactive learning can be off-the-job, where it often takes the form of group discussions to process the messages from lectures or site visits or game-like simulation where many of the Lean principles are demonstrated, or it can be in real life, by attempting a small-scale trial of the principles in the actual work tasks of the company. The five-day "Kaizen Event" format is a popular example of the last of these – a lot of the excitement that is built through these events derives from being involved in the change as it unfolds.

The goal in any educational programme is not only changing people's minds but changing how they see. This means changing their paradigm, since their prevailing paradigm acts as a filter to the incoming stream of information. Changing **what** we see involves changing **how** we see. Many Lean catchphrases touch on this point of making a shift in what and how we perceive: "learning to see" or "seeing through Lean glasses". Opportunities for improvement surround us every day; we need only learn how to identify them instead of passing them blindly by.

Taiichi Ohno, the famous Toyota manager who is credited with developing much of the Toyota Production System that would come to be known as Lean, was said to teach managers to see in the following way: at the beginning of the shift, he would draw a chalk circle in the middle of the factory floor and instruct the student to stand inside the circle for the duration of the shift and observe. After hours of staring at the processes at work, Ohno would come back and ask what the student had learned, and in particular, where the wastes were in the process. This process, though intense, was designed to heighten sensitivity to wastes and to not taking processes in the organization for granted.

While the "Ohno circle" is extreme, learning to see is invaluable when starting a Lean journey. A great way to do this is by combining classroom learning of concepts (like the different types of waste) with forays to the field to search for living examples of those concepts. The lecture sensitizes the participants to what they should be looking for, so that when they revisit the workplace they know so well from being there every day, they are prepared to look at things differently (in this example, searching for examples of waste and areas for improvement).

Many Lean words are borrowed from Japanese, reflecting Lean's heritage at Toyota. One of these is "Gemba". Gemba is the place where value is created for the customer. For a manufacturer, the factory floor. For a hospital, the Gemba is where the patients receive diagnoses and care: consultation and treatment rooms, operating theatres, recovery wards. For a construction company, the Gemba is the work site itself, where the building is being built. The Gemba is of pivotal importance to the company, since the value that is being created therein for the customers is the *raison d'être* of the entire enterprise. Gemba is the source of knowledge and understanding of the true status of the company; it is from Gemba that most opportunities for improvement derive.

Managers have a tendency to become removed from the Gemba, particularly as they rise in the chain of command. Rather than go and observe the issues in context, they rely on second-hand accounts delivered in reports or in conference-room presentations. This distance from the Gemba is a contributing factor to the phenomenon of managers who seem out of touch with reality in their policies and pronouncements; the Lean exhortation "Go to Gemba!" is the countermeasure. As regards education and learning to see, Gemba is the best place to realize the vastness of the waste that exists in any company, but particularly in construction projects. Rather than merely passing through with the typical blind eye and a sense of defeatism, participants who have begun to learn the language and concepts of Lean, when sent to the Gemba, will unearth truckloads of opportunities for improvement. More than any specific improvement idea, the real gain is when eyes are opened to the near-endless possibilities. This is the essence of learning to see.

Sensei

"Sensei" is the Japanese word for "teacher", and in Lean, the sensei acts as guide and educator. Since Lean does entail a shift of paradigms, replete with many roadblocks and frustrations along the way, having someone who has travelled the road before greatly improves the chances of eventual success. James Womack and Daniel Jones,[5] in their book *Lean Thinking* which laid out their description of what Lean was and how to adopt it, urged Lean neophytes to find a mentor who could assist a Lean initiate in their first steps.

At the minimum, a teacher is necessary to provide a basic grounding in the concepts. Even better is someone who can give guidance along the way, particularly if they have experienced the same journey themselves. The goal is not hand-holding but rather gentle suggestion or course correction. The sensei can help with reflection, about what things are working and what could have been done differently, as well as how to improve on the improvement efforts themselves.

Maintaining momentum

As discussed above, an awareness of change management is crucial in shifting "the way things are done" and the way people see. Carefully choosing the right team of early adopters, carefully planning the first improvement projects, carefully crafting the initial messages to the employees are all a part of managing change, and they are critical to making sure that the initial efforts are successful.

Then the real challenge begins: maintaining the initial momentum. People get excited and pumped up about a new initiative, particularly if it shows some quick wins and potential for further benefits. The test is in the longer term, as the organization is challenged to make the new way of thinking and acting a part of the status quo. The difficulty in "continuous improvement" is the "continuous". Rallying the troops to do a big clean-up project of the site is easy; maintaining that same level of cleanliness on a day-to-day basis is much harder.

Ultimately, it comes down to the commitment of the leadership of the organization. Yes, they had the motivation to set out on the journey, but do they have the drive to see it through? Attending the celebratory closing session of the first kaizen event is easy –

but how about the closing session of the fifth, tenth or hundredth? Can the managers change their own work? This is non-trivial, and it is where many well-intentioned Lean implementations stumble. Resistance to change, a proclivity to revert to the old way of how things were always done: these elements will always be present. In order for new habits to take root, they must constantly be reinforced (both positively and negatively, as necessary).

Tidhar's education began with a Lean trek to Japan

Once Gil Geva (Tidhar Group CEO) decided that Lean and BIM were to be the central pillars of the "Growing the Margin" initiative, he realized that people would need to learn not only new skills but also new ways of thinking. The first two steps were to first send Tal Hershkovitz on a Lean study tour to Japan and to thereafter take 15 senior managers on a fact-finding trip to construction companies in Finland and Denmark that had adopted Lean and/or BIM. Later on, formal education was pursued in a Lean boot camp and in study groups.

Tal journeyed to Japan in October 2011, to see and learn about the revolution in productivity known as Lean that began there. The tour group hailed from a diverse range of countries and industries, and it was led by Brad Schmidt, the managing partner and a founding member of Gemba Research (later merged with the Kaizen Institute). Brad grew up in Japan to English-speaking parents, so he is fluent in both English and Japanese, and he utilizes his thorough knowledge of Japanese culture and of Lean practice to help bridge the gaps for his foreign visitors as they learn about the best of Japanese production systems. The tour began with a full day of orientation studies. In addition to visiting a Toyota factory, the group visited eight other companies, among them HOKS Electronics and Daihatsu. Each day concluded with a session to recap the day, process what they had seen and reflect on the lessons learned. Tal described the experience as follows:

> I came back from Japan with two essential ingredients for Tidhar's change: a message and a motive. The message was that the construction industry had to change; the motive was that I was convinced that it was not only possible but that it would also be highly profitable to do so.
>
> The message grew out of an understanding that other industries – automotive, textiles, medical equipment, financial services, etc. – had undergone radical improvements to their productivity, and that only the construction industry had remained oblivious. The aviation and automotive industries, for example, thoroughly adopted 3D parametric solid modelling and Product Lifecycle Management software, yet only a few construction companies adopted BIM – most companies still use 2D CAD. Other industries applied Lean across their internal operations and their supply chains to improve their ability to meet customer needs with less waste, yet only a small number of construction companies in a few countries had begun adopting Lean construction practices. I was convinced that the construction industry was largely anachronistic and had to change.
>
> It was eye-opening to learn that the highly efficient processes we saw on the trip came about as the result of a long series of incremental steps. Each small change was not radical in and of itself, but their eventual combined effect was much greater than

the sum of its parts. While the changes were process-focused, in the organizational systems and the work processes, they had also generated a new culture. What made it all possible was exactly the fact that it was a progression of small steps. Each step was entirely feasible and doable. I knew that we could recreate the same phenomenon in construction, even though it is a relatively conservative industry.

When I came back home and shared my experiences and excitement, people told me that what I learned about manufacturing in Japan was not relevant to construction in Israel. They talked about the differences between product types, the industry structure, work methods, even national culture. Yet I know that at the most basic level there are actually many similarities: inventory is inventory, waste is waste, cycle times are cycle times, buffers are buffers, and safety is safety, across all industries and countries.

One small example: at the HOKS electronics plant on Kyushu Island, we saw an intriguing innovation designed to enhance safety. Machinery operators would forget to put on safety glasses before starting up the machine. The solution? A poke-yoke (mistake-proofing) device [see Figure 3.1]. A worker installed a simple plastic container case right over the "on" switch on each machine that holds a pair of safety glasses. Now, one cannot turn a machine on without first removing the safety glasses from the container; once they're in your hands, it's far more likely that you'll put them on before starting up the machine.

Figure 3.1 A poke-yoke device to ensure use of safety glasses when operating a machine

Note: In this device, a simple plastic container is installed over the "on" switch which holds a pair of safety glasses; one cannot turn the machine on without first removing the safety glasses from their case.

Source: Image courtesy of Tal Hershkovitz.

According to Tal:

> This was a great example of what we in construction need to learn: a mind-set of how to approach problems and find improvements, regardless of the location. The same is true for production flow paradigms such as kanban or CONWIP, and even for BIM. When abstracted as an idea, BIM in construction is essentially the same as 3D parametric modelling in mechanical product design.

Prophetic messages and personal motivations are not enough to change a company. In Tidhar's case the message and motivation that Tal brought back from Japan fell on the fertile ground of Tidhar's culture of pioneering and seeking to excel, and it was nourished by Gil's leadership and the "Growing the Margin" initiative. The message was spread and amplified in management meetings and in staff evenings, and through the initiation of a number of early, small, practical changes.

One of these, implemented very soon after Tal's return, involved remunerating employees at all levels for suggesting improvements. Employees received a token bonus for every serious written improvement suggestion they submitted, a larger amount if the suggestion was judged to have merit by a company improvement committee, and a much larger reward if the suggestion was successfully implemented. The story of this small step and the improvements it generated is told in Chapter 6, "Continuous improvement and respect for people".

Study tour to Finland and Denmark

The second step that Gil Geva took was a study tour in September 2012 with 15 of Tidhar's senior people to see Lean and BIM implementations first-hand in Finland and in Denmark. The team included the senior management, all of the department heads (architectural design, structural engineering, construction engineering, tendering, business development, procurement, information systems) and a number of senior project managers. Rafael Sacks, Vitaliy Priven (a Technion PhD student) and Cheni Kerem, a Lean consultant, accompanied the group.

Gil's explicit goals of the study tour were to show the company's senior managers a set of successful BIM and Lean Construction implementations, to provide them tangible role models and to let them see and learn the state-of-the-art in tools and techniques. In other words, the goal was to show the participants that both Lean and BIM are not only theories but also living, breathing practices that can be successfully implemented. No less important, an implicit goal was to generate a team spirit among the participants, with a positive attitude toward the changes at hand that they could spread within the company. The tour aimed to implement Kotter's second step – "build a guiding coalition" – as well as jump-starting a process of personal paradigm change for each of its participants.

Why Finland and Denmark? Construction companies in Finland and in Denmark had been early adopters of both Lean and BIM. The earliest deep use of 3D modelling technology in the construction industry (what later became known as BIM) was in the structural steel industry, and the earliest and arguably the most sophisticated software for this purpose was Tekla Structures. Tekla, a structural engineering and building construction IT service company, was established in Helsinki in the 1960s. Its 3D steel modelling software became a world leader first for shop drawings and fabrication, then for structural steel engineering, and later for precast concrete and cast-in-place reinforced concrete. The Finnish government had promoted the use of BIM through commercial incentives, funding of research and publication of appropriate standards. World-leading Finnish BIM start-up companies, such as Solibri and Jotne EPM, benefited from the positive climate. Thus the AEC industry in Finland had already been working with BIM for some years before it was called BIM, and by 2012, when Tidhar planned its tour, BIM was in regular use by a number of construction companies.[6]

While Tekla and other companies were promoting the use of BIM in Finland, Finnish academics were prominent in rethinking scheduling and production in construction. This led to development of location-based planning and control for use in non-linear construction projects and was implemented in the DynaControl software (an extension of the DynaRoads software that was common for linear construction).[7] But perhaps of most significance was Professor Lauri Koskela's role in development of Lean Construction as a recognized paradigm for managing design and construction. While on a sabbatical from the Finnish research agency (VTT) and a visiting scholar at CIFE, Stanford University in 1992, he authored the seminal report *Application of the New Production Philosophy to Construction* mentioned above.[8] This work was later consolidated in his doctoral thesis, published by VTT in 2000.[9] The first conference of the International Group for Lean Construction was held in Espoo, Finland, in 1993, and a branch of the Lean Construction Institute was established in Finland in 2008. Finland has long been at the forefront of Lean Construction thinking and practice.

All of these developments contributed to the exposure of people in the Finnish construction industry to the ideas of Lean Construction. The industry adopted many practices. Skanska Finland adopted a range of Lean Construction practices and established a "Lean and BIM" department. The Last Planner® System was made standard on all of its projects, and BIM was utilized for a range of design, construction planning and construction control activities. Skanska's story of Lean and BIM adoption is told in Chapter 5, "The Skanska Finland way".

Denmark has a strong record of Lean Construction adoption with a unique twist – strong support from the Danish Federation of Building, Construction and Wood Workers Unions (known as BAT).[10] Sven Bertelsen, an engineer with a passion for improving the productivity of construction and an appreciation for the importance of flow in construction processes and operations, led the preparation of a government report in 1999 that laid the foundations for the adoption of Lean by two of the three major construction companies (MT Højgaard and NCC) with full support from BAT. The trade union federation had some 125,000 members out of a total labour construction force of some 150,000. The federation recognized that MT Højgaard's implementation of the Last Planner® System on some 30 construction projects by 2002 had resulted in higher wages and enhanced safety. More importantly, it had not reduced employment. BAT was instrumental in supporting the establishment of the Lean Construction Institute of Denmark in 2002 and introduced Lean Construction education into its vocational training courses. The result was a strong record of Lean Construction experience in Denmark, with development of practice and theory.

Despite the differences between the industrial relations in the construction industries of Denmark and Israel – the latter being typical of many countries where construction labour is not organized and labour-only subcontractors are prevalent – Denmark was a natural destination for Tidhar's people to observe Lean Construction in action.

Tidhar's Lean and BIM study tour

The team prepared for the tour with a series of lectures on Lean and BIM with Professor Sacks and his students. The topics covered the fundamentals, introducing the ideas of value and waste and the technological background to BIM. Working through Koskela's "transformation, flow and value" (TFV) theory provided a basis for thinking about how production flows in construction. Exposure to the wide range of architectural and engineering simulations that could be run on a BIM model opened eyes to potential uses. Learning about the economic and cultural contexts of the construction industry in Finland and in Denmark, and discussion of the differences they were likely to observe when compared with their local construction industry, helped set expectations and prepared people to ask intelligent questions.

Table 3.1 The syllabus outline for Tidhar's preparation course for its Lean Construction study tour

Session	*Session content*	*Homework*
	Preliminary reading	Read Hopp and Spearman *Factory Physics*, Chapter 7 (7.1 to 7.4)
1	Principles of Lean production management: value stream; flow production concepts (batch size, WIP, cycle time, throughput, Little's Law); push and pull; flow; waste. Koskela's seven flows and uncertainty game (90 min.)	Read Koskela *Making-Do – the Eighth Category of Waste.* Read Court, Pasquire, Gibb and Bower *Transforming Traditional Construction into a Modern Process of Assembly Using Construction Physics*
2	Variability in production (60 min.) Parade of Trades game (15 min.) Discussion (30 min.) Introductions to Skanska; NCC; Lemminkäinen; Tekla; Enemærke & Petersen; MT Højgaard; E. Pihl: presentations by course participants	Read Tommelein et al. *Parade Game – Impact of Work Flow Variability* Read Howell *The Oops Game: How Much Planning Is Enough*
3	Last Planner® System (60 min.) OOPs game (30 min.) 5S game (30 min.)	Read *Last Planner® Workbook* (pp. 1–34) Read Bertelsen *Construction Physics*
4	Tidhar presentations – review and critique	Teams of two participants are each required to prepare a presentation about one of the host companies. One team prepares a presentation of Tidhar for the hosts

The four-day tour began at Skanska Finland's new headquarters building in Manskun Rasti,[11] Helsinki and ended at E. Pihl & Søn headquarters in Lyngby, Denmark. All in all, the team visited six construction project sites, five construction company headquarters, two BIM and virtual reality display rooms, a university construction management department, a prefabricated wood panel production plant, a construction trade union lecture room and some excellent restaurants.

Figure 3.2 Tidhar senior staff visit Skanska Finland: (above) viewing virtual models at the Skanska BIM and Virtual Reality centre with 3D glasses; and (below) visiting a residential construction site, October 2012

Lessons learned from the study tour to Finland and Denmark

Two weeks after their return from the tour, the participants gathered for an extended debriefing session. Here they shared their impressions, reflecting on what they had seen and how they understood what they had experienced. Their comments revealed not only what they had learned, but also what actions they felt were needed at Tidhar going forward, now that they could see their current practices in a new light. Box 3.1 relates some of that discussion.

Box 3.1 Tidhar Finland and Denmark study tour debriefing session, 24 September 2012

Ilan Nachman, the head of purchasing, began:

> I was most impressed by the extent and detail of the MEP BIM models for all the buildings, from Skanska's HQ, to the Pakkalarinne apartment complex, the Nova Nartis pharmaceutical HQ in Copenhagen, and the Aitio Business Park. I was even more impressed by the impact it had on the construction itself. We spend so much time coordinating the ducts with the cable trays and all the piping on our projects, here it seems to be all sorted out before the installation starts. Modelling can really save time and money on site. When we start the Dawn Tower project next year, we must model all the systems for all 56 storeys – basements, commercial, offices and residential. I'm convinced we can do it.

Ilan continued, describing his changing attitude to BIM through the trip:

> On the first day I was optimistic. Then over the next two days I began to feel that no, there is no way we can achieve this in Tidhar. But on the last day, by the time we got to E. Pihl & Son, I became convinced that not only can we do it, we can do it really well. It can give us some major benefits: accurate quantity take-offs will let us negotiate with subcontractors with more confidence, less uncertainty. When the subs have accurate quantities, they will also be able to give reliable cost estimates, and that will reduce risk in our estimates. We'll save money. We need to set-up a modelling department with clear protocols for modelling all aspects of our activities: property development, contracting, construction planning, interfacing with designers. And since it needs to be done in-house, we'll be building a major competitive asset. Of course, a major barrier to setting this up is a lack of good and highly-skilled people, for us and for our designers and subcontractors.

Amir Putievsky, a senior project manager, pointed to the lack of six-week "look-ahead" planning at Tidhar:

> Our current work quantity forecasts are not detailed enough and we have very poor coordination between site supervisors and subcontractors. We

> don't involve our subs in planning the work in the way we saw. I've instructed my site supervisors at the Carasso project to begin collaborative weekly work planning immediately. I'm not sure exactly how to do it, but with four 22-storey buildings and some 192 apartments in the finishing stages I can see how badly we need it.

This led to discussion of the Last Planner® System, which they had seen in action at the Pakkalarinne apartment complex and at the Nova Nartis pharmaceutical HQ. The consensus was that they needed to train the site supervisors and the subcontractors, and to invest time and effort not only to initiate it but to sustain it.

Zohar Raz, head of the design management group, praised the study tour highly. He felt that it hadn't given him any new information per se, but it had sharpened his understanding:

> I feel that the main thing the trip achieved was to put us all, the company's leadership team, on the same page with respect to our partnership in the Lean/BIM effort. It really helped me to explain to others in the company something I've been trying to explain for some time: that BIM requires a serious investment, but it really improves the quality of construction. It touches everyone in the company (except maybe the accountants). I was really influenced by their approach to Lean design and the way the BIM modelling is managed for flow and value. When we started CU [a large commercial and office development] two months ago, we weren't aware of how much BIM could help but we started modelling the MEP systems anyway. Now I see that it has to be much more carefully managed, and we have to catch up fast.

Tal Hershkovitz, the CEO of the construction division, showed everyone his notebook from the trip. He had a system for marking up his own ideas, drawing stars next to the good ones. This time, he said, his notes were full of "double star" ideas":

> Here are some things we can do in the short term (i.e. right away):
>
> - Start using visual management in site offices – safety posters, making things visible, signposting, using notice boards to show progress to everyone, etc.
> - Set up a training course for Last Planner®, write a company guide for implementing it, and do it.
> - Prepare and plan the modelling procedures for the Dawn Tower project. We need to plan the design stage thoroughly.
> - 3D is not our goal. Our goals must be 4D first and then 5D.

Nadav Galai, the VP for business development, spoke about his surprise and delight at NCC's Aitio Business Park project. They had been working for just six months, and the eight-storey building was almost complete. The elevators were installed, the site was clean and organized, workers seemed to follow safety rules by the book, quality was good and they were already planning to dismantle the tower crane. Perhaps, he surmised, this rapid progress was partly a result of the fact that the BIM model was completely detailed and thoroughly coordinated at least three months before work began.

Finally, Gil, the company CEO, summed up his impressions. From his perspective, the main achievement was that the team now shared a common story and a common vision. He put it this way:

> Before we went, I was frustrated because I could see the potential future state of the company, but I couldn't explain it in a way that would get people behind the idea and the changes needed to achieve it. Now I feel that there is a core group of 15 people who share the vision, and that's a force that can change the whole organization of 400 people. I understand how frustrated Rafael must have felt, knowing for years how building design and construction could be improved, but without partners in the Israeli construction industry. Over the course of the past two years, I've slowly begun to feel a similar frustration, but now I feel like we have new partners with whom we **can** make a difference.

He went on to lay out some of the next steps, as he saw them:

- Increased attention and effort, including bigger management teams, for specifying and standardizing Tidhar's main product, the "Tidhar Apartment".
- Build up BIM modelling capability, through the steps of 3D modelling, then 4D and all the way to 5D.
- Design and planning of the Dawn Tower project using a "big room" set-up with workstations for all the architects, engineers, consultants and Tidhar's own construction planners at the project site.
- Initiate a company-wide training program, "Tidhar School", with support from academia, and encourage graduate students to pursue research in collaboration with us.

Gil emphasized that he would make the necessary resources – staff, equipment, technology, office space, etc. – available. *"I expect all of you, and for that matter anyone else in the company, to approach me for resources where needed, to ask my help in motivating design-service providers and suppliers, to provide technology: whatever is reasonably needed to promote Lean and BIM."*

Toward the end of the meeting, Guy Frumer, the manager of the contract bidding department, felt he had to challenge everyone:

> You guys are hypocrites. Look at the preparation for this meeting. Instructions for us to prepare the A3 page analysis of our lessons learned were

distributed too late, so only the most pious among us prepared them. The meeting itself has been run quite differently from what had been planned, and that's symptomatic of everything we do. We still have the mentality of the "division commander on the hill", continually improvising and adapting our actions to the changes we're faced with, rather than the mentality of a production planner.

Just look at what's happening at the Dawn Tower right now: we have design holds on two major slurry walls and on the façades, but we're already excavating. There's no replacement for solid management, and neither BIM nor Lean can fix everything.

Tidhar's Lean boot camp

As the next step in their journey, Tidhar's management decided to undertake a six-day training programme in the form of a "Lean boot camp" for managers from every level. The goal was to have an intense introduction to Lean to jump-start the Lean implementation. The highlight of the programme was two full days of work in the Gemba, during which the managers had to work as professional trade workers. By getting out of their offices to where value was actually being created for their customers, the managers would gain new insights into concepts like value and waste.

What is a Lean boot camp?

A Lean Leadership boot camp is a special training programme designed to transform traditional leadership behaviours into behaviours centred on Lean principles.[12] While many Lean initiatives are based on adopting the right tools, the purpose of the Lean Leadership boot camp is to train leaders in the organization in the deeper level of principles, allowing them to better implement the right tools, and more importantly, coach others in the principles and the use of the tools. The workshop is based on a mix of lectures, discussions, simulation games and on-site experience to ensure a profound understanding of the principles. In order to maintain the positive momentum generated by the workshop, each participant is required to make his own improvement plan laying out a project he will undertake following the workshop, with detailed milestones to be achieved in 30, 60 and 90 days. Follow-up meetings are set in advance to ensure that the full plan-do-check-act (PDCA) cycle is carried out.

Boot camp programme

The boot camp programme consisted of six days in total, spread over the course of three weeks (two days each week). The six days were divided as follows:

- The first two days were dedicated to learning Lean principles such as waste and the Lean paradigms. During these days, to transition from classroom lectures into hands-on learning, the participants took part in the LEAPCON™ simulation (see Box 3.2). At the end of the first day, a Gemba walk was scheduled, in which the group visited one of Tidhar's construction sites to see if they could find real-life examples of what they had been learning about in the classroom.

- The second week was dedicated entirely to working at the Gemba. Each of the managers was out in the field, working shoulder to shoulder with the front-line workforce. Examples of the assigned trades: formwork, rebar fixing, plastering, tiling, warehouse, electrical and plumbing finishes.
- The last two days were divided between learning a variety of Lean tools and preparing personal improvement work plans to implement them. Among the tools were 5S, SOE (sequence of events), Last Planner® System, visual management and Takt. To practice the SOE tool, the participants again went to the Gemba, where they were divided into three groups. Each group prepared detailed SOEs in order to begin understanding the work as it was performed so that they could develop standardized work sequences.

To ensure the full attention of all participants and to make sure they were disconnected from their daily project management chores, the classroom portion of the boot camp was held in a conference room away from the projects and the company headquarters. All participants had to deposit their computers and smartphones in a box for the duration of the programme (except during breaks and lunch).

Box 3.2 The LEAPCON™ Simulation Game

The Technion "Lean Apartment Construction" (LEAPCON™) simulation game demonstrates the impacts of changes to a production system according to Lean Construction principles on the outcomes of execution of the interior finishing works in apartments with client-specific customization in multi-storey residential buildings. It is a live simulation game in which participants play the roles of subcontractor trade crews, general contractor (GC) staff and clients, and build the apartments using Lego® bricks. The game is played in two rounds – first using a traditional construction management approach and second using a Lean-inspired production system design.

The game focuses on the variability in timing and content of client design changes and its impact on the flow of production that the trade crews experience. The phenomenon of apartment buyers requesting tailored customizations of their apartments – custom kitchens, bathrooms, partition walls, HVAC systems, flooring and other decorative options, etc. – is common throughout the world. These changes introduce uncertainty and disrupt the regular flow of work because general contractors find it difficult to control the clients. The timing and content of design changes is unpredictable.

This scenario is ripe for Lean rethinking, particularly in terms of the timing of the way in which the work is structured and scheduled. The LEAPCON™ simulation illustrates three interventions: pull, single-unit batching and work restructuring in multi-tasking teams. Each can be tried separately, but most facilitators suffice with two rounds – one using a traditional approach to work planning and control, and a second that applies all three interventions together.

In the simulation, four players represent the trades required to execute the finishing works in each of 32 apartments, four on each of eight floors of an imaginary building. The eight floors are represented by drawings that are taped to tables spread around the room. Initially, the apartments are all to be built according to a standard design which has four work packages: flooring, partitions, HVAC ducts and false ceilings. The work packages are modelled as four steps in building an apartment from Lego® bricks, as illustrated below. However, once play begins, a player who represents the clients randomly draws apartment numbers and change orders at fixed time intervals, and delivers them to the project manager so that she can instruct the trades where to go next and what to do there. The design changes are minor, but they often require a trade to return to an apartment where their work was previously completed, which introduces the wastes of rework and waiting (because such events disrupt the flow of the other trades). Once apartments are complete, a second player representing the clients is called to receive the completed apartments, and records the time of delivery and any quality defects (deviations from plan).

Figure 3.3 Lego models representing apartment interior finishing works in the LEAPCON™ game

Each round is limited to 11 minutes, which would be marginally possible if each player spent 15 seconds on each apartment and flow was perfectly stable. In reality, players usually manage some 8–10 apartments using a traditional "bottom-up" sequence of flow of the trades within the building (i.e. starting on the first floor and working their way up to the eighth floor, with each trade exclusively occupying a whole floor during its work on that floor). Play is lively, with the GC's staff struggling to optimize the sequence of work for the trades as design changes are delivered.

Results are computed at the end of the round. With $1,500 paid for every apartment delivered and $1,000 charged to the GC for every apartment on which work was begun – even if not completed – the GC usually has a negative cash flow in the order of $9,000 at the end of the first round. This reflects accumulation of the WIP inventory that is not controlled, as the flooring trade is able to progress more quickly than the other trades, generating a stock of unfinished apartments that the other trades cannot complete because the following trade, partitions, has a slower work rate and is affected more by the design changes. The partitions builder is the bottleneck in the process.

Between rounds, participants discuss the process they have experienced. The facilitator raises a range of questions: what was the project management team's experience of the game? Were they kept busy? Did they feel that they were in control of what was being done? Were they able to keep to the predetermined construction plan? Where was the bottleneck in the production system? Why are cycle times long and WIP high? And, finally, do the players have any suggestions for improving the process?

In the second round, the basic ground rules are maintained, but the Lean interventions are introduced as changes in the work plan and procedures:

- Batch sizes are reduced to single piece flow by allowing each subsequent trade to move into apartments as soon as the preceding trade is done, instead of waiting for the preceding to finish the entire floor. The effective batch size of four apartments that was implicit in the practice of completing whole floors – a condition imposed in real projects by trades that try to optimize their productivity by accumulating big batches of work – is reduced to a batch size of one.
- Pull is implemented by sequencing the work to follow the sequence of delivery of the design change information. This is possible because the structure of the building is assumed to be complete before the finishing works begin. Pull means that apartments are worked on only when their preconditions are complete – in this case, the design information is final and not subject to change. The effect is that rework is no longer needed.
- Cell production, or multi-skilling, is simulated by allowing each of the four trade crews to perform all of the four trades' work types. Effectively, this means that each worker can start and complete all of the work for a single apartment. This balances the workload and removes most of the waste of waiting.

The result in almost every case is that the same four pairs of hands manage to produce almost twice as much value in the same period of time. On average, teams compete 16 apartments and have a positive cash flow of $6,000. WIP is reduced to a maximum of four, because the system is now a CONWIP (controlled work in progress) system (trades complete each apartment before commencing another, which limits WIP to the number of trades).

Unfortunately, it is not possible to distinguish the separate influence of each of these interventions with the two rounds of the game described above. However, an enterprising facilitator can elect to introduce an additional round, in which one of the changes is left out. For example, if participants express the view that multi-skilling is impractical, an additional round can be played in which subcontractors return to their specializations, but the sequence of work is still pulled by the clients' change orders and single piece flow is maintained. Another alternative is to allow the HVAC trade to support the partitions trade, which evens out the workload.

A deeper understanding of the way the simulation works is afforded by a discrete-event computer simulation of the game prepared by Alberto Esquenazi, a former graduate student at the Technion. Using the computer simulation, a range of permutations of interventions could be tested and the game could be run through to completion of all of the apartments.[13] It turns out that reducing the batch size and pull are the main sources of improvement – the batch size reduction is the main contributor to better cash flow, and pull reduces the wastes of rework and waiting.

The LEAPCON™ game was inspired by the "Airplane Game" simulation, a well-known Lean educational tool. The adaptation to the construction industry and the focus on client changes provide a powerful hands-on educational tool for engineers and managers to learn the concepts of Lean thinking within the context of their own industry. It has been used at universities and companies in 36 countries.[14]

Tidhar used the game in their Lean boot camp numerous times. At one point, a group of senior managers asked to try the game in a third round. In the spirit of continuous improvement, they wanted to see how far they could go in the 11 minutes allowed by introducing additional improvements. They implemented three elements of the 5S methodology that they had learned in the boot camp lessons: *Sort* – removing all extraneous objects from the playing area; *Standardize* – each trade worker specialized in a particular type of design option; *Set in order* – pre-arranging the Lego® blocks in piles of the right size in appropriate locations around the play area. Naturally, they also benefited from the learning curve, given that this was their third round. The result was that they managed to complete 28 apartments, an all-time record for the LEAPCON™ game.

The three rounds were recorded on video and made available within the company. One of the construction engineers, who was not a member of this group, wrote an email to the boot camp facilitators:

> Well done! It seems that once the planning is thorough and the production system is well designed, there is no need for a "superstar" project manager who knows how to improvise and solve problems on the fly. The team has learned to plan slowly in order to build quickly. Imagine what we could achieve if we designed our systems and planned our work thoroughly for the kitchen fit-out, the door and window installations, the marble window sills and surfaces, and the other finishing trades. Please show the videos of the first and third rounds to Gil [Gil Geva, Tidhar's CEO], they are great examples of the "before" and "after" of the learning process. The major difference that one sees right away is how much less shouting and improvisation there is in the third round. It's a real paradigm change from the way we select project managers these days – we look for superstars who are expert at fire-fighting; we should be looking for planners and thinkers.

Learning to see wastes

Lean thinking is different from other operational excellence methodologies in that the focus is on creating more value for the customer by eliminating waste. For most managers this is not trivial, since they are used to focusing their improvement efforts on the value-adding activities, and not on the wastes inherent in the support activities that are done between the value-adding activities. In order to change the boot camp participants' points of view, they were first introduced to the notion that in most activities only 10 per cent of the resources are consumed in value-adding work; the rest (90 per cent) is waste. The participants then learned about the different types of waste that Lean thinking has identified (covered at length in Chapter 8, "The waste of non-value-adding work"). In small groups, the participants were charged with identifying examples for each of the types of waste from their daily work process. To further cement their understanding of the concepts, the first Gemba walk was dedicated to finding examples of each of the types of waste in the actual work being carried out on site. The visit concluded in a group meeting to share the examples they had gathered and summarize the findings and lessons learned. Although the participants were highly trained professionals who had lived and breathed construction over the course of decades, most of them only realized the huge amounts of waste inherent in the work processes during the Gemba walk. To put this in terms of the concepts discussed earlier in the chapter, they were sensitized during the boot camp training sessions to the Lean paradigm, and this gave them the ability to take a fresh look at the work environment they knew so well. They were able to begin seeing the wastes.

Working in the Gemba

Bringing ten managers to work in the Gemba required a great deal of preparation, both on the part of the participants themselves and by the workers and the managers who hosted them. The preparation of the participants included sessions on what mindset they should approach the experience with from a Lean point of view and technical classes explaining the type of work each of them needed to perform. The emphasis was on the request that they act as a simple worker and try not to be judgemental during the two working days. Each boot camp participant received the name of his Gemba manager, the name of the worker he was assigned to work with and the working hours for the particular crew.

To prepare the hosting managers, a one-hour meeting was conducted for the management team of each project that the boot camp participants would visit. During the meeting, the purpose of the boot camp was explained, as were some basic Lean principles regarding giving respect and continuous improvement (not all of the hosting project managers and site superintendents were familiar with the Lean methodology and principles). The hosting site superintendents and the crew leaders were asked to treat the boot camp participants as regular workers with no special privileges. They were instructed to give the participants a safety briefing, sign them up for special equipment as necessary to their trade, and explain the daily site routines and the actual task they needed to perform. The hosting workers were told that the participants were going to work with them as an extra pair of hands. They were asked to teach them the basic skills required for the job and supervise them during the work execution. The workers were also instructed not to give the boot camp participants any special treatment during the time they spent with them. They were to do everything the crew did: the work itself, breaks and clean-up of their workplace at the end of the day.

To understand the site routine and to learn as many details about the work as possible, the participants spent two full days with the same crew doing the same work. They were instructed not to take any notes during the day (though they were encouraged to do so during breaks), so the working crews would not feel as if they were under observation. The participants were also asked not to give any suggestions or try to change the way the crews executed their work.

At the end of the second day of working in the Gemba, the group held a three-hour summary meeting during which each participant shared his experience and presented two or three case studies regarding wastes. They used different visual aids, such as actual tools, materials, charts and photographs. The case studies did not necessarily have a direct connection to the presenter's regular daily work; the idea was to share the observations and findings around the group so others could decide whether to include them in their personal work plans. All of this was geared to begin teaching the participants to "see". Some of the examples of the observations will be discussed in Chapter 9, "Learning to see".

Personal experience and learning

During the boot camp summary meeting, each participant shared his personal experience throughout the camp. Here are a few quotes:

> "The atmosphere and the dialog contributed to the learning experience and allowed me to gain a deeper understanding of what Lean Construction is and how it can help us improve."
>
> "I think the programme was amazing. I enjoyed it very much. We came, listened to lectures and put on different 'glasses' only to see, a week later, the things that have always been right in front of us in a different light."
>
> "You realize that if you're not in the field, things will never come up. It's part of respecting people; if you're in the Gemba, people will relate to you and follow your leadership."
>
> "The boot camp programme generated a buzz throughout the entire organization."
>
> "It was powerful and overwhelming, and the reverberations were felt throughout the entire company like the ripples from a stone thrown into a puddle. The two days we spent in the field were a total blast."
>
> "I had been hearing 'Lean this' and 'Lean that' for a long time, but only today can I say that I'm starting to understand the meaning of Lean."

Modern management revolves heavily around IT dashboards, reports and charts. This results in a decision-making process that is usually performed far away from the place where the actual work is being carried out. These decision-making processes often lack the proper information needed for a fully informed decision, information that can only be found in the Gemba. The boot camp experience showed that for construction professionals, going to the Gemba is a powerful concept and should be a standard management practice. None of the wastes observed during the time spent in the Gemba had previously been reported, and they would have remained unidentified if not for the Gemba session.

The most powerful observation is that the true magnitude of wastes is likely much greater than just the examples that surfaced during the boot camp, discovered over the course of two days in the smattering of assigned locations.

As a practice, managers should train themselves and the people that they work with to go to the Gemba to see, ask why, and show respect. The capability to go to the Gemba, to look at a process and to ask the right questions that will make the worker or the foreman think about the problem in a new way, is a genuine skill that can only be developed over time and with much practical experience.

As noted above, the two days spent by the participants of the boot camp in the Gemba had a ripple effect throughout the company. The fact that managers were "really" interested in the actual work, and were willing to invest their time in order to learn about it, made many people in the company feel that their work was important. Workers and foremen, who were used to working alone, felt empowered by the realization that senior managers were asking sincere questions about how they did their work.

In summing up their experience at the boot camp, several managers said that even though they had heard and learned about Lean in the past, they had not been able to make the cognitive connection between Lean and their own daily work. Only after learning to see through the training in the boot camp did they understand the meaning and full potential of continuous improvement and eliminating waste.

Personal projects

At the end of the boot camp, each participant was asked to prepare an A3 report to define a project that he or she could undertake personally to improve some aspect of their own work, implementing Lean tools based on the principles they had learned. Each improvement plan was structured using the A3 method, including choosing a topic for improvement, going to the Gemba to see and learn, performing a root-cause analysis, testing possible countermeasures and setting up follow-up checkups to ascertain whether the countermeasures had indeed addressed the root cause that had been identified. Each participant had to choose the right Lean tool for investigating the current condition and implementing a new process. Among the tools were value-stream mapping (VSM), SOE (sequence of events), the Last Planner® System, 5S (sort, set in order, shine, standardize and sustain) and others.

After 90 days, they were asked to report back as part of the boot camp debriefing, using a standardized form which asked six basic questions:

- What was the problem you confronted?
- What was the waste involved?
- What was your process improvement idea?
- How did you implement it?
- What did you achieve?
- What did you learn from the process?

This fixed format provided a structure from which to capture the learning. A wide variety of problems and potential solutions were presented by the meeting participants, such as:

- removal of waste and rework in the formwork design process;
- education of work supervisors for independent action and decision-making by the project manager;
- an improved process for managing the repair and supply of electric hand tools;
- reducing waste of formwork tie "butterfly" hooks;
- visual management on site.

All participants were very satisfied, both with the process improvement efforts and with the way the meeting was run. However, when the participants enquired whether or not there would be a structured continuation process of Lean coaching, and whether or not the review meetings would continue, the organizers informed them that no follow-up had been planned. Instead, their hope was that since the boot camp participants had been given the skills to improve and there was already enthusiasm for Lean within the company at large, the individuals would take the initiative and carry out more improvement efforts on their own. However, in retrospect, it is possible to see that without any sort of formal framework, the day-to-day pressures of running the company overcame the momentum that the boot camp had created. This lesson can be seen at other companies as well: without management follow-through (in creating the frameworks etc.), sustained improvement efforts (even those begun with as much enthusiasm as the boot camp created) are likely to fail.

BIM training

The trip to Finland and Denmark, in which the management team was able to see how BIM operated as a crucial element of the host companies' day-to-day operations, proved pivotal for Tidhar. The scales were tipped in favour of rolling out BIM at Tidhar; seeing implementations by companies not that different from their own gave them the confidence that they were capable of implementing it and a clear vision of the benefits. Upon returning to Israel, it was decided that like for Lean, a formal educational programme was needed to teach Tidhar's employees and suppliers what they would need to know about BIM in order to adapt to a new way of doing business. Ronen Barak, in his role of R&D director with responsibility for Lean and BIM, along with the BIM "model managers", ran a number of training sessions in which various functions participated: design managers, client change designers, engineering consultants, structural team members and others.

Reading *The Toyota Way*

In April 2014, Ronen Barak felt that the Lean journey was getting off track. He was upset at the slow pace of the implementation, and he was disappointed by the lacklustre results. The Lean journey had been underway since late 2011, but the organization was not where he felt it could be. As contrasted with the big changes that had been undertaken in bringing BIM to the company, there hadn't been parallel Lean achievements. For example, the attempt to make a fundamental change in how quickly the structural works progressed (improving the traditional "four storeys a month" to a new record of eight a month) had failed. The shortcomings included:

- not carrying initiatives through to their conclusion;
- not using Lean tools on an ongoing basis, including 5S, SOE etc.;
- no "kaizen events" (rapid improvement workshops);
- not performing enough VSM of the current state of chosen processes, as well as not creating a vision for the future state of the process under study;
- not carrying out the annual plan for Lean activities, which had included many of the preceding items.

Ronen prepared a presentation for the "Growing the Margin" steering committee, in which he outlined his observations. In particular, he pointed out that there were not enough people involved in the Lean efforts beyond the few select managers who were part of the steering committee. Even the people who had participated in the study tour to Finland had not been sufficiently engaged in the day-to-day Lean activities. He felt that he needed people who would help him "make things happen" in terms of Lean, who would help carry out the annual Lean plan, implement the tools, help build the culture, maintain the improvement processes in the long term and in general be supportive of the efforts. In the ensuing discussion, the steering committee came to the realization that what was lacking was sufficient knowledge about Lean. As a result of the meeting, Gil, Tidhar Group CEO, decided that the best course of action would be to use the forum of the monthly management meetings as a study group for furthering the organization's Lean knowledge, with the goal of changing the mindset of these managers who were largely responsible for how the company functioned.

To that end, Tidhar purchased copies of Jeffrey Liker's book *The Toyota Way* for each member of the management team. In the book, Professor Liker outlines 14 principles that underpin Toyota's approach to business, starting with philosophy, then covering process, people and problem solving.

Each one of the principles is discussed in a separate chapter. At Tidhar, each chapter was assigned to a different manager. Over the course of 14 weekly meetings, each principle was presented by the manager who had been assigned to prepare it. Putting his money where his mouth was, Gil Geva undertook to prepare and to present the first chapter, advocating the first principle, which states: "*Base your management decisions on a long-term philosophy, even at the expense of short-term financial goals.*"

Each presentation focused on three topics:

- An explanation of the principle, as it was written in the chapter. Here, the presenter was urged not to judge or add any personal slant to the material, but focus instead on teaching the other participants what Toyota had done.
- Three concrete examples from Tidhar's experience in which they had acted in accordance with the principle (whether or not they were conscious of their actions being in line with the principle).

- Three concrete examples from the company in which they had acted in ways that were at odds with the principle. In addition, the manager had to explain how they could have acted in each scenario in order to not contradict the Toyota principles (how things could have been done differently).

This process of presenting the principles and identifying examples was one of learning coupled with reflection. In Lean, reflection (or *hansei*, as it is called in Japanese) is strongly associated with the "check" stage in the PDCA cycle. Only by looking back and being self-aware of the actions after the fact can we generate the self-knowledge that is sufficient to influence behaviour going forward. Identifying examples in which they had worked against the principles was not done to chastise or punish, but rather to try to understand what forces were at work that had lead the processes astray. On the flip side, finding the positive examples where they had worked in harmony with the Lean/Toyota principles helped to highlight ways in which the prevailing culture and practices at Tidhar (as a company that had always prided itself in its focus on excellence and had taken time to develop clear statements of its vision and its values) were already aligned with Lean. Here, by "putting a point on it" and explicitly recognizing the examples, the goal was to take pride where it was due, which can help in the long term to continue modelling the desired behaviour.

For example, in the presentation of Principle 5 (*Build a culture of stopping to fix problems, to get quality right the first time*), Chaim Kirshon, the head of construction engineering, brought up the example of the Aviv Gardens project in Lod (mentioned in in Chapter 1, "Introduction – Growing the Margin"). For one of the high-rise apartment buildings, it was discovered that the sewer line had been improperly designed. Rather than push onwards, the decision was made to stop the project until the design team could sort the problem out and develop a countermeasure. Only then did work resume. Despite the fact that fixing the problem was very hard, the decision proved to be a good one, when compared to another parallel building that had a similar problem but did not stop their work. Ultimately, the rework that was required to undo the error caused this second project to take much longer than the one in which they had indeed stopped and fixed the problem.

For an example of a case in which Tidhar had not worked in accordance with this principle, Chaim brought up the all-too-common scenario in which subcontractors kept on working, even if the next work area was not fully completed by the preceding tradesmen. Since they are paid when work is done, they have a strong impetus to finish whatever and wherever they can get some portion of the work done, as quickly as possible. "Quality" often takes a back seat. Tidhar is not capable of effectively stopping the work in this situation. In order to address this issue, Chaim suggested that the payment terms and incentives would need to be changed. Another tack would be to try to develop a culture of "working on the same team", despite the fact that each subcontractor has his own groups of workers that have no contractual connection to the others. And in fact much of the Last Planner® System, which will be discussed in Chapter 17, is geared to just this end: helping the subcontractors organize as a team with a common goal, seeing how each part interacts with the whole of the project.

The reactions of the participants were mostly positive. Some of the managers reached out to Ronen and asked to meet with him one-on-one to get further explanations of the material in the book. Ronen relates that the conversations that grew out of these sessions were very thought provoking. One of the managers invited Ronen to come to his department and give a presentation to his subordinates about Lean. This lead to a "mini-kaizen event" in the finance department on invoice authorizations, and a visit of the finance department to one of the construction sites to see the Lean efforts first-hand. Gil, the CEO of the group, decided to do a VSM of the business development process, which had far-reaching impacts on the company, as a result of *The Toyota Way* sessions.

The last presentation to the management took place in November 2014. The legacy in Tidhar is a greater familiarity with the concepts, to the point at which they have entered the daily vernacular. At a recent all-hands meeting of the company, the presentations were focused around themes like respect, value, waste and the customer. These are all themes at the core of Lean thinking.

Lessons learned

Education is a key component of any successful Lean journey. Lean involves many new concepts and behaviours, which often go against the years of experience and training workers and managers have gained. This means that their very paradigms must change in order for Lean to succeed, which can only take place in an active and mindful Lean education programme.

Tidhar's multi-phased educational journey is a good example of a layered approach that included different participants, learning styles (from the experiential to the international) and messages crafted for each stage of their journey. Much of the groundwork for Tidhar's successes with Lean was set during this first crucial step. A one-time lecture would not have done the trick.

This touches on another important point, which is particularly relevant at the beginning stages of Lean implementation: people's innate tendencies are often to settle back into their own habits, even after intense experiences like the trip to Finland and Denmark or the boot camp. In order to maintain the momentum, the organization's management must make sustained and active efforts to keep Lean at the fore. A good example of this is with the follow-up after the boot camp; there wasn't really any. As a result, improvement projects got pushed to the back burner, since no one in the senior organizational hierarchy made a point of checking in about their progress and no one rewarded successes.

Notes

1 IPD stands for "integrated project delivery". IPD is a construction contracting approach that calls for all main parties – owner, designers and builders – to collaborate closely from the start of a project within a contractual arrangement that closely aligns their commercial interests. This includes gain- and pain-sharing, open books, co-located design and construction teams and extensive use of BIM.
2 Koskela (1992).
3 Kotter (1996).
4 Hoeft and Pryor (2016).
5 Womack and Jones (2003).
6 Notable among them were Skanska, NCC and Lemminkäinen.
7 This software was later developed into Graphisoft Control and was the basis for VICO (Virtual

Construction), which merges BIM and location-based management and control. VICO is now a part of Trimble.

8 Koskela (1992).

9 Koskela (2000).

10 An excellent review of the role of the Danish Construction Trade Union (BAT) in the adoption of Lean Construction can be found in Chapter 7 of "Collaboration on industrial change in construction" by Christian Koch (2007).

11 Skanska's headquarters building in Manskun Rasti, Helsinki, is a monument to Lean and BIM implementation in construction. The story is told in Chapter 5, "The Skanska Finland way".

12 Some of the material describing Tidhar's boot camp was presented to the International Group for Lean Construction annual conference at Fortaleza, Brazil, in July 2013. The authors are indebted to Cheni Kerem and Vitaliy Priven, two of the authors of the original paper (accessible via http://iglc.net/Papers/Details/917), for their insights and their permission to reprint material from that paper.

13 Sacks et al. (2007).

14 For more details see http://sacks.net.technion.ac.il/research/lean-construction/technion-leap con-management-simulation-game/.

4 The Dawn Tower

Project profile

Figure 4.1 View of the Dawn Tower. The photograph was taken a short time before completion of the structural works

Project description

The Dawn Tower project, located in Giva'atayim, Israel, was developed by Tidhar Ltd and a private investment group (Sufrin Ltd) whose members were the individual owners of apartments or offices in the building. The group was formed to collectively purchase the property and was directly involved in the whole life cycle of project development. The tower is part of a new urban precinct with another two residential buildings (136 units) and an underground access tunnel.

The tower has 57 storeys of mixed use areas:

- commercial (2,300 m^2);
- offices (49,000 m^2);
- residential (156 units);
- five levels of underground parking (1,250 parking spaces).

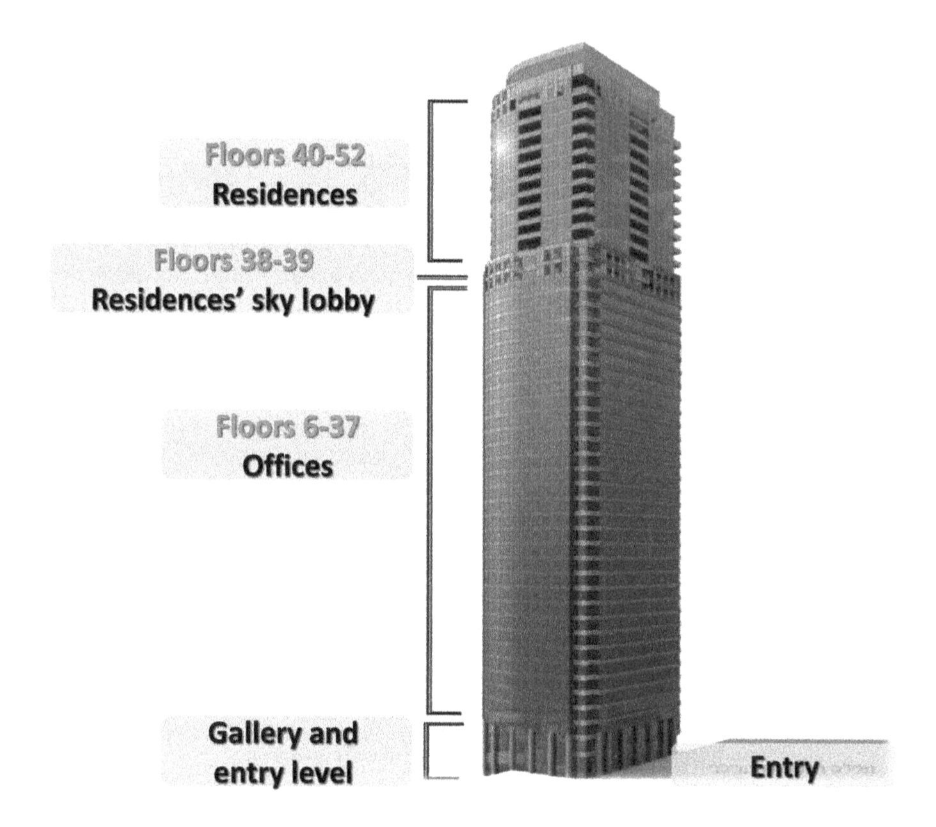

Figure 4.2 Tower layout

Construction methods

The tower was built of reinforced concrete with self-climbing forms for the core, a self-climbing concrete pump boom and three luffing jib tower cranes. Columns and beams were poured in situ, and the slabs were composed of hollow-core precast panels. The foundation was a concrete raft and a 1.2 metre thick transition slab was used to change the structural grid at the interface between the office tower and the residential tower above it.

Tidhar's project team included a project manager, three zone managers, four construction engineers, seven site superintendents, a building systems engineer and a logistics supervisor.

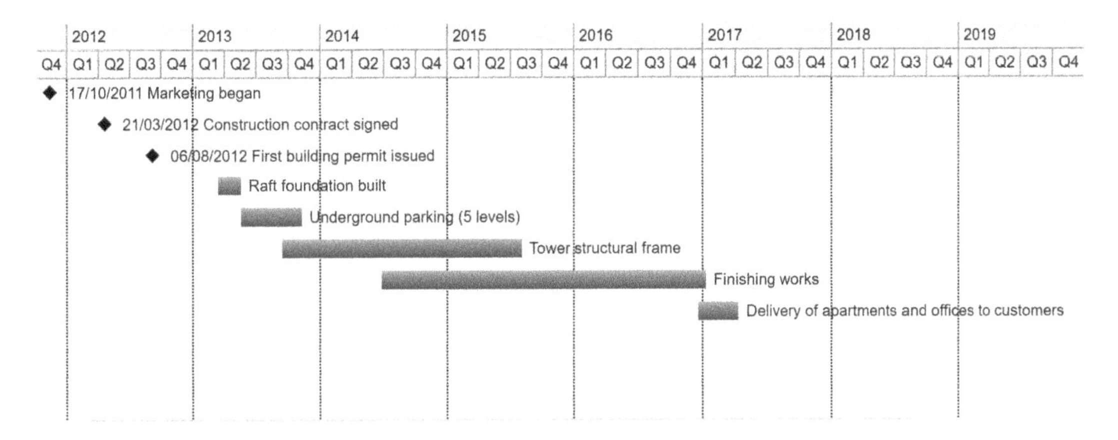

Figure 4.3 Dawn Tower project timeline

Application of Lean and BIM

The project was not modelled as a whole in a single BIM model, but BIM tools were used for site organization and system coordination in strategic areas (structural cores, underground parking floors, mechanical, electrical and plumbing [MEP] systems, etc.). The integrated model was prepared and maintained by Tidhar staff using Revit and Navisworks software. The Last Planner® System was used for some of the building system works in the basement floors.

5 The Skanska Finland way

Pioneering companies that have adopted Lean and BIM in an integrated fashion have redefined the level of service in their market, be they real estate developers, general contractors or subcontractors.

The first company that Tidhar's management team visited during its learning tour of Finland and Denmark was Skanska Finland. Skanska was known for its pioneering work as a construction contracting company in the areas of both Lean and BIM through the efforts of two of its key people, Jan Elfving and Ilkka Romo. They had not only worked hard to implement these innovations, they had also taken active roles in presenting their work to international audiences at conferences and in the press. Here we tell the story of Skanska's experience as an early adopter of both Lean and BIM.

Skanska is a leading international construction company with headquarters in Sweden. The company was founded some 130 years ago. Over the course of its history, the company expanded its market throughout the world, undertaking a wide variety of projects, including some which turned out not to have had a positive impact on the bottom line.

With the economic downturn of the early 2000s having a global impact, Skanska decided to rethink its "bigger is better" approach. While the benefits of running such a large operation (financial flexibility and a wider breadth of knowledge) could not be ignored, Skanska management saw a need to consolidate their operations and focus on a home-market strategy. This entailed a pull-out from Africa, Russia and Asia, standardization of existing processes and a focus on their operations in their home markets, primarily in Europe.

In this chapter, we tell the story of Skanska's Finland branch on their change journey. We will see how their implementation of BIM and adoption of Lean Construction principles was instrumental in allowing them to adjust and be successful in the changing environment of the construction industry.

The start of BIM and Lean at Skanska

From 2001 to 2005, the Confederation of Finnish Construction Industries sponsored a programme designed to encourage the industry to integrate BIM technology and practices. Under the initiative, called "Pro IT", representatives of the participating design offices, construction contractors and software companies developed national BIM design and production protocols. Pro IT was led by Ilkka Romo, a civil engineer by training and a development manager at the Confederation.

Skanska Finland was one of the member companies of the Pro IT consortium. In 2006, when Pro IT ended, and with it the government's funding, many companies ditched the new BIM procedures and reverted to their old practices. Skanska Finland was one of the few who saw the potential in the practices and persisted. They took the opportunity to hire Ilkka and establish a BIM team to help institute company-wide change. This was largely seen as an R&D project, but it was a crucial decision in the company's pursuit to remain among the leaders of the industry. Ilkka was to be responsible for coordination of BIM development in Skanska's Nordic offices, as well as in Skanska UK.

Two years earlier, in 2004, Skanska Finland's CEO, Juha Hetemaki, recruited a former Skanska employee named Jan Elfving, who had worked for Skanska in the 1990s. Jan had recently completed a PhD in construction management, with a focus on Lean Construction, with Professor Iris Tommelein at the University of California, Berkeley. His academic knowledge of Lean Construction, combined with his previous experience at Skanska, made him an ideal candidate to lead the company's implementation of Lean. Juha called Jan and asked him to come back to Skanska; when Jan asked what for, Juha responded, "Just come and do here what you did at Berkeley!"

In 2005, Jan and his team started collecting data on the existing processes at the company. In addition to interviewing the company's own employees, they approached suppliers, warehouse managers and others who were involved in the process. Working in this way, they hoped to identify the problems plaguing production on site and to find their root causes.

At the time, Skanska was building the athlete's village for the European Track and Field Championships to be held in Helsinki. According to Jan, this was a "red project", which the team was finding very stressful: "A 'red project' was a slang term for a project where the schedule was tighter than usual." Jan's team measured (1) the reliability of commitments in supply, planning and construction and (2) the quantity of work in progress, for precast concrete elements in design, in fabrication and waiting for installation. Among the problems they identified were:

- Production was highly unreliable, as reflected in the reliability of commitments. Contractors weren't meeting their goals as planned, which resulted in delays in subsequent work, resulting in waves of delays.
- There were large inventories of WIP. Precast elements, beginning with the design phase and continuing until installation, were being produced in an ad hoc order. This resulted in high average waiting times during the element production process.

The investigation team proposed many solutions to tackle these and other issues in the supply chains. Among those were systems designed to track production and WIP, such as inserting radio-frequency identification (RFID) chips in order to track materials and installing cameras at storage sites. However, the root cause of the production reliability problem was not that materials were arriving late or missing: it was that the workers on site were not being put in the ideal position to complete their tasks. When they were asked to present their weekly goals and actual production rates using a simple A4 sheet of paper and some coloured pens, the results were astounding. Among the 150 residential, infrastructure and commercial projects that were analysed, the per cent plan

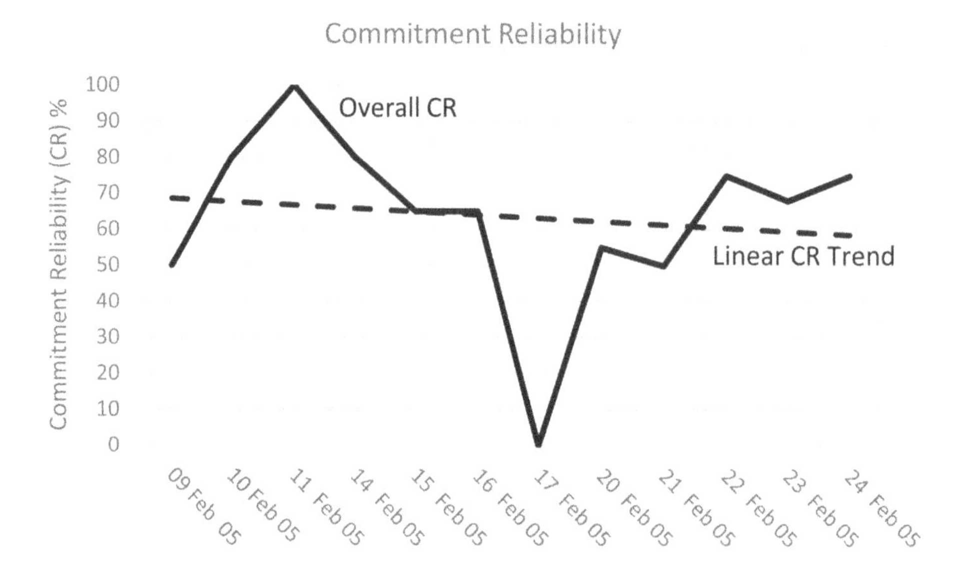

Figure 5.1 Commitment reliability over a two-week sample in 2005

Note: The degree to which commitments were met fluctuated widely.

Source: Image courtesy of Skanska Finland.

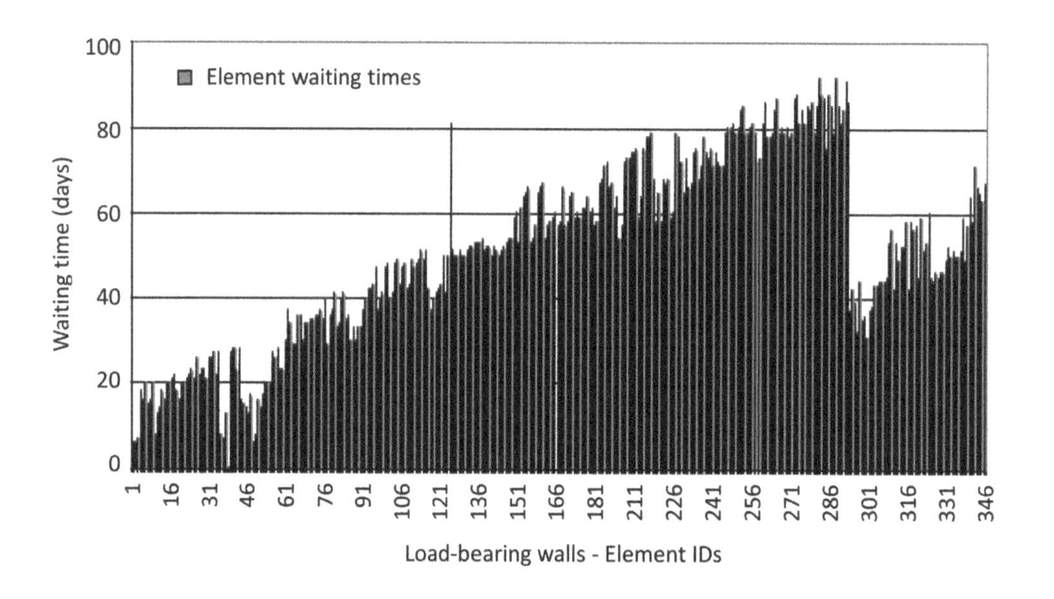

Figure 5.2 Records of the average waiting times for load-bearing walls during the production process

Source: Image courtesy of Skanska, Finland.

complete (or "PPC") was reported to be between 63 and 67 per cent. Throughout the company, less than two thirds of the targets planned for each week were being achieved!

With this data in hand, Jan drew the initial phase of the Lean project to a close. He could now identify the root problem leading to the poor reliability and work delays: improper planning and preparation of the work. Therefore, in addition to the adoption of BIM, Skanska Finland decided to start implementation of Lean Construction principles. The first step was to use the Last Planner® System (LPS) to prepare the work plans and schedules and to control the work itself.

When Jan proposed rolling out Lean Construction and LPS at the company's biannual R&D meeting, the idea was greeted with both enthusiasm and scepticism: "Sounds like a great idea, but can it work here?" Yet even for those who didn't truly understand the process, the low cost of implementing Lean coupled with its high potential for gains was a persuasive argument. With the blessings of the board in hand, Jan invited Glenn Ballard from the Lean Construction Institute (LCI) in the US, the inventor of LPS, to assist with the development of an LPS process suitable to Skanska's context and needs.

Lean and BIM implementation begins

When BIM and Lean were first introduced at Skanska, most people perceived the two initiatives as relevant for the managers in the head office. BIM was a tool for IT specialists, but not practical for the typical worker at the job site. Therefore, there was very little impact at this point in the company's day-to-day operations. This, however, was about to change.

The team developed and implemented LPS in stages. Initially, they chose three pilot projects, one from each of Skanska's key operations (residential, commercial and infrastructure). The Lean teams worked closely with the whole project command chain. They explained to employees how to properly define a daily task, and together they developed reverse phase scheduling, look-ahead plans and weekly work plans.

At the same time, Ilkka and his team were implementing BIM modelling on their first project, one of Skanska's residential development projects. At this early stage, they only modelled the architectural elements, using ArchiCAD software. In the ensuing projects, structural elements were modelled using Tekla, and finally in 2008 they completed their first fully modelled project, including all building systems.

One of the biggest challenges Ilkka and his team encountered with the implementation of BIM was the additional costs incurred. This was because their choices of outsourced design firms were limited to the very few capable of modelling in 3D. These design offices were generally the best in their fields and their level of compensation reflected this. In the early years, many people in the company and among its consultants perceived that BIM would cost more than traditional design using CAD, and that the fruits of BIM implementation would not be reaped for several years. If it were not for the full support of the CEO and upper management, it is quite possible that the programmes would have been scrapped.

However, the first steps of Lean implementation were proceeding well. Jan and his team made sure to spend a lot of time at the construction projects where LPS was being rolled out, attending the weekly work planning meetings. At the meetings, the team representatives paid attention not only to the words that were spoken, but also to the group dynamics and to the conversations between the superintendents, the foremen and the front-line workers. They encouraged the superintendents to elicit from the foremen all information needed to remove any constraints that might hinder production, rather than simply demanding performance. Superintendents at Skanska learnt that just changing the tone of their request could make a difference. For example, a simple request of "We need to do this pour on Thursday" changed to "What do we need to do in order to pour on Thursday?" would in turn change the tone of the discussion, leading to a greater chance that the pour would take place on the desired date.

Jan's team also paid attention to training superintendents to complete the cycle of commitment within the formal project hierarchy: when a superintendent approved a foreman's request, he was required to relay this information explicitly to the project manager and to request and obtain a clear cooperation commitment from him as well.

While there was the occasional hiccup, overall the subcontractors' foremen appreciated this new approach. They felt freer to speak their minds and to explicitly request the things they needed, such as use of the crane, exclusive access to a particular area on the site or additional materials. This dynamic also helped create dialogue at the subsequent week's PPC review. When tasks were not completed as planned the root cause of the problem could be investigated. Was information not being properly relayed? Had the contractor overestimated his needs? Questions like these opened the door for continuous improvement.

Yet despite all of the team's efforts, communication, a crucial aspect of collaboration, was still proving to be the biggest hurdle in implementation of LPS. The Lean team had to pay a lot of attention to this issue, as communication skills are typically not taught at professional or vocational schools for construction workers. Traditionally, situations would arise where a superintendent, foreman or contractor was afraid or embarrassed to admit to his peers that he did not know something he was expected to know. The legacy of this confrontational dynamic could be felt during the LPS roll-out, when one contractor was asked what he needed to complete his task. "You're the superintendent, you should know!" was his sharp response. It was at moments like these that the Lean team would step in and help coach those involved to ask not just the right questions, but also to ask the questions in the right manner to get more productive answers. To improve employees' communication skills, the Lean team set up a training programme that included classes, videos and simulated scenarios.

Over the course of approximately one year, the team refined all the aspects of LPS within the pilot projects. Once LPS was being implemented correctly and was working smoothly, it was spread to the other 10 to 20 projects in each region. LPS implementation was expanded under the careful supervision of the Lean team and with the support of the educational resources developed during the pilot projects.

Skanska Finland HQ – An opportunity for Lean and BIM

In 2009, a unique opportunity arose for implementing BIM. Skanska's leadership decided that their new head office building, a 28,228 m^2, eight-storey office block in Manskun Rasti, north of Helsinki's city centre, would be a 100 per cent BIM project. As this was the first commercial project to be executed at Skanska in this manner, many from the commercial construction division were sceptical. Ilkka and his team, utilizing their experience from the residential projects, worked hard with the commercial division's designers to support BIM implementation.

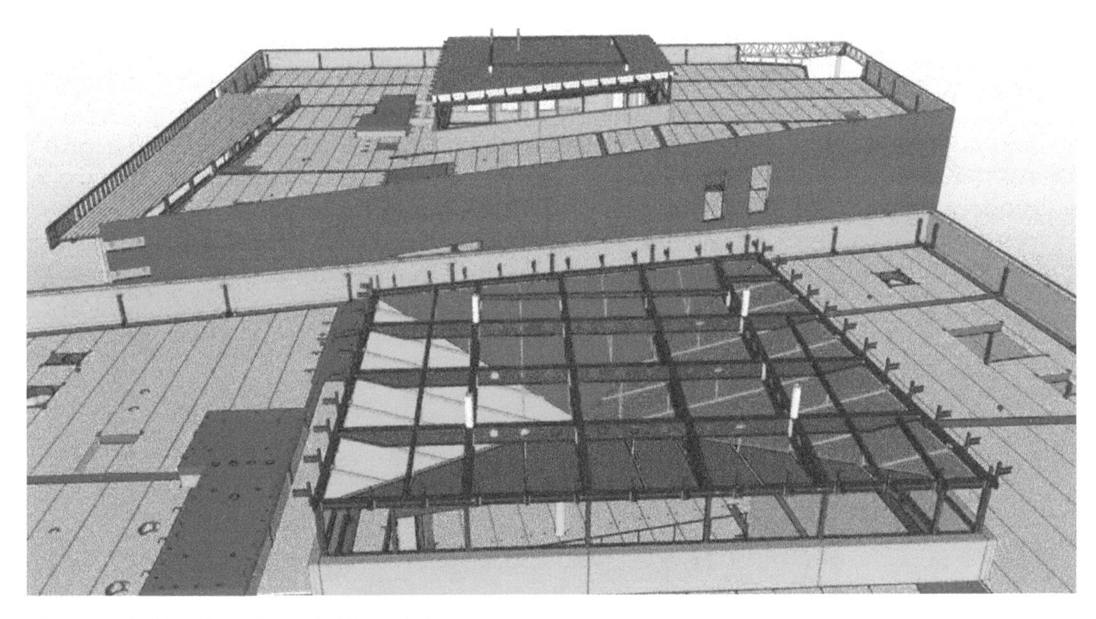

Figure 5.3 Manskun Rasti BIM model
Source: Image courtesy of Skanska, Finland.

The hard work and effort started to pay off. As with the Lean implementation, site managers were very sceptical about this new technology at first. With the head office project, they were able to experience the benefits first-hand. The quality of their work was of a higher standard. This was due in part to the ability to better understand the design. Many workers had trouble visualizing 2D drawings, yet BIM afforded all members of the team an opportunity to look at and "feel" the building prior to its construction.

The successes of the project did not go unnoticed. Skanska won the 2011 Tekla Global BIM Awards "outstanding BIM model" category for the Manskun Rasti project.[1] This was to be just the first of many, as Skanska has been recognized numerous times since for its excellence in BIM.

Box 5.1 MEP modules – a BIM-dependent Lean solution

Skanska's headquarters building in Manskun Rasti, Helsinki, is a monument to Lean and BIM implementation in construction. Design began with a laser-scan of the existing site and a complete model of the existing pipes and tunnels below ground. The design team used an open BIM approach with architectural, geotechnical, structural and MEP models exchanged using the Industry Foundation Classes (IFC) standard. As many as seven structural designers were able to work concurrently on the Tekla structures model, which included all structural steel, precast and cast-in-place reinforced concrete details, reinforcement, etc. Site safety and logistics were also managed using the model.

Lean practices included use of LPS for production control and VSM analyses to improve production methods. For example, the traditional approach to installation of MEP systems was recognized to be wasteful, with a great deal of waste: time wasted in coordinating among the trades, ergonomically challenging working conditions (installation of pipes, cable trays and ducts into a compact ceiling space) and frequent clashes among physical system parts due to incomplete design coordination. Skanska significantly improved this by leveraging BIM to design, model, prefabricate off site, deliver and install one-piece modular ceiling units containing all of the MEP systems. The modules, 5.6 metres in length with all MEP systems incorporated within, were prefabricated in a factory in Tallinn, Estonia, and shipped across the Baltic Sea in custom container-size frames for installation at the site. No problems were encountered during the installation on-site, thanks in large part to the accurate BIM models that were used.

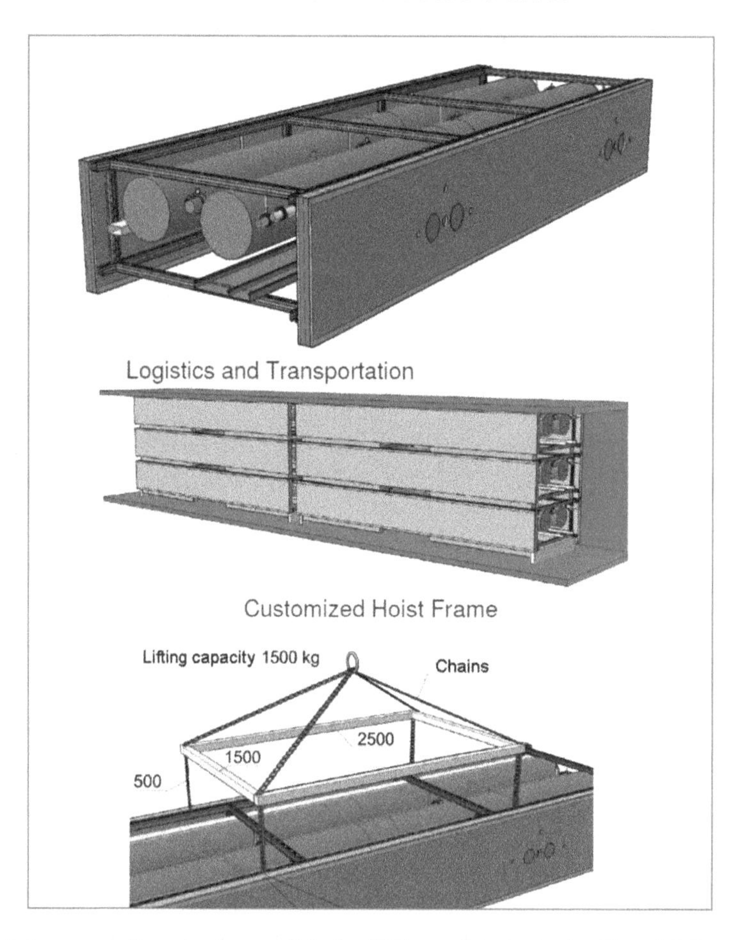

Figure 5.4 A Lean solution made possible by BIM: prefabricated ceiling MEP module for Skanska Finland's headquarters building in Manskun Rasti, Helsinki

Note: From top to bottom: a single module, six modules packaged in a standard shipping container and rig for lifting modules. Site labour was reduced to 5 hours per module, versus the previous 60 man-hours for an equivalent amount of work.

Source: Images courtesy of Skanska Finland.

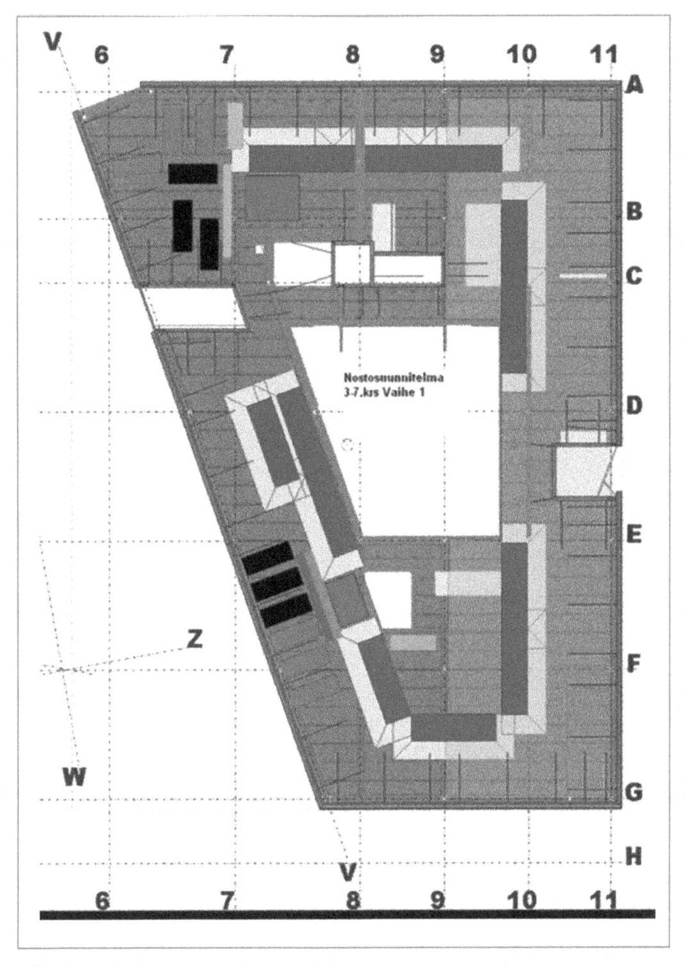

Figure 5.5 Layout of MEP modules on a floor of Skanska Finland's headquarters building in Manskun Rasti, Helsinki

Note: The modules are shown in blue.

Source: Image courtesy of Skanska Finland.

Each module included ducts with thermal insulation, cooling pipes, valves, thermal insulation on the pipes, electrical cable trays, sprinkler pipes and gypsum board corridor walls with sound insulation. The modules were designed and detailed in BIM to a degree of detail that allowed for prefabrication. The result was improved productivity and fewer hours of work on site with a concomitant reduction in exposure of workers to hazardous conditions, better economy and improved constructability. Each module required approximately five work hours to install and connect to its neighbour, but each module saved some 60 work hours – the labour allocation that would have been required to install the same systems using conventional methods.

BIM and Lean are introduced as mainstream practices

In 2009, after the initial introduction of LPS at Skanska's lower levels, many projects were being modelled in BIM, and employees were starting to feel the benefits. It was time to expand the introduction of Lean philosophy and BIM to the upper management.

Ilkka established a BIM competency centre. The centre, which functioned for three years, employed some 10 people who worked to train Skanska employees in proper BIM methods. At any given time, one or two representatives among the trainees were from other Skanska offices, including the UK, US, Poland and Sweden, making it a truly global initiative.

In the Lean area, Jan conducted workshops with the management team and regional management, each of which lasted about half a day. First the basic philosophies of Lean were introduced, followed by results from LPS implementation in the field. The communication hurdles were discussed extensively and, as with the field workers, role-play-simulated scenarios were reviewed.

While Lean was proving to be profitable for Skanska's bottom line, it required a huge shift in the standard approach to scheduling on the part of the superintendents. The shift to reverse phase scheduling could be an extremely difficult cultural change for someone with many years of experience. Whereas before the superintendent dictated what was to be done, the LPS method required input from all those involved, and at times, it required workers to challenge one another. In the eyes of the Skanska management, it was important that the change be natural and gradual and not revolutionary. Therefore, management made sure to commit sufficient support resources in order to help facilitate the project teams, and managers were very careful to address problems posed by employees who didn't take to the new Lean approach and who might have discouraged others. Jan enlisted senior management support for this issue, and they set a policy that demanded that all employees must, as a minimum, support the initiative by adopting a positive attitude. The statement "I am not going to learn this because I am retiring soon, but you younger guys must learn to work this way", was considered a minimum requirement. A superintendent who did not implement the process, but encouraged the rest of the team to make the change, kept his position. In such cases – which were the exception – one of the foremen would run the weekly work planning (WWP) meetings instead of the site superintendent.

In parallel to the implementation of LPS and Lean, the company had started preparing constructability reports and productivity analyses with the 3D models. With the information from BIM, they performed clash detection and simulated production. They simulated different construction schedules using Tekla and VICO Control, in order to identify potential risks and optimal sequences. This was originally seen purely as a BIM exercise, but over time it was recognized as an integral part of the LPS make-ready process.

Inferior communication skills, as mentioned before, were a big barrier to implementing Lean Construction methods, but BIM models began to fill some of this gap. Today, most of the site engineers fulfil the role of information specialists, using the model to provide information for everyone on site. Jan observed that this role was reminiscent of the way in which people would approach site engineers to help them access information from Excel sheets when that technology was new. The increased access to information improved the communication among all of the people in the project.

A BIM solution to a Lean problem

LPS was a big part of Skanska's implementation of Lean principles, but it was not the only change undertaken. An example can be seen from Skanska's development of a production tracking system integrated into the BIM system to streamline processes and reduce waste in the production process of the precast concrete elements. One of the lessons from building the athletes' village for the European Track and Field Championships in 2005 was that there was an extremely high amount of WIP: most of the pieces were in the production process simultaneously. The cycle time (CT), or the amount of time a single element is in production, was between 60 and 100 days! This was certainly not desirable for a precast element. Jan explained:

> The root cause of this phenomenon was found to be the large batches in which the shop drawings were prepared and the pieces were produced. The structural engineer prepared all of the shop drawings for a whole floor at a time, the manufacturer produced all the similar elements in one batch (to minimize the amount of form work setups they had to perform), and so the site team installed the elements floor by floor.

Drawings waited for approval, elements waited for production and prepared elements waited on site. Too many elements waiting on site increases the potential for damage by equipment, or as is the case in tight construction environments, elements need to be moved in order to access different areas, wasting time and resources, and further increasing the potential for damage. From an outsider's perspective, this may have looked like any normal construction site, but Jan and his team easily identified a great deal of waste. A solution had to be found, and the sooner the better.

In this case, as many construction teams in today's technological construction age are learning, the answer could be found in BIM. The team had access to complete BIM models for Skanska's projects, and had only to tap into the resources already available to find a solution. A thorough analysis using a construction simulation showed that elements were not being erected in the order that had been planned.

With this information, Jan developed a countermeasure: they would make the process visible in real time, by applying RFID tags to the pre-cast elements, stamps to all drawings and plans and colour coding for the BIM model to visualize each element's real-time status. With this system in place, everyone would know exactly where everything was in the process, and more importantly, they could prioritize the needs of the project. The idea, in essence, was to create a pull system, a system which makes certain that elements are ready exactly when they are needed, without waste and attendant costly overheads. The system proved to be a huge success, and it was implemented for windows, doors and many other items produced for the project.

In addition to streamlining the production process, the BIM model was very useful for ordering and controlling delivery of materials. Materials had been delivered to sites just once a week. Yet, because the model was broken down into locations, the items could be placed precisely in the location where they were needed for that week. This enabled the sites to operate a "Just-in-time" (JIT) delivery system. Figure 5.6 is an example of the way in which deliveries of pallets of interior finishing materials could be planned, so that crane operators and superintendents could direct them to the right location, avoiding double handling, and without occupying the space needed for other activities, such as the slab formwork shoring shown in the figure.

Note that this is strikingly similar to Tidhar's practice, begun at the Rosh Ha'ayin project, where the BIM model was used to control the delivery of autoclaved concrete blocks into floors before pouring the slab above (see Chapter 8, "The waste of non-value-adding work"). This similarity is not entirely surprising, for two reasons – it is a relatively obvious way to use a BIM to achieve a Lean Construction goal, and recall that Tidhar's senior staff visited Skanska Finland in 2012. Either great minds think alike or Tidhar managers learned from Skanska, or both. In any event, the solution is both elegant and productive.

Once the results of the initiative to use BIM and other technologies for supply chain monitoring for precast concrete construction on four projects in Finland proved to be so positive, Skanska decided to implement the approach at another major project, the Meadowlands Stadium project in New Jersey, USA. The stadium, with a budget of $1.6 billion and designed to seat 82,500 people, was to be one of the largest football stadiums ever built. Here too, the new RFID tracking system was a success.

And yet, despite that success, Skanska did not use the system again. Why would such a successful system with proven results be abandoned? Increased productivity from the workers, less waste and more streamlined production: it seemed to be too good to be true. In Jan's words:

> The system was too cumbersome, and required a lot of support. We needed a solution that could be maintained by the standard site office personnel, without additional dedicated resources on site. The Meadowlands was big enough to justify the extra people with specialized training, but most other projects are not.

For the vast majority of Skanska's projects, the need for infrastructure support, RFID technicians and IT support on the site was too demanding and the cost of the extra staff was more overhead than the company was willing or able to maintain for smaller

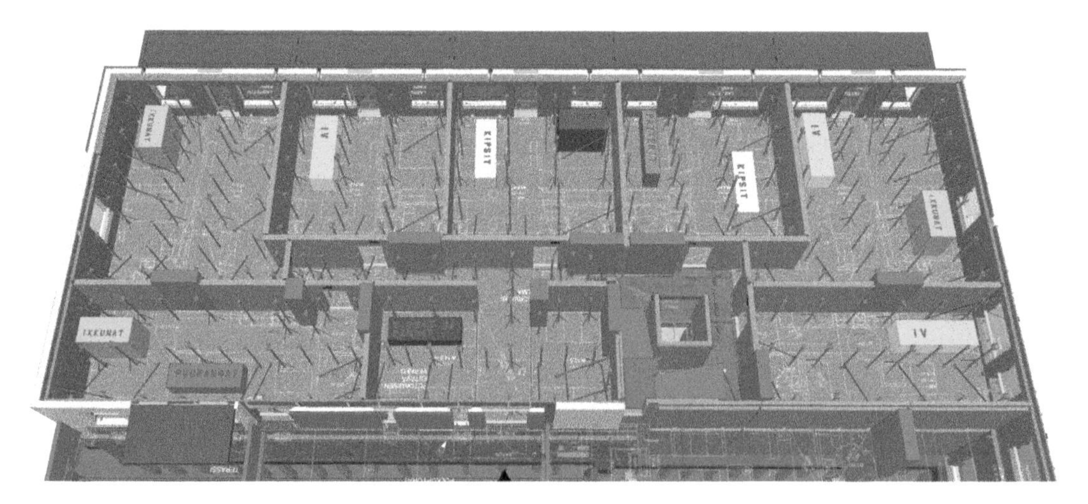

Figure 5.6 A Tekla model used for just-in-time delivery of pallets of material, avoiding the wastes of double handling and of obstructing other activities

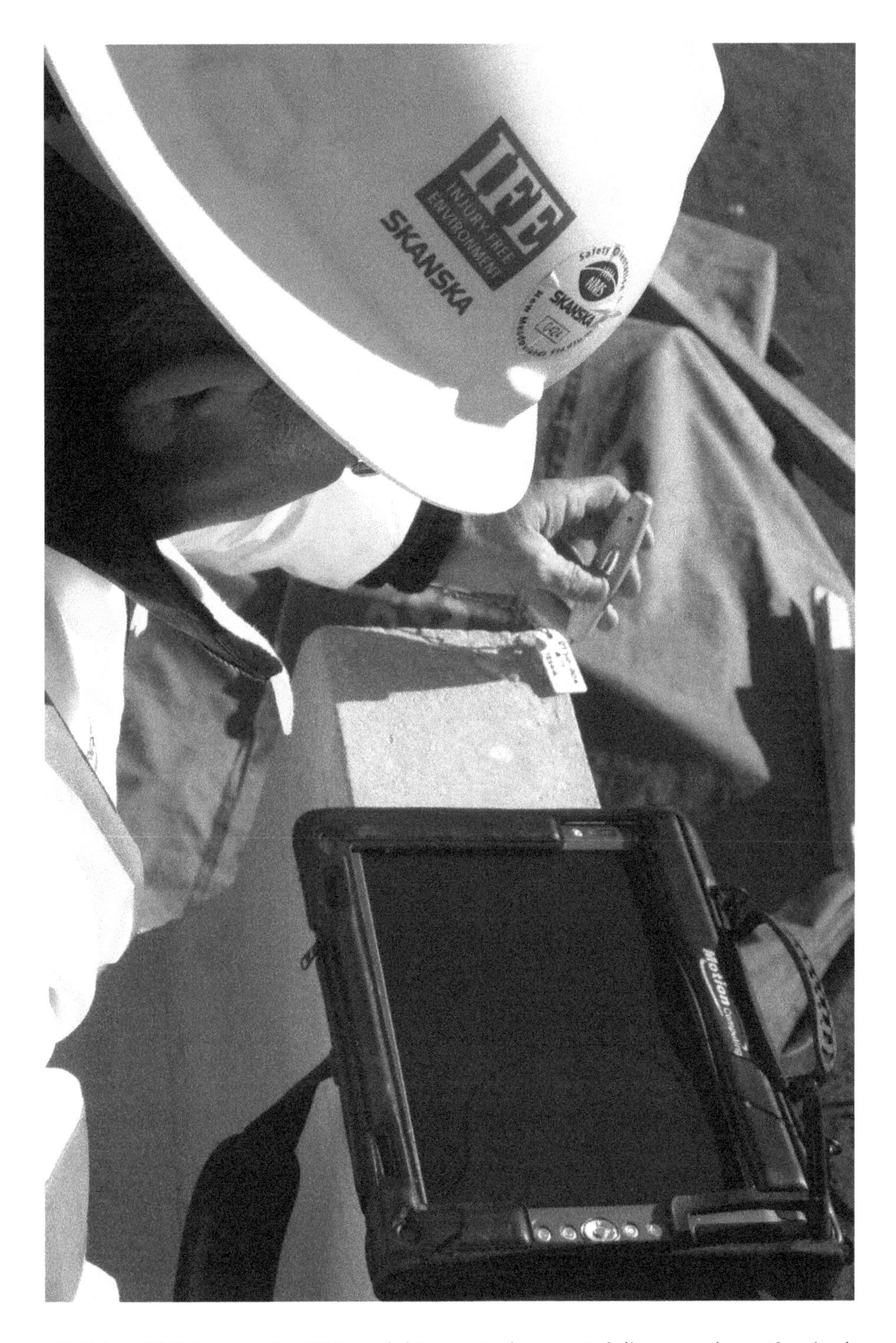

Figure 5.7 Using RFID tags and a BIM model to control precast delivery and erection in the Meadowlands Stadium project

Source: Image courtesy of Skanska.

projects. Skanska had about 150 projects in Finland alone and over 1,000 across the entire Nordic region, with an average project size of €3 to €4 million. Projects of this size typically employ only one or two superintendents. It was not possible to train existing site personnel to use the system and, equally, not enough people with the right skills could be hired. Solutions which required large inputs of skilled resources could neither be launched nor sustained for the majority of projects, even if they were fundamentally economically justifiable. Integration of Lean and BIM using RFID tag tracking for supply chain monitoring was no longer implemented, although Skanska maintained the use of BIM for logistics and JIT materials deliveries.

LPS, in contrast, with its very modest needs for training and investment, and its ease of implementation, was a much more sustainable solution for small- to medium-sized projects. After a relatively short guidance period, a foreman or superintendent could take control and run the WWP with ease. However, there was an issue here too: how could LPS be made simple enough so that more than one or two people could use it as a personal tool? The problem was one of access and availability of the information. How could a large team see the current weekly work plan and collaborate on it, given that they were not in the site office all the time? This meant that some of the Lean solutions that Skanska implemented were not necessarily useful for large, complex projects. This problem is beginning to be solved with online and mobile apps dedicated to Lean planning and control, such as Visilean, Touchplan and other services.

"Steady state" BIM and Lean

As of 2016, Skanska had a BIM support centre with a permanent staff of five to six employees who supported an average of 80 projects at any given time. These BIM and Lean support staff are responsible for training Skanska's site employees in the company's version of LPS, for providing back office support and for supporting clash detection on the BIM models. In other words, they are responsible for implementing the "Skanska Way" of doing things across the company's projects. The centre is constantly working to innovate and to improve the BIM and Lean processes, often through developmental partnerships with outside organizations. For example, Skanska works with the construction software subsidiaries of Trimble Inc. (an international software and navigation hardware company) to integrate BIM models with supply chain management using more sustainable technologies than those available at the time of the Meadowlands project. Employees of the centre are also tasked with preparing Skanska's BIM standard, as this is seen to be a crucial aspect for streamlining BIM implementation.

The centre is evidence of the increasing in-house technological sophistication required for a construction company, as construction matches pace with the increasing computerization of other production industries. Building with BIM is quite different to building without it; it encourages the growth of stronger companies, with more head office resources. These resources are then better able to establish and sustain their specific ways of operating than smaller companies might. As such they are increasingly able to develop Lean Construction practices in parallel, provided they use standardization as a platform for continuous improvement and work hard to avoid the pitfall of becoming set in their ways.

As of 2017, Skanska modelled all of its building projects, although only about 90 per cent of the scope of the project is modelled. The remaining 10 per cent consists mainly of the earthworks and landscaping aspects of the projects, for which BIM software solutions are still cumbersome and quantities are imprecise. Use of BIM has resulted in a significant drop in the amount of paper on site, as engineers and superintendents become familiar with working with tablets. Sales teams benefit too, as the 3D models have helped their clients better visualize their buildings.

Box 5.2 Skanska's Sara Troberg on BIM for standardization

Sara Troberg earned her MSc degree from Aalto University in 2015. She began working for Skanska in 2007 at construction sites, and became a full-time member of Ilkka's BIM centre team. The company encourages its employees to study advanced degrees in parallel with their day-to-day work.

At any given time, Sara is involved in about 20 construction projects, at various stages of design, development and construction. She uses her knowledge and experience of Skanska's field operations to improve the BIM experience for site workers. In her MSc thesis, she wrote: "As Building Information Modelling (BIM) becomes more common, there are even more possibilities in terms of standardization. BIM can be used in creating new parametrical and object-based standard solutions that could help architects and engineers in the design process." This idea is being put into practice at Skanska.

Some of the advantages of the standardization of BIM practices are seen directly in the fact that whereas previously the BIM models would need to be upgraded from a design model to a site model, the BIM centre staff are increasingly able to make the model "site-ready" with very little effort. Designers traditionally did not understand how to properly specify the components needed for construction. Using BIM standards provided early in a project's design phase by the Skanska BIM centre, the designers' BIM models tend to be much better aligned to the needs and practices of construction. Many of them prepared parametric component libraries according to Skanska's standards. Members of the BIM support centre help guide the designers in implementing Skanska's BIM standards.

Lessons learned

When implementing change in a company there are always challenges. A successful company will learn from these in order to improve its processes, improve implementation and learn where to better focus its resources next time. The introduction of Lean Construction and BIM in Skanska Finland certainly had its share of challenges.

With the implementation of Lean, Jan Elfving's team saw how much the human and social aspects affect the change. The need to pay attention to and to improve communication skills, and the need to help employees identify the problems in their production systems, were identified as crucial for motivating employees to become change agents instead of obstructing change. Perception was key to success. When implementing the change to Lean, complex technical jargon proved to be a big obstacle. Lean Construction was initially seen as some foreign system, with a foreign language, foreign to the employees' culture. Translating the terms to more understandable words proved highly

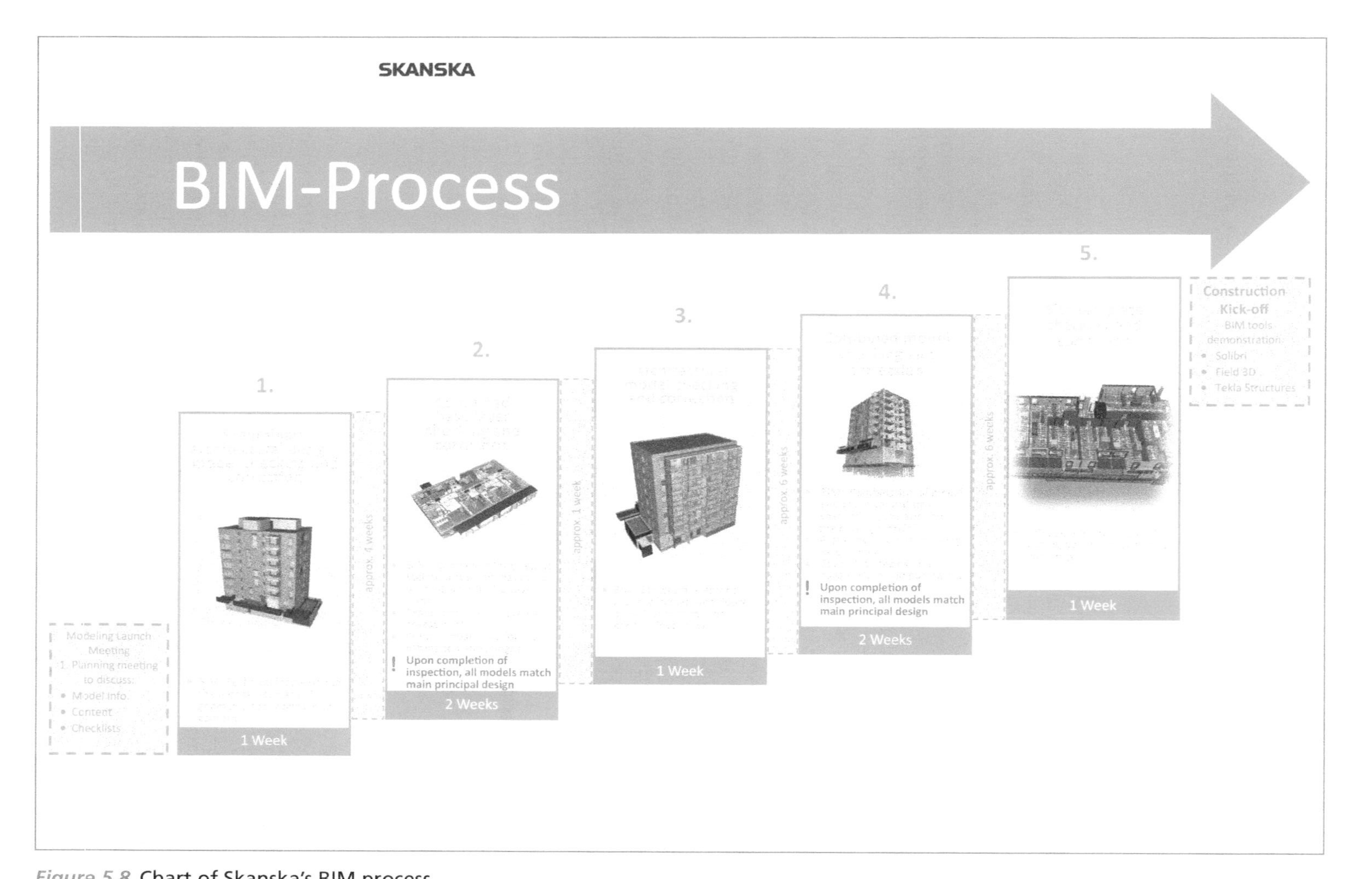

Figure 5.8 **Chart of Skanska's BIM process**

Source: Image courtesy of Skanska Finland.

successful in overcoming the cultural gap. For example, the Last Planner® System was translated as "Productivity Tools" in Finnish.

Interestingly, at the start, many people held the opinion that Lean Construction was only appropriate for large and complex projects. Skanska's implementation team therefore focused particularly on small- and medium-sized projects. Over time, this tactic gave people the mistaken impression that Lean Construction was in fact intended primarily for small- and medium-sized projects. However, with the passage of time and the acceptance of Lean Construction as an integral part of Skanska's culture, this issue faded.

Another important lesson learned – which like many is obvious when seen in retrospect – is that the scope of technological change is limited by the skills and abilities of the people available in the organization, and in this case on the construction sites. Skanska learned a major lesson from the implementation of the RFID tag system for supply chain management: if the technology is not robust, familiar and transparent, small- and medium-sized construction sites find it difficult to sustain and use effectively. Management processes dependent upon technological systems requiring large set-up and maintenance investments for every new project, although economically justifiable, may ultimately be abandoned by site staff.

Similarly, when implementing such large changes to its processes, a company must actively support the process. The company must supply the resources for employees to learn the new system, inform them prior to its implementation and support their day-to-day use. In some cases, rewarding those who excel can improve morale and motivate others to adopt the new systems as well. Understanding that employees will fail, identifying the most common problem areas and supplying the necessary support – such as the Lean Construction team and the BIM centre – was a big part of Skanska's success.

In this sense, Lean Construction, and some aspects of BIM, are endothermic processes – i.e. they need continuous application of energy in order to continue running, even if they are net producers of energy. Only exothermic processes (processes that supply their own energy to maintain momentum), such as the use of BIM by designers who profit directly from the productivity savings, are naturally sustainable for construction contractors too.

Note

1 For a thorough explanation of the project, see https://youtu.be/M9hJIr7immU.

6 Continuous improvement and respect for people

> The Lean approach stands on two pillars: continuous improvement and respect for people. Remove either and the entire façade comes falling down.

Upon CEO Tal Hershkovitz's return from his trip to Japan (see Chapter 3, "Education and motivation"), he implemented an employee suggestion programme. Hershkovitz was inspired by seeing the suggestions that had been put in place by front-line employees in the companies he visited, and even more so by hearing the employees explain their improvements to the visiting group. The initiative meshed well with the Tidhar culture that rewards innovation and prizes equality, and in which people at all levels feel free to express their opinions and ideas. Within the first three years, employees submitted more than 800 suggestions, of which about 100 were successfully implemented. The suggestions ranged from the mundane to the grandiose.

One example of a small suggestion with a big impact was the creation of a call system for exterior work elevators. These elevators are installed on the outside of a building while it is under construction, allowing access to the higher floors for workers and materials before the internal elevator is installed. The elevators require an operator to control them, pushing the buttons to drive the elevator up and down the building and open the lift gates. The problem was that there was no way for the operator to know when someone was waiting on a higher floor to be picked up and returned to ground level. One of Tidhar's employees suggested using a wireless call system that is used in restaurants. Each table has a button they can push, which relays a signal to a wrist-worn device worn by the server. By putting the call boxes on each floor next to the elevator, and giving the wrist device to the elevator operator, they created a stand-alone elevator call system. The cost of this off-the-shelf system was insignificant relative to the cost of the workers' time, and Tidhar adopted it immediately to great effect.

A larger improvement that grew out of the suggestion system was the creation of a "standard" for the layout of site offices and "big rooms" used by the design teams (see Chapter 14, "BIM in the big room" to learn more about the design process). By identifying the current best layout for each of these critical office spaces (the layout that is most conducive to productive interactions among team members) and applying it in future projects, Tidhar could carry their learning and successes from site to site and project to project. As they learned more and further refined the designs, this too could be captured in the evolving "best practice" standard.

If run correctly, employee suggestion programmes like these can have a powerful effect on a company, since by giving workers a platform to contribute their ideas, the implicit message is that the company is interested in and wants to hear what employees think. This may seem trivial, but in many companies, the opposite message is broadcast through management's behaviour. All too frequently, the underlying belief is that employees' ability to contribute their ideas is tied directly to pay scale; the engineer is the one who solves problems, while the front-line worker is paid mainly for manual or rote labour.

And yet, this approach (besides the belittling undercurrents it reflects) represents a loss for the company, since it is not tapping into the brainpower that came along with those many sets of hands. Yes, the engineer may have specialized training, but he or she does not perform the work day after day. It is the worker who has this expertise and deep understanding of the process, gained by hour after hour on the front line of value creation for the company. And in most companies, there are many more workers than engineers, meaning that the effect is multiplied. If only there was a way to capture and utilize that understanding and inherent creativity, and refocus it towards improvement.

Lean offers a way: the twin principles of "continuous improvement" and "respect for people".[1] The two complement and reinforce one another. Like many terms in Lean, they have very specific meanings (for example, here "respect" differs from its colloquial usage), which can make them difficult to comprehend and implement successfully. Yet if the two are successfully implemented, the organization can exploit improvements to improve productivity. This chapter discusses the background of continuous improvement and respect for people, what they mean, the barriers they face in the field of construction and what Tidhar has done to implement them.

The two pillars at Toyota

Just like many other elements of Lean, the concepts of continuous improvement and respect for people were developed by the Toyota Motor Company. In 2001, Toyota was expanding so rapidly and so globally that they felt they could no longer trust their oral tradition to ensure the spread of their values and management approach to new factories and offices. They created a document known as the *Toyota Way 2001* to assist in this effort. In this document, Toyota described what it believed were the two pillars of the "Toyota Way" that had led to their global success: continuous improvement and respect for people. Today, it is part of the company's public image, as shown on Toyota's corporate website (see Figure 6.1).

As can be seen in Figure 6.1, Toyota portrays continuous improvement and respect for people as two interlocking elements of equal weight that together comprise the "Way" that they operate. Continuous improvement is usually what most people get excited about in Lean – how to use one of the many tools from the Lean toolbox to slash waste in the organization while creating more value.

In Toyota's diagram, respect for people is depicted as no smaller than continuous improvement, and in truth it is no less important. But the most important is neither the former nor the latter; it is their intersection portrayed on the graph. Only the combination of continuous improvement and respect for people allows Lean (and all the improvements it brings) to be sustained and thrive.

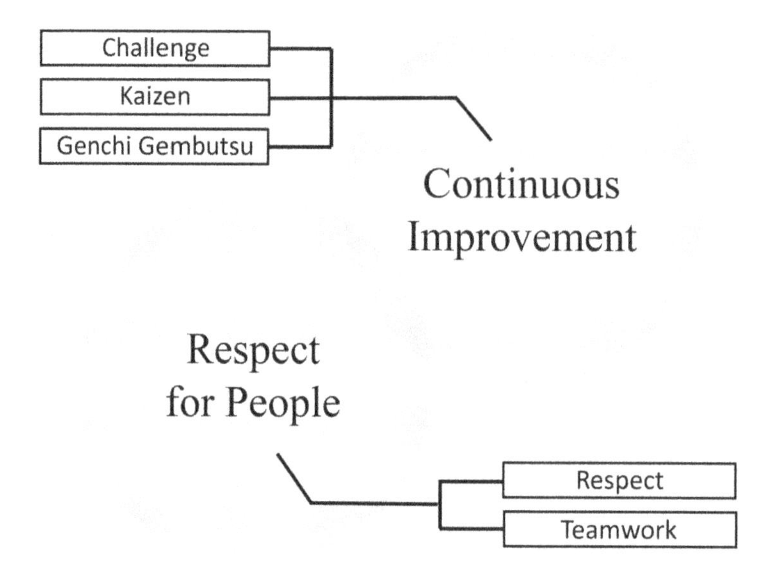

Figure 6.1 Continuous improvement and respect for people
Source: www.toyota-global.com/company/history_of_toyota/75years/data/conditions/philosophy/toyotaway2001.html.

Without respect for people, Lean is only continuous improvement, which tends to focus very quickly on the technical side. This has been called "Fake Lean", a name that evokes the situation in which management pursues a strategy of continuous improvement without including respect for people, with results that are short lived and grate upon the workforce. Toyota, by contrast, proudly proclaims: "We make people before we make cars." This statement shows their commitment to respect for people as a guiding principle. Likewise, other companies that have successfully implemented Lean consistently state that their achievements would not have been possible without sustained employee engagement and support at all levels of the organization.

The "secret sauce" of true Lean is not the use of tools to achieve short-term, point improvements (as impressive as they may be). Rather, it is creating an organizational culture and climate in which improvements to the work methods and processes (the way the work is done) are being made every single day, by every single member of the organization, in every area of the organization. This is what Masaaki Imai, the author of the seminal works *Kaizen* and *Gemba Kaizen*, has attempted to convey in his definition of the word *kaizen*: "Everyday improvement, everybody improvement, everywhere improvement". The most successful Lean organizations are not the ones with the largest "kaizen promotion office" or the most elaborate Lean posters; true success comes from creating an organizational culture and organizational climate in which "improvement" is a daily responsibility for everyone.

For a 150-person company, this would mean 150 pairs of eyes actively looking for and capable of identifying wastes and the corresponding opportunities for improvement in the processes with which they are intimately familiar, 150 hearts knowing that their contributions will be respected and valued, and thus motivated to make those improvements, 150 brains puzzling out the wastes identified in order to develop countermeasures, and 150 pairs of hands to pick up the pieces if the planned countermeasure fails to work out and additional work is needed to improve further.

The way to get to this ideal state – where continuous improvement is indeed continuous (and not just another short-lived corporate initiative) – is through respect for people.

What is respect for people, and why is it important for Lean success? Respect for people as it relates to employees includes respecting the innate ability of every human being to identify waste and develop creative ways of improving. It means respecting their contributions and simultaneously challenging them to always be improving their problem-solving skills. It means asking people not only to put out fires but also to prevent future flare-ups from igniting. Giving them the time and resources to experiment with countermeasures, even if it means allowing the experiment to fail (albeit in a controlled manner). And it means providing the training and leadership necessary to both provide the skills to recognize waste and develop countermeasures that are in line with the company's overall objectives.

When understood in this fashion, it becomes clear why respect for people is crucial to long-term Lean success: continuous improvement that is truly continuous and ongoing can only survive in an atmosphere where respect for people is being practised. Respect for people creates the fertile ground that allows continuous improvement to flourish. Companies that see the true potential of Lean are those that are constantly investing in the problem-solving abilities of their workforce at all levels and that make time for "daily kaizen", thereby challenging them not only to do their jobs but also to be responsible for improving them. Mike Rother, author of *Learning to See* and *Toyota Kata*, has claimed that the Lean tools of continuous improvement are in fact no more than structured frameworks for developing people and improving their problem-solving capability. A true Lean implementation is more about growing people than throwing out all of the inventory in the organization or finding a few point examples of waste to remove with fanfare.

Box 6.1 An example of continuous improvement at Tidhar

One example of pursuing continuous improvement at Tidhar centred around the construction of the niche for the main electrical switchboard in the apartments they built. The main switchboard for an apartment is typically located right inside the entrance door to the apartment, housed in a small compartment in a partition wall that is built primarily for that purpose. All the electrical circuits in the apartment go through the board, which means the compartment has a large number of electrical conduit pipes routed to it.

To reduce the amount of work needed to cut grooves in the finished concrete walls and ceilings for the conduit pipes, many of them are placed during the structural work and cast into the concrete. They are routed through the rebar inside the formwork so that they will be encased in the concrete ready to accept the electrical wires. When the electrician arrives to complete the wiring and fixtures in the apartment, his first step is to use a hammer drill to break the thin layer of concrete covering the conduit ends so that he can access them. In the case of the electrical compartment niche, he then has to route new pipe to the partition wall, and cut grooves in the wall to accept the vertical sections of conduit to connect to the panel. Finally, after the electrical work is completed, the grooves and the conduit must be covered over with a filler material to create a smooth surface to accept plaster and/or paint.

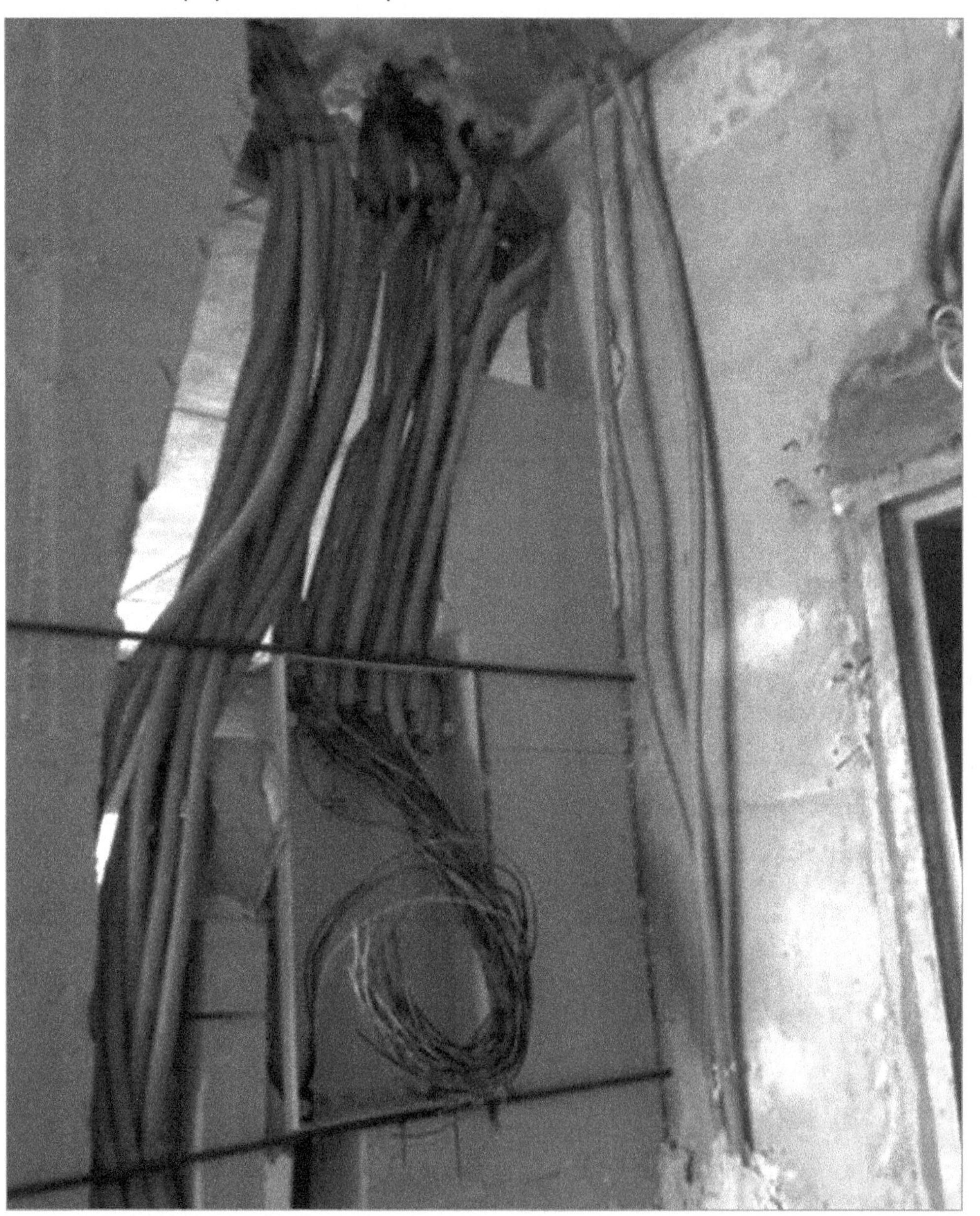

Figure 6.2 Cutting into a block partition wall to install the conduits for an electrical switchboard. Little remains of the blocks to constitute the wall

At Tidhar, the partition wall for the niches was typically built of gypsum blocks 16 to 20 cm thick. But what would often happen is that due to the large number of grooves cut, coming in from different directions for all the conduit terminuses, very little of the block partition material would remain intact. The partition resembled Swiss cheese, full of holes, and structural integrity was compromised. With enough cement and plaster, eventually a cosmetically passable wall could be reconstructed, but the rework needed was clearly a waste.

Once Tidhar had started their Lean journey, they decided to try to tackle this recurrent problem. A group of experts was convened to try to think up a solution to the problem. The idea they came up with was a floor-to-ceiling steel frame that would support the electrical distribution board compartment box and provide the structural integrity that the cut-up gypsum blocks could not. A prototype was even produced to test out the concept. However, ultimately it was decided that the steel frame was too complicated to install and too costly to purchase for it to be a viable solution.

The next step in the thinking process of the improvement team was to realize that, in essence, the problem was caused by a lack of dimensional tolerance. When the electrician placed the conduit pipes in the concrete formwork, the locations were not very accurate. This meant that when they were chiselled out of the concrete later in the process, the locations weren't exactly over the gypsum-block partition wall. This led to creative routing on the part of the electrician, which consumed most of the block. The obvious solution would be to try to extend the grooves in the ceiling to bring the conduit right over the location of the wall, but steel-reinforced cast-in-place concrete is very difficult and time consuming to cut grooves into. So, the next solution idea was to place a pour stop (a block of Styrofoam) into the formwork before pouring the concrete in the general vicinity of the partition wall. The conduits could be routed to this Styrofoam block, and then once the partition wall was built, accurate grooves could be cut into the Styrofoam to bring the pipes to the wall in a more organized fashion, thus preserving a lot more of the wall.

At this point, while up on top of the building under construction, experimenting with the placement of the Styrofoam block, one of the team members asked a simple question: "why is there a problem with dimensional tolerances? Why did the electrician in charge of placing the conduit pipes in the formwork prior to the concrete casting have so much difficulty in 'hitting the mark' in terms of where the partition wall for the electrical panel would be located?" Luckily, there was just such an electrician in the vicinity, so the team asked him. The answer amazed the team of experts; he said that on the blueprints he was given, which showed the electrical plan sketched out over the general lines of the apartment, there were no dimensions given for the location of the partition wall. The designer who had created the drawings put those dimensions on the "architectural" plans that were given to the masons in charge of actually building the partition walls. So, the electrician had to guess more or less where the wall was located based on the limited information he had been given. The team asked him if simply providing him with the right measurements on the drawings would be enough for him to be able to place the conduits in the right location, and he said that yes, it would be no problem. Thus, the ultimate solution to the original problem was no more difficult than instructing the designers who generated the electrical plans to put the required dimensions on the correct blueprints. This was done and the problem was solved. The need for rework on every panel that had been a permanent feature of every apartment disappeared.

This story, while having a happy ending, is a good example of a problem-solving approach that is less than efficient. After the problem was identified, the response was to put together a group of experts and have them try to solve the problem from the comfort of a meeting room in an office. No representative of either the electrician who had to route the cables and the conduit both before and after the concrete casting was included in the team, nor was a representative of the masonry crews who built the partition walls. It was only late in the process that the team got out to the construction site itself, to the Gemba where the value was actually being created for the customer and where the problem was also being created. As discussed in earlier chapters, the Gemba is the source of knowledge, and it is the place where problems are best (and most effectively) solved.

Toyota Chairman Fujio Cho would always instruct his managers to "go to Gemba". When they asked what they were to do once they got there, his response was concise and threefold: "Go see. Ask 'why?'. Show respect." And in this case at Tidhar, that was exactly the way they were able to (eventually) solve the problem. They went to the Gemba to see the issues in the complex environment in which they were created. This allowed them to grasp the problem in much greater fidelity than they ever could have done in a conceptual discussion in the conference room. They showed respect to the electrician who was doing the work by asking him "Why?", being open to hear his input and learn from his experience and his expertise. Once they had done all three – gone to see, asked "why?" and shown respect – the problem unravelled more or less of its own accord, and the countermeasure became exquisitely obvious and simple. No elaborate steel frames, no need to add Styrofoam blocks to their already-complicated building processes. This is the combined power of continuous improvement and respect for people when properly applied.

Difficulties implementing continuous improvement and respect for people in construction

The fact that the marked majority of all Lean implementations (in construction and beyond) fail suggests that there are barriers to successful implementation that are not specific to construction. They include being overly enamoured with continuous improvement as well as not fully understanding respect for people and/or underestimating its importance to the long-term success of Lean.

Beyond those initial barriers, we can consider difficulties that are specific to construction, which will additionally need to be overcome for the enlightened Lean Construction implementer. The subject of "construction peculiarities" that differentiate construction from other production industries has received much attention, with three main peculiarities identified at the level of construction projects: site production (the fact that each project must be built *in situ* in a new location), one-of-a-kind products (each product is typically different than any other in terms of styling and geometry), and temporary organizations (each project is typically composed of a new collection of subcontractors and consultants, different from the last).

Of these, the last is the biggest potential barrier to continuous improvement and respect for people in Lean Construction, since the temporary nature of each project tends to cause the parties to focus on short-term outcomes and seek to optimize at the level of the project. Site production and one-of-a-kind products do not offer the same challenge, since they do not assume the same level of worker transience. The first of the 14 Toyota Way principles identified by Jeffrey Liker, the same one Gil Geva presented to the management forum at Tidhar, is "*Base your management decisions on a long-term philosophy, even at the expense of short-term financial goals.*" For no other area is this more relevant than continuous improvement and respect for people; making an investment in developing people and their problem-solving skills requires the organizational self-discipline to maintain a degree of focus beyond the event horizon at the outer bounds of the current project. In addition, since subcontractors who are not directly employed by the GC perform the majority of actual work on the building, it will be an uphill battle to make the case (business or otherwise) for developing the same front-line workers who will be gone soon after their contribution to the project is complete. The assumption that subcontracting labour is cheap and transient (and often comprised of foreign workers) leads to an unwillingness to invest in developing their skills.

The question then arises: how can continuous improvement and respect for people be applied despite those difficulties? Is there any light at the end of the tunnel?

First, as with any Lean implementation in any company in any industry, Lean aspirants in construction companies must begin the work within their own four walls. Even the smallest architecture, engineering and construction (AEC) company can teach all of its people to see waste and develop countermeasures to address it, while empowering them to make changes in the work processes.

For those not yet ready to commit time each day for every employee to work on improvements, an employee suggestion programme may be a more viable first step. The emphases in making a programme like Tidhar's a success, and ones that will reinforce respect for people and continuous improvement, are small-scale changes, within the employee's sphere of influence, that do not necessarily require large capital outlays, for which approval to begin a trial can be rapidly granted, and that the employee is directly involved in trying out. More important than establishing financial objectives for the programme is aiming to get everyone contributing, with coaching by direct managers as necessary (and not using coercion by any means). The opposite scenario, in which suggestions are placed in a locked box, reviewed infrequently by management and implemented by a third party (typically an engineering or maintenance function), does not constitute respect for people and thus will not reinforce Lean efforts.

At the same time, only so much can be done in-house (though it is possible to do quite a lot over the course of years as people grow and develop). A lot of waste may be "locked in" at the design phase, and therefore any company that is only involved with executing plans developed by others may be limited. Likewise, for GCs, their ability to impact the work methods of the subcontractors who actually perform the work may be limited. Thus a typical progression, once Lean has started to become "the way things are done" within the company, is to start reaching out to key suppliers (and in construction, designers and subcontractors are key suppliers) and beginning to work with them to teach some of what has been learned and begin jointly implementing in order to find mutually beneficial improvements.

Another tack entirely would be to work through local trade unions, spreading Lean thinking and Lean training horizontally through the local industry, as was done in Denmark. The Federation of Building, Construction and Wood Workers Unions has embraced Lean Construction, seeking to make it the industry standard. In this way, Lean understanding can diffuse horizontally rather than requiring one company to invest in what are perceived as "here-today, gone-tomorrow" subcontractors.

Two key Lean Construction tools share elements of respect for people, even if they don't mention it explicitly: LPS (discussed in Chapter 17) and integrated project delivery (IPD). The success of LPS is likely due in part to its inherent combination of continuous improvement and respect for people – the very act of involving subcontractors in the process of planning engages their mental skills and asks them to take an active part in improving the process of construction. Thus respect is being shown for their creative and cognitive abilities, which is key in fostering continuous improvement. IPD also highlights the importance of respect for people, since it creates an atmosphere where the interests of the collaborating parties are aligned. This allows for more investment of energy in finding solutions and improvements that are beneficial for the project. By creating a win–win atmosphere, respect is indeed being shown for all stakeholders: owner, design professionals, contractors.

FastCap

In many ways, making continuous improvement a part of the daily practice for everyone in the company *is* how respect for people is practised. Take for example Paul Akers, the founder and president of FastCap, a company that produces products for contractors and the woodworking industry. Akers relates how the message of respect for people finally hit home for him.[2] His company was a number of years into their Lean journey and they had made great strides. But Akers felt that whenever he was not physically present to push the improvements along, the company made no progress. During one of his study missions to Japan, he had the opportunity to meet a VP from Lexus. Akers asked the executive to tell him what the most important thing was for Toyota. The response he received echoed Toyota's declared mission of building people before it builds cars. "Our number one concern is how to build our people and how to build a culture of continuous improvement." Reinvigorated, Akers returned home to introduce respect for people to a company that had been steeped only in continuous improvement up to that point. Today, FastCap employees spend the first hour of every day of work making improvements. Akers asks that they make no more than a two-second improvement to one of their work processes each day, since he knows that it is consistency of improvement that will over time lead to a competitive advantage (*kaizen*, Japanese for "incremental change"), not a few "home runs" hit intermittently (*kaikaku*, Japanese for "radical change"). When his employees are stumped for ideas (as often happens when people are put on the spot and ask to come up with suggestions), he asks them a simple question: "What bugs you?" In other words, what are the parts of your job and the processes you carry out that always seem to get in the way or cause problems. Very often, these "pain points" are also sources of waste if not waste themselves. Removing them has a double benefit: the process will run more smoothly, increasing productivity for the company, and the employee will have a tangible sense that Akers is interested in helping them make their job less frustrating and easier to perform. Even better, the focus is on working with the employees so that they are the ones developing the solutions to those things that bug them, so they are taking ownership for the improvements and developing their problem-solving skills. They are also being given the trust of the company to make changes to the process, which is a great responsibility not often bestowed upon front-line workers. This only serves to further develop them and increase their sense of responsibility.

After spending an hour making a small improvement, the second hour of each day at FastCap is also spent in developing people, with an all-hands stand-up meeting to review the core values and metrics as well as share improvement ideas. When other business leaders are aghast to find that two hours of every day are spent in apparently non-productive work, Akers responds: "In only six hours, my people can outperform anyone else working eight but not taking the time to improve."

Akers is a shining exemplar, but the theme of harnessing the creative power of all employees through respect for people runs through all the stories of the most successful Lean implementations. And as Akers has shown, there need not be grandiose changes every day; an improvement by each person that shaves no more than two seconds off a process will suffice, as long as one is made every single day.

Box 6.2 House brooms replace street-sweeping brooms

Tidhar continues to take a holistic view in improving their production processes, understanding how product design and the various parts of the production system interact in ways that are sometimes ignored. If they can be put together with mindfulness, it is possible to create synergy that leads to greater systems optimization far beyond any least-cost-piece-rate local optimization.

An example of this became apparent in a project in which Tidhar experimented with ways to improve construction flow through changes in the product design (even if the changes involve methods that are supposedly more expensive than the current methods when viewed in a purely per-metre light). In this project, two major changes were made. First, instead of routing the electrical and plumbing supply lines through the floor and then filling with gravel, these lines are routed through the ceilings and walls (including some areas of drop ceilings), with the flooring tiles glued to a self-levelling concrete screed floor and no gravel. Second, instead of gypsum or aerated concrete blocks, the interior partition walls are constructed of drywall (gypsum boards hung on lightweight metal framing). The "wet" trades that create lots of the dirt and rubble in the workspace (plaster, masonry, floor fill) have been minimized or outright eliminated.

As a result, the site is much cleaner throughout the finishing works phase. In fact, the workers responsible for keeping the job site clean asked the site superintendent to supply new brooms. Their standard issue street-sweeping brooms were well suited to cleaning up blocks of concrete and dried bulk materials that characterized the older methods. But now, with all of that bulk debris gone, there wasn't the same level of dirt. The coarse street brooms failed to clean the smooth screed floors of the fine dust that remained. They needed indoor brooms! This both reflects how much cleaner and more organized the site was, and how each improvement leads to additional improvements, which ideally come from the workers themselves.

How small changes add up

Lean neophytes do not always see how the sort of small improvements that mark a successful marriage of continuous improvement and respect for people (particularly when they are no more than two-second improvements) can lead to the significant bottom-line improvements that Lean promises. But by building a culture where every person, every day, is making a two-second improvement, larger improvements are inevitable, since people will be turned on, motivated and experienced in problem solving by the time larger opportunities present themselves. This is the concept of "turning the flywheel of improvement" writ both on the micro (two-second inputs each time) and macro (getting to a pace of improvement that allows the organization to outpace their competition). Art Byrne, who led the now-famous Lean implementation at the Wiremold company, explains how small improvements in reducing set-up time (SMED or single-minute exchange of die) are actually a strategic move for the company: by reducing the time required to change over from one product to another (set-up), it is possible to reduce the batch size of the products being produced.[3] A reduction in batch size means that the lead time of any given product is reduced. Thus the company will be able to respond more quickly to customer requests and changes in demand than their competitors and gain more market share as a result. This is the essence of *kaizen*: small changes that are made consistently, accumulating over time to lead to big improvements.

In the construction sphere, despite the peculiar barriers present in the industry, it is also possible to gain a competitive edge from the sustained application of continuous improvement. But this can only happen when respect for people is present, so the two must be implemented together if either is to survive. Bob Emiliani, who documented the Lean journey at Wiremold, has suggested that respect for people is a practice that defies simple verbal definition; it must be implemented in order to have its full effect be understood.

Problem solving at Tidhar and the PDCA cycle

Walter Shewhart developed (and W. Edwards Deming popularized) the "plan-do-check-act" (PDCA) cycle as a management method for creating continuous improvement.[4] The cycle breaks down the steps of the scientific method into specific phases that can be implemented. First, a plan is developed, including a hypothesis of what the outcome will be. Next, the plan is carried out ("do"). Then the results are studied to see if indeed the hypothesis was correct ("check"). Finally, the insights gained from the "check" stage are reintegrated into the process during "act". Then the cycle is complete and a new one begins. The scientific approach aids in effectively solving problems which enables improvement.

Problem solving and continuous improvement do not end once a countermeasure to a problem has been applied; it is essential to carry the PDCA cycle through to its end, not just rush off to the next problem. Chapter 8, "The waste of non-value-adding work", tells the story of how Tidhar identified significant wastes in the process of building interior partitions out of gypsum blocks. Many opportunities to improve were identified and plans were created for how to address them. Yet despite the wonderful plan that the team developed for the gypsum blocks, things have not always gone accordingly. More than anything else, there has been a lack of "check".

Part of the problem is that the plan did not create internal checks to get any deviations that might arise back online. For example, the individual responsible for placing the pallets of blocks on each floor prior to pouring the slab on top of the floor, one of the key improvements that was made to the process, is the superintendent for the structural works. This superintendent is primarily tasked with building the reinforced concrete frame of the building. He decides what work the crane will do at any point and he is primarily judged based on the volume of concrete poured and the skyward progression of the building according to schedule. While he is technically also responsible for placing the pallets of blocks on each floor before the ceiling slab is poured, no mechanisms of accountability were created to ensure this responsibility is carried out. And thus, just as the management truism opines, what is not measured does not always get done. If the full quantity of pallets of gypsum blocks have not arrived on site by the time the ceiling is ready to be closed, in general, the superintendent for the structural works would proceed with the pour rather than wait for the pallets to arrive and thus delay the progress of the structure. Rather than the full complement of 50 pallets, for example, this means that maybe only 40 pallets will make it to the storey since that was what was on hand before it came time to close the ceiling. This has been a recurring problem for Tidhar.

The person who is charged with using the human resources in effectively and efficiently constructing the partition walls (who thus cares very much that all of the pallets arrive in the proper location before the ceiling is poured) is the superintendent for the interior finishing work. But he has no say over the work priorities of the crane operator and thus has no direct ability to ensure that the pallets are properly and timely placed.

One possible solution to this problem would be to add a formal check that the pallets had been properly placed somewhere in the process. One candidate inflection point would be during the inspection by the site engineer that takes place before the concrete is poured. The change could be as simple as adding a line to the latter's checklist before he signs off. Alternatively, the structure superintendent could have a measurement of his "pallet-placing performance" added to the parameters that he is measured on, instead of only measuring his performance exclusively according to the quantity of concrete the crew pours each month (as is currently the prevailing practice).

Another possibility would be to sidestep the issue of needing to put the blocks into the floor before the structural concrete work could proceed and find a way of efficiently bringing the pallets of blocks to the floor after the slab above the floor had been poured. There is already a solution in use, since many items (flooring tiles, fixtures, etc.) are delivered long after the slab has been poured: a temporary steel cantilever platform, attached to an existing balcony. However, these platforms are not considered very safe. Another alternative would be to use a "magic arm" extension for a tower crane, which can deliver a pallet to the interior of a building's floor without the need for a surface that is exposed from above (see Figure 6.3). Whatever the exact method chosen, it could allow the two steps to be decoupled and would have the further benefit of reducing the waiting time of the blocks on the floor until the framers were ready to begin.

One could go even further and remove the need to build masonry block walls altogether, by replacing them with a different construction technology. As mentioned in Box 6.2, Tidhar did this in a later project, in which all of the interior partitions were built using the drywall method, obviating the need to deliver pallets during the structural phase entirely.

Tidhar's challenges

But perhaps the deeper issue at work is that the team (and in turn the company) has forgone "continuous" improvement, hoping to replace it with point-based improvement. Yes, during their analysis of the gypsum-block processes, the team unearthed large wastes, and the countermeasures they designed to address them were admirable. But assuming that their work of improvement was done as soon as the team completed their assignment was an assumption that does not jibe with the complex dynamics in any workplace. The very reason continuous improvement has been developed to be continuous is because any countermeasure, no matter how well thought out, will always have unforeseen ramifications as the changes ripple through the system. Some will be predictable, like people's resistance to changes that are foisted upon them. But others may be less obvious beforehand, and therein lies the need to be constantly adapting and improving. Any "team" carrying out a time-limited "project" by its very nature cannot be engaged in continuous improvement. Like Masaaki Imai said, true *kaizen* must be done every day, by everyone in the company, everywhere in the company. A large part of this pervasiveness is required in order to address the additional problems that arise from the changes that were made in solving yesterday's problems.

This is Tidhar's challenge, but it is the challenge of any organization that seeks to implement Lean, respect for people and continuous improvement: improving not only today but tomorrow, and the next day, and the next, and the next, ad infinitum. On the one hand, this long-range view can be daunting or even insurmountable (leading to a tendency to give up without ever trying). But on the other hand, as Toyota and FastCap have shown, each step along the way, including the first, needs only be very small. Just as in eating an elephant, continuous improvement and respect for people need only be undertaken one bite at a time.

Lessons learned

- For Lean to take root in an organization's culture, it must combine both continuous improvement and respect for people.
- Respect for people means helping them learn, grow and develop their problem-solving skills, as well as empowering them to make improvements to the work processes.
- By definition, continuous improvement is never-ending; each solution implemented is only the platform to reach for the next.

Figure 6.3 **Delivering pallets of building materials in the interior of a floor using the "magic arm" extension for a tower crane**
Source: Images courtesy of Eitan Leibovitz, Contex Global Technology Experts.[5]

- In most organizations, the opportunities for improvement are numerous; by engaging the workforce in improvement, great strides forward can be made. Management need only ask.
- In construction, there are barriers to implementing both continuous improvement and respect for people, but with dedication they can be overcome.

Notes

1 Much of this chapter draws on a paper presented at the 2016 International Group for Lean Construction Conference. See Korb (2016).
2 For details of the Two-Second Lean approach, see Akers (2011).
3 The story is told in Emiliani (2008).
4 More detail about the development of the PDCA cycle can be found in Deming (1982).
5 See http://contex-products.com/about-us/.

7 Tidhar on the Park, Yavne

Project profile

Figure 7.1 General view of Tidhar on the Park in Yavne
Note: The "Weather Vane" buildings are the tall towers on the right.

Project description

Tidhar on the Park is located in the small town of Yavne, Israel. The project is a residential neighbourhood with 32 buildings and 981 apartments. It was built in six stages, each consisting of two to ten buildings, as shown in Figure 7.2. The architecture, structure and construction methods used were common in Israel, especially for high-density neighbourhoods. The buildings were between 6 and 22 storeys tall, with natural stone cladding and aluminium framed windows. Each storey had three to five apartments, each ranging from two to four bedrooms (60 to 110 m^2). The parking spaces for the smaller buildings were provided above ground; parking for the two taller "Weather Vane" towers was provided in two underground parking storeys. Between the buildings, the company developed a park with gardens, playgrounds and a small stream.

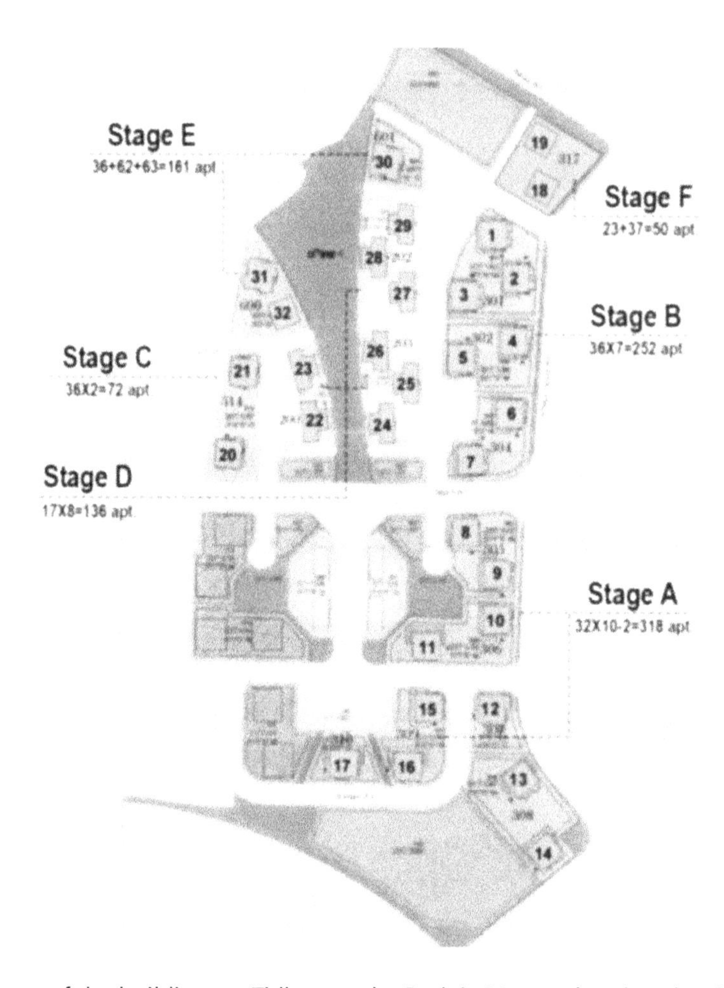

Figure 7.2 Layout of the buildings at Tidhar on the Park in Yavne, showing the six stages in which the project was built

This was the first Tidhar project on which Lean and BIM tools were used intensively, including LPS, 5S and complete modelling of the buildings' systems for clash detection and construction coordination.

Construction methods

The structures were cast-in-place reinforced concrete frames with a central core consisting of two staircases and two elevator shafts. Local building regulations require that at least one room in each apartment must function as a private reinforced security room. The exterior walls are also part of the load-bearing structural frame and were built using the "Baranovich" method. In this method, stone cladding pieces are set in exterior metal formwork panels in a ground-level workshop. The forms are hoisted to the needed storey and concrete is poured *in situ,* with the cladding already attached. The method is explained in detail in Chapter 18, "Raising planning resolution – The four-day floors".

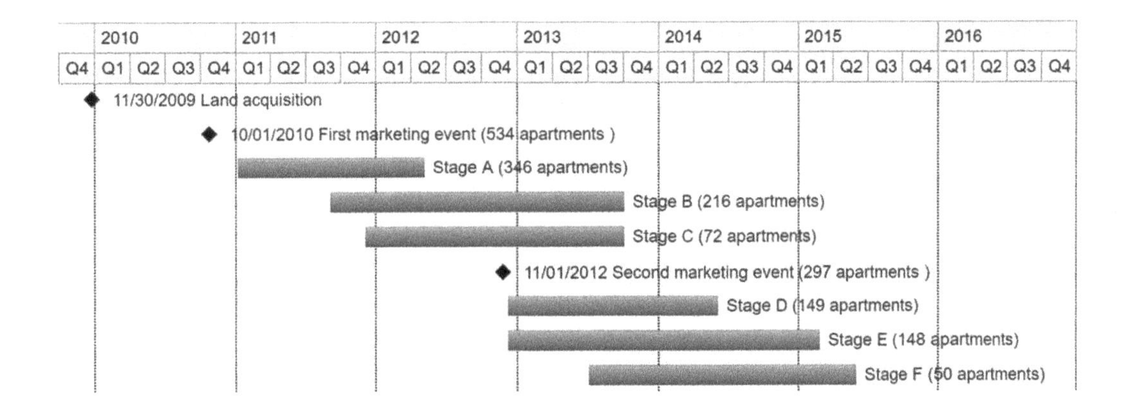

Figure 7.3 The timeline of Tidhar on the Park, Yavne

The interior finishing work included gypsum block partitions, tile flooring and plaster as the finishing layer of the walls. Israeli construction law dictates that all MEP systems must be within the apartment volume, which means that all drainage and sewage pipes must be placed above the concrete slab. To accomplish this, a 12 cm thick layer of gravel substrate is provided between the reinforced concrete slabs and the flooring tiles.

Tidhar's project team included four project managers, six construction engineers, 20 site superintendents, a logistics manager and a BIM modeller.

Application of Lean and BIM

Two of the buildings in Stage A (buildings 10 and 11) were the location of the company's first attempt to implement the LPS Weekly Work Planning meetings. The pilot proved to be a great success (see Chapter 17, "The Last Planner® System"), and as a result the company standardized the LPS procedures and implemented them throughout its portfolio of projects.

During Stage C of this project, the site superintendents for the structural work on buildings 20 and 21 made the first attempt to apply the 5S method, creating an exemplary site in terms of order, cleanliness, visual management and organization of the layout. This effort eventually led to a 25 per cent productivity gain in building the cast-in-place reinforced concrete structure.

The first complete multidisciplinary BIM model used in Tidhar was created for building 30 (Stage E). The architectural model was created by Pivko Architects. This was the practice's first experience with BIM; Gil Geva had persuaded them to work with BIM: "I invited Ilan Pivko to embark with me on a BIM journey, and I promised him a 'safety net' and the support of our own BIM experts." Pivko Architects' successful adoption led directly to Tidhar awarding them the contract for the Rosh Ha'ayin project (see Chapter 13, "Tidhar on the Park, Rosh Ha'ayin"). All of the MEP models were created by Tidhar's own modeller. This model set the standards for both the product and the process for designing in a BIM environment.

Figure 7.4 Layout of the buildings at Tidhar on the Park in Yavne

8 The waste of non-value-adding work

Non-value-adding work is a major part of the waste in construction. Finding a better way to build, in which workers are relieved of these aspects of their work, can significantly improve labour productivity.

Waste is one of the central concepts of Lean thinking. Waste is any expenditure of resources (time, money, energy) that does not bring value to the customer. The customer pays the company money to obtain something that he or she values; the company only exists to the extent that it can continue providing the value that the customer is willing to pay for, thereby injecting cash into the company. Any activity that does not directly provide value to the customer detracts from both the short- and long-term success and survival of the company – these activities are called "*non-value-adding work*".

Seeing the waste of non-value-adding work

How is it possible to distinguish value-adding from non-value-adding work? How can we tell if someone is working hard, even efficiently, and yet not adding value? During the early days of Tidhar's Lean adoption, Gil Geva often used the example of the work of building interior partition walls to explain to his management team what non-value-adding work was. The partition walls were built from 50 cm × 60 cm × 10 cm gypsum blocks of two kinds: regular blocks for most of the walls, and water-resistant blocks for the bathroom walls. Using these big, relatively lightweight blocks is considered to be a highly productive solution when compared with regular concrete blocks or clay bricks, for three reasons: the gypsum blocks are smooth and do not require a finishing layer of plaster before they can be painted; they have interlocking tongues and grooves that ensure a smooth, planar surface; and their size and low density mean that it is easier for the block wall builders to work with them.

At one of Tidhar's biannual off-site retreats, in which all project managers and site engineers take part, Gil explained how he had begun to see things differently:

> We've been either stupid, blind or asleep on the job, maybe all three. Just think about our process for building interior partition walls. In every apartment that we build, we have some area of partition walls, let's call it X m^2 of interior partitions. First, we buy the blocks, let's say Y m^2 worth per apartment. Obviously, Y is

significantly more than X. So the first question is: where is all that excess going? I'll tell you where – we throw it away! Something is wrong if we throw away something we paid good money for. To make matters worse, think of how much handling goes into that excess: we buy it, we transport it to the site, we lift it by crane into the building, we cut it up, we crane it back out of the building, and then we truck it to a landfill. Not only do we pay for every step on the way, we're not exactly good citizens in terms of sustaining the environment. Seems to me we could save ourselves a lot of money if we just buy the excess blocks and ship them directly to the landfill, without all the work in between. But that's not even the half of the waste – let's look carefully at the whole process.

The blocks are delivered to site by the truckload. The truck driver shows up at the site and unloads the pallets wherever he can find space. Then, typically, we move the pallets a few times around the site because they're invariably in the way of some other activity before finally lifting them up with the tower crane to the floor that's under construction. We pay people for their time and effort to do all that moving, but we still don't have any partitions built yet. Does the customer who bought the apartment care how many times we move their partitions around the site? No, not at all. And yet they end up paying for it all, since it's all in the price they eventually pay.

Figure 8.1 Pallets of gypsum blocks for partition walls unloaded haphazardly on the site

And then the blocks begin their wait for the builders. It takes some time, and it's not a very peaceful wait. When the concrete formwork and shoring are stripped from the ceiling slab above the pallets, the beams fall on the top row of blocks. The blocks don't survive very well – the top layer of the pallet are the first casualties that are headed for the landfill. More money down the drain.

After a few weeks, the builders arrive. They open up the pallets and distribute the blocks for each wall. But the top blocks on the top pallet are 2.20 m above the floor, and each weighs about 17 kg, which while light as blocks go, is still pretty heavy to repeatedly take down from that high! Workers have to climb up on something to get them down. It takes a lot of sweat. Next, they sort them out, cart them around the floor to the different apartments and partitions. We're paying for their time – and still we have no walls!

And finally, they begin building. This is the good part – they're finally doing something of value. It doesn't last long. Once they get to the end of the first row, the blocks are too big. And near the ceiling, there is almost always 10 cm to 20 cm left to fill. How do you fill a 20 cm row with a 60 cm block? You have to cut it! And that's what they do. They cut the blocks for the ends of the walls where you need a half block on every other row, and when you meet a door opening, and also for sections that are less than 50 cm long, and for the top row, and so on. There is a lot of cutting, and a lot of offcut pieces.

Figure 8.2 Block offcut pieces waiting for disposal

> And where do the offcuts go? You think the workers collect them and reuse them in other places? Think again. We're so smart, we learned a long time ago to subcontract the block work and pay the builders for finished product, for every square metre they build, not for the time they spend working. Do we pay them to collect blocks and save material? No. If they do collect them, will they build more square metres per hour? No. Why would they? Almost all the offcuts go to scrap, into the dumpster on site. We pay a lot of money for every dumpster load that is trucked away, and we even pay the city for the landfill service!
>
> So why am I telling you all this? It turns out that the time the builders spend actually building walls is only about **one third of their time**. If we could get rid of all the carrying, measuring, waiting, cutting, clearing away offcuts, rework, etc., in theory we have the potential to triple their productivity! That's the waste of non-value-adding work. The excess material I told you about, the stuff that goes to the landfill – that's waste for sure, and it's easy to see and to understand – but the cost of this wasted material pales in comparison with the wasted time of the builders.

There were murmurs across the room, as some of the engineers and site superintendents shifted uneasily in their seats. One of them spoke out:

> Gil, I understand about the material we throw away, I realize that's waste. On my site, we pay the builders a small bonus to collect and reuse as much as they can, and that saves us money for sure. But we don't pay for their wasted time, we pay per square metre of wall built, no matter how long it takes the subcontractor. So why should we care? Why should we invest our time and effort to do anything about it?

Gil smiled. This was the question he had been waiting for:

> If you were a builder, where would you rather work? For a contractor who paid you NIS100 per square metre, and where you average about 200 square metres per month, or for a contractor who paid you NIS80 per square metre, but where you average about 300 square metres per month? What about as the contractor: would you rather pay NIS100 per square metre, or NIS80? And would you rather have the subcontractor teams continue taking a week and a half per floor on average, or have the work be done in one week? This is where our opportunity is. This is how we're going to beat our competitors.

Gil himself had recently read about this example in a research report. Four years earlier, Professor Sacks's team at the Technion had undertaken a value-stream analysis of the process stream of the gypsum blocks, with the goal of quantifying the waste inherent in the current state process, as part of a study funded by the Ministry of Housing and Construction. The research report was available to all, but industry had largely ignored it.

Measuring the waste of non-value-adding work

The researchers used the standard VSM approach: follow a product from its manufacture in a plant, through storage, transport to site, installation and disposal, and to its final use by the end customer. The site observations were carried out, coincidentally, at a construction site where Tidhar was the general contractor. Research assistants observed the builders continuously over a number of weeks, recording their activity every five minutes. Using this work study method, they collected data defining the times required for a number of cycles of each operation. The results were eye-opening. The data showed that the builders were engaged in building block on block for some 32 per cent of their time. As can be seen in Figure 8.3, the rest of the time was devoted to non-value-adding activities, particularly to cutting blocks (24 per cent), waiting and rework (11 per cent), cleaning up (10 per cent) and dealing with design changes (7 per cent).

The researchers also looked into the material waste and compared the equivalent costs of the different wastes. They compiled a BIM model of the partition walls in a typical three-bedroom apartment and used it to assess how many unique block shapes and how much cutting was required. The total net area of the partitions was 71.4 m^2. Their construction required 314 individual pieces and 197 cutting operations were needed. The offcuts that could not be reused somewhere else and thus had to be disposed of were the equivalent of a further 14.6 m^2, making a total requirement of 86.0 m^2. Each pallet contains 44 blocks, for a total area of 44×0.5 m $\times$ 0.6 m = 13.2 m^2. This means that seven pallets (92.4 m^2) are required per apartment at the minimum, to provide the 86.0 m^2.

The result was that the difference between the gross amount delivered and the net amount remaining in the walls in each apartment was 21.0 m^2, an excess of 22.7 per cent. This amount, which is destined to become scrap that must be disposed of, includes the offcuts, blocks broken during transport and storage. To flip the numbers around, one could also say that the cost of materials and their handling was 29 per cent higher than what the customer valued and should have been willing to pay for (92.4 m^2/71.4 m^2 = 1.29).

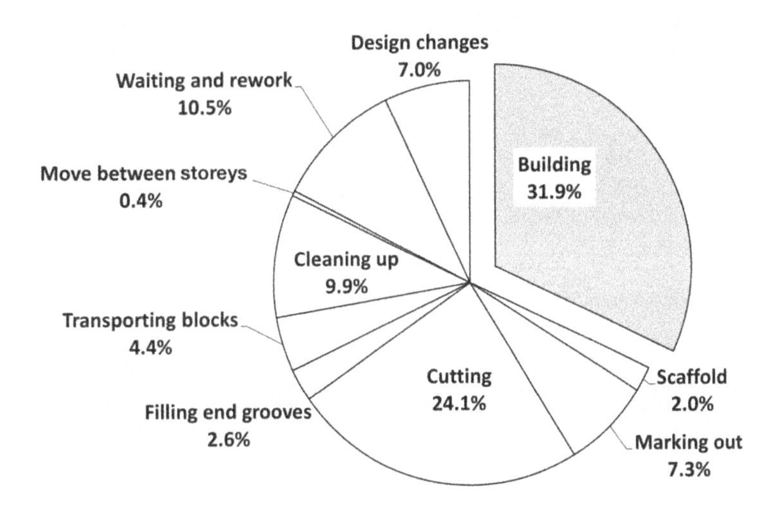

Figure 8.3 A breakdown of the time spent by builders when building gypsum block partitions, showing value-adding activity in grey and non-value-adding activity in white

Source: Sacks et al. (2008).

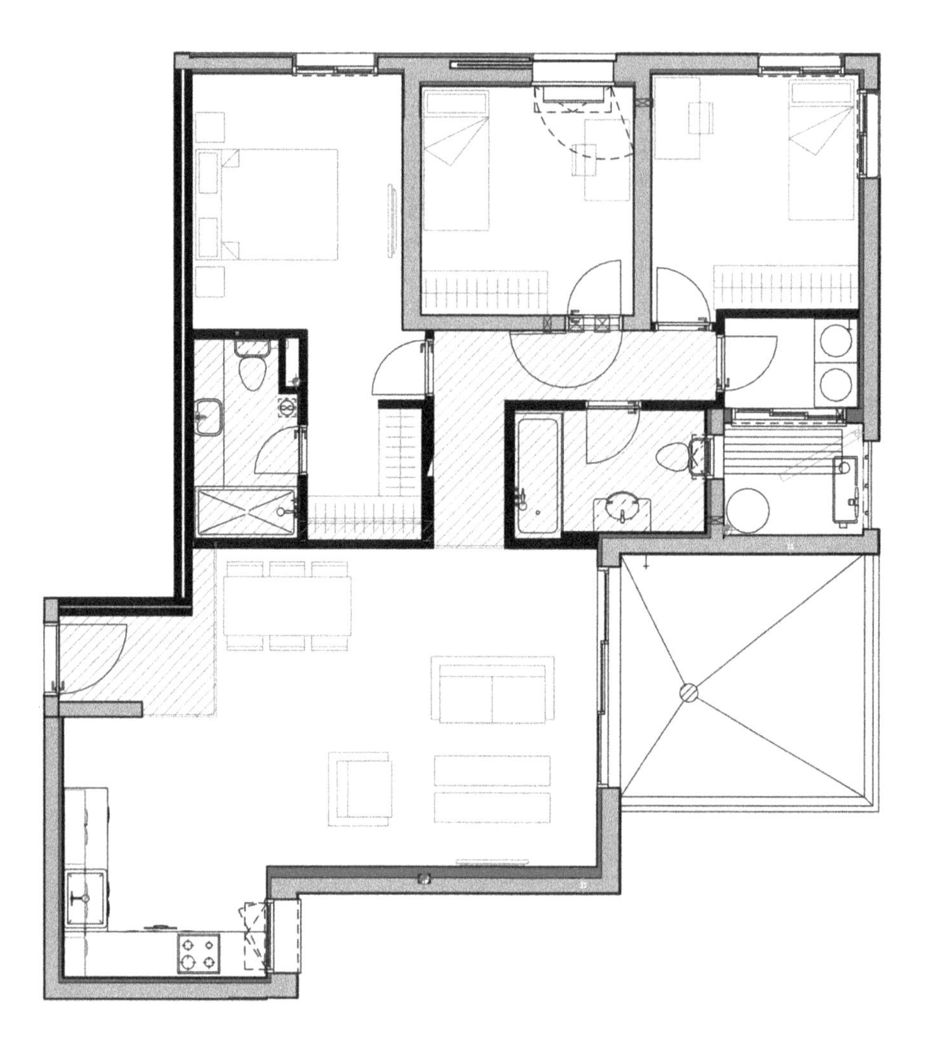

Figure 8.4 The floor plan of a typical three-bedroom apartment, showing the block partition walls with thick black lines

Note: The exterior walls are made of cast-in-place concrete with a natural stone finish and an internal drywall thermal insulation layer. Source: Sacks et al. (2008).

These results were presented to industry audiences on numerous occasions, along with the broader literature available on Lean research and thinking. Senior construction managers from different companies would listen politely, ask questions and agree that the problems exist, but none took any action within their companies to effect change, in either this or in any other production methods. As discussed in Chapter 2, "False starts", in many ways the organizations were not ready to see this waste, in the sense that many of the issues that must be resolved to reduce non-value-adding work are systemic and require broad action to have any lasting effect. The report ultimately served as just another witness to the inertia of production systems that have evolved within the

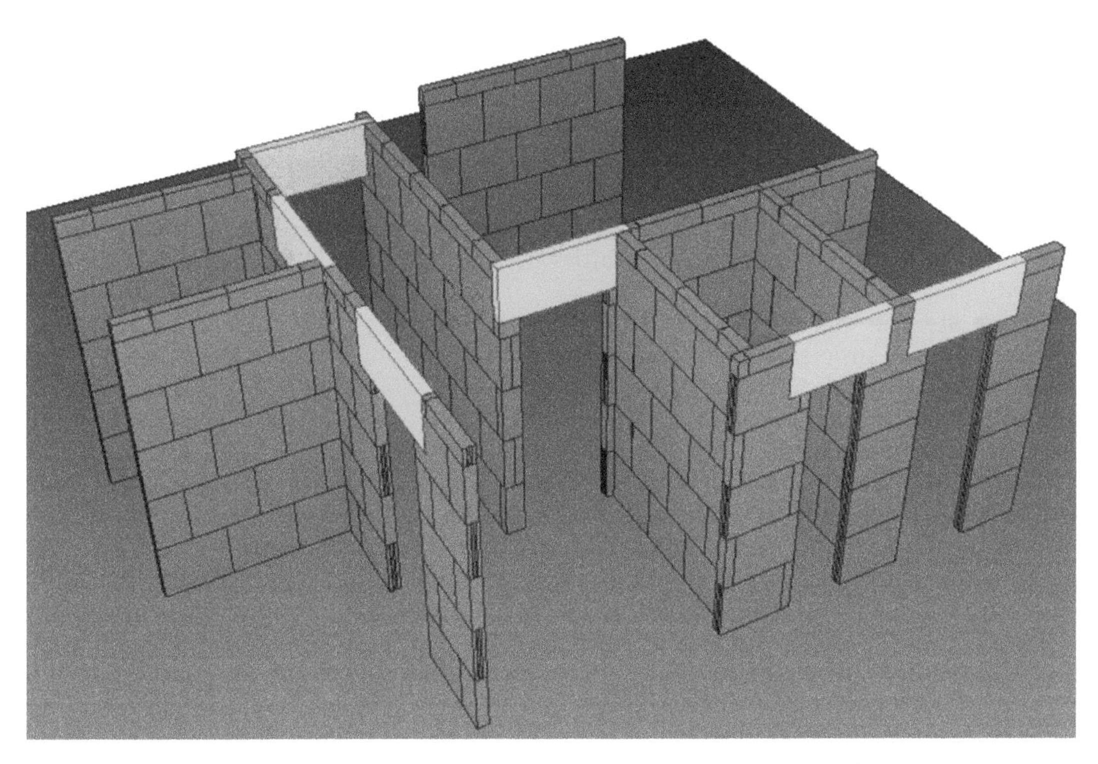

Figure 8.5 A perspective view of the block partitions in the typical three-bedroom apartment shown in Figure 8.4

Note: The model was prepared using Tekla Structures software.

Source: Sacks et al. (2008).

economic context of the highly fragmented construction industry, seemingly beyond the reach of any one individual engineer or company. The pressing concerns of today (dealing with the urgent rather than the important), the tendency to see projects as islands disconnected from one another, the pervasive practice of transferring risk by subcontracting work to the lowest level possible (and thereby reducing construction management to contract administration devoid of production management), appear to discourage people from confronting the challenge.

The report was shelved, but it was not forgotten. When the Tidhar team kicked off their Lean effort, they invited Professor Sacks to give a series of presentations. With ready listeners, the report, prepared four years earlier about one of their own projects, was now far more significant than it had been earlier when presented to unprepared audiences.

What is the waste of non-value-adding work?

The concept of waste (defined as any activity or consumption of resources that does not add value to the end customer) is key to Lean and the philosophy of *continuous improvement*. It is intertwined with the notion of *respect for people* in the company since it gives people ways of thinking, seeing and communicating with each other that all lead to additional value being created with fewer resources being consumed. When people in the organization have a shared vocabulary of waste, a shared understanding of the need to work continuously to minimize and eradicate each waste found, and a shared commitment not to blame individuals, they can begin to work collaboratively to improve. Each waste identified represents an opportunity to improve, thus increasing the efficiency and effectiveness of the organization as it creates value.

In order to help people identify the various forms of waste around them, Toyota identified three categories of waste, which have similar names in Japanese: *Muda*, *Mura* and *Muri*.

Muda are the wastes inherent in the work processes. These are the traditional impediments to efficiency identified in the field of industrial engineering from as far back as Frank and Lillian Gilbreth and their studies of the wastes in bricklaying.[1] Toyota identified seven canonical types of Muda, which further help focus the energies and the eyes of employees, which we discuss below.

Mura is the waste of unevenness. Any peak or trough in the workload represents either the overburdening of resources (this is the Muri that will be mentioned next) or lack of full resource utilization, and sometimes both. Slow and steady wins the race not only for the proverbial tortoise competing against the hare, but also in production; while constant, even demand and production are more boring than firefighting and racing to meet deadlines, they are much less prone to errors and waste. Construction is rife with examples of unevenness (like any industry, but writ even larger given the cyclical nature of the work of building a building), and all of the unevenness represents waste and opportunities for improvement.

Muri is the waste of overburden. Any resource, be it human or otherwise, has a limit to how much it can deal with at one time. It may be possible to exceed those limits for short periods of time, but long term, the behaviour is unsustainable. Machines begin to break down and people begin succumbing to stress. Overworking does not demonstrate respect for people, and it leads to increased errors, which in turn cause defects.

Muri and Mura are often given less emphasis than Muda in all its forms, but they are no less important and no less wasteful.

The types of Muda

Taiichi Ohno identified seven main types of process wastes that can occur. Just like the concept of waste in general, the point of this typology is to help people learn to see as they look for examples of each type. There may be examples of waste that do not fit neatly into any of these categories, and more often than not there is overlap between the types (or one may lead naturally to another).

The seven types can be remembered in the mnemonic TIMWOOD, as follows:

Transportation: the waste of moving raw materials around on their journey to becoming finished products. In practice, it is infrequent that value is added while the material is being moved (ready-mixed concrete is a wonderful exception that strongly illustrates the rule), which means that almost all movements of materials are waste. One way to track the more egregious wastes of transportation is to create what's called a "spaghetti map" – trace the path that the material takes on its entire journey through the process. This can often produce eye-opening results.

Inventory: any material or products that are sitting by themselves, not being worked upon, are by definition not having value added to them. And this is what inventory represents: a stockpile of waste. In financial terms, inventory must be purchased (requiring an outlay of cash), but until it is sold as a finished product, the purchase price is not recovered. Thus, a company that has a large inventory has much more cash tied up than one that has managed to reduce its inventory, and cash flow is a primary concern of most construction companies. Improved cash flow is one of the major benefits for companies that begin seriously implementing Lean. Inventory is one of the easiest wastes to find since it stands out to any observer who has been sensitized to see it as waste. In construction, all of the piles of raw materials are clearly inventory, but also any unfinished building, unit or apartment is also a form of inventory. If the project has repeating sub-units (like an apartment building), then the number of units being worked on at any point in time can be considered a form of WIP inventory.

Motion: the waste of movement of workers around the job site that does not add value. This operates at both a macro level (for example, if the worker repeatedly has to walk from one area to another to retrieve tools or materials) and at a micro level (including ergonomic issues like having to reach up too high to retrieve a product from a shelf). Transportation is about materials while motion is about people, but both have to do with moving around without adding value. Here too, spaghetti maps can be helpful aids; one organization found that in the course of a typical shift, a factory worker was walking over three miles to fetch tools and materials from different parts of the building.

Waiting: if a person or machine has been employed to perform a job but is prevented from doing so since not all of the prerequisites are in place (i.e. the raw materials have not arrived, equipment has broken down, a quality problem is being investigated upstream), then that person's or machine's time is wasted.

Overproduction: performing a task before it is required by the next step in the process and/or processing a larger quantity than is required by that next step in the process. Taiichi Ohno cited overproduction as the worst waste of all, since it is overproduction that causes inventory. Inventory then must be stored somewhere, which requires transporting the material to the storage location. When the workers come to retrieve it, they incur wasted motion. Also, while the inventory is being built up, there is a greater possibility that defects will go unnoticed by the downstream processes, which means that more will have to be reworked or discarded once the defect is eventually unearthed. Not to mention that the inventory can be damaged by external factors while it is waiting around (precipitation, wayward forklifts, changes in demand). All of this arises because of overproduction. The concept of overproduction being a waste flies in the face of traditional efficiency thinking, and requires a paradigm shift to fully appreciate. Traditionally, each step in the process tries to pump out product as fast as it can, with no regard to the readiness of the next step in the process to accept the output. If any latter process is not ready, then technically any production done by the former process is overproduction, which is a waste.

Over-processing: putting additional effort into a product beyond what is desired and valued by the customer. Sometimes companies try to outdo themselves in terms of the quality and number of features that their product offers. But if those features and quality are not what the customer requires (and not what he or she would want to pay for), then they are waste.

Defects: any quality issues that lead to a product being discarded. In addition, beyond the part that gets thrown in the trash, all the processing and transportation steps that led up to the point where the item is rejected are implicitly also wasted. They too are all being thrown in the trash. The waste of rework is included in this category. Rework does not result in the whole product being thrown out, but it does require handling and processing the product much more than required by the customer.

In the West, employees are sometimes put off by the use of the term waste, particularly when they are informed that a very large proportion of what they do in their jobs adds no value to the customer. Where the respect for people aspect is weak, people may misinterpret the assertion as a personal attack, perhaps fearing that if most of their work is waste, they are of little value and may be superfluous to the company. To soften the message, an alternative concept was added to reduce the threat level. Typically, this involves drawing a distinction between wastes that can easily be removed and wastes that are more entrenched. The term "supporting activities" refers to activities that do not add value, but are considered essential or unavoidable. For example, transporting materials to the work face does not change their form, fit or function (the three value-adding operations), but it is essential. Some people call these "Type 2 Muda", and the other non-value-adding activities are called "Type 1 Muda". But euphemizing does not change the fact that either the activity adds value to the customer or it does not, and if it does not, it is waste, pure and simple. The compromise in language can cause additional problems as team members argue about whether a particular example of waste is "required" or not. Ultimately these arguments are pointless, since waste is waste – it is only a question of how much the system must be changed to extricate it.

The problem is almost always in the processes, not the people. If anyone is at fault, it is the management for crafting an organization and its processes that contain so much inherent waste. Just because an activity is understood to be waste (in the sense that it adds no value to the customer), that does not mean that the organization is ready to stop doing it immediately and promptly dismiss the personnel who are employed to carry it out. That would be a clear violation of the respect for people principle (successful Lean organizations maintain a "blame-free" approach to problem solving, and the best among them make a commitment that no one will be fired as a result of process improvements). Similarly, despite Lean pointing a clear path towards zero inventory, no Lean enthusiast would begin the improvement process at a company by first throwing out all the inventory. The system and the processes are not yet ready for such a drastic change.

Likewise with waste: some wastes can be eliminated immediately, with no harm done to anyone or anything. But most wastes are there because they fulfil some real or imagined internal need or to cover up some shortcoming elsewhere in the process. If the underlying issues are not addressed first, the waste cannot be extricated or discarded. This does not mean that it is not waste, and there is no need to sugar-coat that fact by calling it by another name – doing so might work at cross purposes to the desire to improve. If a worker is cognizant of what is waste day after day, that awareness can be the spark for ideas for improvement. The aggravation of having to repeat the waste over and over should only fan the flames of motivation to improve. The best thing management can do in this case is to allow the changes to be made and experiments to be carried out, and not block improvement efforts once people are keen to improve.

Tidhar begins to address its block problem

One of the first Lean activities undertaken by Tidhar was VSM. The exercise aimed to compile a shared understanding of the current state as the team learned to see the opportunities around them. The opportunities for improvement were apparent, particularly since the team had been sensitized to look for the various types of waste, but mapping was needed as a basis for identifying and defining the specific steps in the proposed future states. Two production processes were chosen as the initial projects: tile flooring and gypsum block partition construction.

The gypsum block team had eight members. They represented the full spectrum of people related to the task at hand, starting from block builders and site supervisors, to the quality inspector who could represent customer value, all the way up to the CEO of the construction division. The members were chosen not only to represent all of the interests, but also for their personal communication skills, creativity, and teamwork abilities. The members included:

- a block-builder subcontractor who had many years of experience leading crews building partition walls;
- a site superintendent specializing in interior finishing work;
- a site construction engineer;
- a quality inspector;
- a construction engineer from the head office controls department;
- a regional project manager;
- the CEO of Tidhar's construction division;
- an external Lean consultant.

The team got to work. They began by observing the process, much in the way that the research team had done earlier, but with closer attention to detail. Their results were similar to those shown in the pie chart in Figure 8.3, but with one striking difference: their proportion of value-adding work was just 10 per cent, significantly less than the 32 per cent measured by the Technion researchers. The discrepancy arose because they distinguished between the time spent placing a new block on a wall and the time spent levelling that block using a spirit level, which occupied 10 per cent (considered value adding) and 23 per cent (considered non-value-adding) of the total time, respectively. Is the time spent levelling the blocks waste? This is essential for achieving value – no one wants a wall that is not straight and level – but viewing it in this way raises the question of whether it could be done in a better way. Regarding the remaining 67–68 per cent there was no question – it could all be considered non-value-adding.

An early observation related to the way that the blocks were transported to each floor of the building in pallets. Apartment buildings with reinforced concrete structural frames, like most other buildings, are built from the ground up. The construction of each storey of the building has two basic steps: build the vertical elements (walls and columns) and then the horizontal elements (ceiling slabs). Both steps consist of sub-cycles of building formwork (moulds), installing steel reinforcement and any electrical or plumbing conduits that may be needed, pouring the concrete itself, curing and stripping the formwork. In the time after each slab is poured and before the next slab's formwork is put in place, there is a short window of opportunity in which tower cranes can lift materials from the ground and place them in the building for further construction work after the structure is completed. This is the way that the pallets of gypsum blocks are traditionally brought up to the apartments, since the only alternative – using temporary steel balcony platforms to receive deliveries and then transporting the pallets horizontally through the storey – requires much more effort and expense and is far less safe.

As one of the engineers on the team put it:

> At least we do one thing right: given the difficulty of delivering pallets of blocks via crane into the apartments once the structural frame is complete, and only then distributing them within the building storeys by hand, we hoist them before the ceiling slab of each storey is built and spread them out on the poured floor slab. Have any of you watched what actually happens? The crane guide has the operator set them down wherever he can find room between the column and wall starter bars, so they're usually stacked two pallets high.

What the work team noticed was that when the cranes placed the pallets on the slabs, they were always stacked one on top of the other. "Noticed", of course, is not an exact description – all of the team members were experienced enough to know that this was common practice. What the team did was to begin zooming out from this local optimization to look at the implications on the wider system of construction of the interior partition walls with the gypsum blocks.

The drawbacks of stacking the pallets were twofold. First, the act of stacking itself increased the load on the blocks, which meant that a higher percentage of the blocks were broken before they were even unwrapped (Gil's statement at the beginning of the chapter goes as far as implying that the entire top layers of the pallet had to be scrapped). Since the pallets were stacked up almost to the ceiling, when the forms of the ceiling slab were struck, there was a tendency for the formwork to fall onto the pallets of blocks when their supports were removed.

Figure 8.6 From left to right: pallets stored within a building storey with two levels, a special electrical saw used for cutting gypsum blocks and a wheelbarrow full of offcut pieces

The builders typically arrived on the floor after the ceiling slab was "closed", and they were faced with an array of pallets stacked almost to that ceiling. In order to unload the pallets and begin the work of building the walls, they had to start from the top layer of blocks. But with the blocks weighing 17 kg and the top of the stack at 2.2 m, there was no way a worker on the floor could unload the top layer. This meant that the real first task at any site for these builders was to build makeshift work platforms so they could climb up and reach the top layer of blocks. They would lower the blocks to the platform, then get down off the platform themselves and move the blocks to the floor. Given the height and the need to descend the platform, it was easy to drop a block. Once dropped, a block would be broken and have to be scrapped. But the more insidious wastes, according to Gil, were those surrounding the work of actually building the walls: "To begin with, they mark out the locations for the partitions, which of course depends on them having the updated drawings. Mostly we succeed with that, although sometimes the designs change after they build – but that's a whole other story".[2]

In the task of actually constructing the partition walls, there was more waste. The blocks are each 50 × 60 × 10 cm, but a completed wall was almost never an exact multiple of that size. This meant that some of the blocks had to be cut to size to fit the finished wall. Construction blueprints very often do not go down to the level of detail about which blocks should be in which configuration in a partition wall; typically, the location and the overall dimensions of the wall are specified and builders are left to their own devices to figure out how to "best" put the wall together out of the blocks at hand.

This lack of standardized work means that each builder could have a different method for constructing a wall (which blocks need to be cut where), and each builder's choices might vary from apartment to apartment. These factors virtually guaranteed that a large percentage of raw material was wasted. The cut-off fragments mostly found their way to the dumpster, even though some of them could have been used. This was in part due to a divergence of interests between the general contractor and the framing subcontractor: the GC pays for the blocks, including those destined for disposal, and would like to see as much reuse of off-cuts as possible, whereas the subcontractor is paid by square metre of wall assembled and thus does not want to spend time sorting and reusing off-cuts – easier to throw them out and take a fresh block each time.

Although the builders were supposed to be spending most of their time building walls, they spent a lot of time moving blocks around. Since the crane operator would place pallets of blocks wherever he could find room on the storey, blocks for partitions in apartment A could be located in apartments B or C. If a particular type of special block (lintels, for example) ran out, it would have to be fetched from a different storey.

Of course, all the waste blocks – those broken in delivery, discarded off-cuts, excess blocks that were part of the buffer stock – had to be removed from the building, loaded on a truck and carted to the landfill. An engineer commented:

> And here's another nice thing – all the partitions are built, but there's still work to do. The clever grooves and tongues on the edges of the blocks? Where they're exposed, at openings and at the ends of some partitions, they have to be either cut off or filled up. Another few hours of work per floor that we pay for.

Continuing to observe the partitions after they were ostensibly finished revealed an additional waste. When electrical, plumbing and HVAC workers arrived, in turn, to install their systems, they always began by cutting grooves and holes into the partitions. Large rectangular openings had to be cut to allow penetration of air ducts. As we saw in Chapter 6, "Continuous improvement and respect for people", fitting an apartment's main electrical switchboard meant not only cutting a recess to mount the plastic box, but also cutting blocks above and below to allow cables to feed into the box. A great deal of rubble resulted and the walls were often so damaged that builders had to return after the systems workers had left to patch and repair them. This is the waste of defects, which incurs motion, transport and rework, and interrupts the flow of the builders' work in other locations.

Clearly there were many opportunities for process improvement. As the team learned to see the wastes, the processing steps and delays that failed to add value to the customer stood out in stark relief. For the customer, value is added when the raw materials are changed in such a way as to process them toward the finished product. The three "Fs" of form, fit and function, are the three things that can be changed while adding value. In this case, value is added for the customer when the gypsum block is set into place as part of the partition wall (changing form). A certain minimum number of cuts of some of the blocks (assuming there is no possible way that blocks could be provided in exactly the size necessary – more on that later) can be considered value adding, since they change the form of the raw material in the direction of the finished product.

According to the Lean paradigm, everything else in the process was waste. Using the seven types of waste identified by Taiichi Ohno as a rubric, some of the examples above can be categorized as follows:

Transportation: all of the movement of the physical products, in this case the blocks. From the delivery truck to somewhere on the site, from there to the crane staging area, then up to the storey under construction. From there to the proper apartment, then finally into the partition wall and finally the transport of scrap material following the same trajectory in reverse.

Inventory: the large number of pallets that come off the truck, placed all together on each storey and stored, sometimes for weeks, until the builders arrive.

Motion: the motion of the builders as they climb up on the makeshift platforms, cart the blocks around the floor and dispose of the off-cuts.

Waiting: the blocks have to wait on the site after delivery to be lifted up to the storey under construction, then even longer while the slab forms are assembled, the rebar is laid, the slab is poured, the shoring is removed and the forms are stripped before it is even possible to begin the work. Even then, there may be additional delays before the builders arrive.

Overproduction: the symptoms of overproduction are inventory and waiting, so it is clear from this example that bringing the pallets to the site and to the floor so early was overproduction – a prior step in the process (bringing the pallets to the storey) that is carried out before or in a larger quantity than required by the next (building the partition walls). Building walls without the large openings needed for ducts is also overproduction, in the sense that areas of wall are built that are not needed. It leads to rework (cutting holes) and if damage is caused, it interrupts the flow of the builders' work elsewhere, with the added wastes of motion and transportation of rubble and new blocks needed to repair damage.

Over-processing: the additional cuts of the gypsum blocks beyond the absolute minimum required to construct the wall represent more processing of the product than required by the customer, and thus are a waste. Where off-cut pieces are stored to be reused, the waste of over-processing manifests itself in the collection, sorting, storing (inventory) and searching to retrieve a useful off-cut piece when needed.

Defects: blocks that are broken along the way and are never used and then discarded. If any of the finished walls are not fully plumb, square or planar, they must be repaired.

The team takes action

The first step was to abolish the practice of stacking pallet upon pallet when placing the blocks on the storey. But there was a problem: twice as many stacks of blocks meant twice as much floor space taken up by pallets, which meant that the operation required more detailed planning of exactly where each pallet would go. The team prepared a procedure for planning the quantities and the delivery locations for the pallets using the BIM model for the typical storeys. They prepared the necessary graphic families and a view template with a quantity take-off schedule for project production planners[3] to use for marking out exactly where each pallet should go, leaving a buffer around the exterior walls so that the tie rods from the forms could be removed, and making sure none of the pallets interfered with the planned position of the shoring posts for the slab formwork. Also, floor space where partitions would eventually go was kept clear, so the finishing work could proceed without the need to move the pallets first.

Figure 8.7 shows a plan and an isometric view of the layout provided for the crane operator and signaller to facilitate their work. Visual details of this kind go a long way to avoiding errors and to making the work less cognitively onerous for people who work under time pressure. This level of resolution had the added benefit of allowing the right number of pallets to be placed in each apartment to further cut down on transportation and motion. To understand the scale of the waste, consider that the team had made a detailed assessment of the extra work of moving pallets of blocks as a result of inaccurate deliveries. Moving the pallets around within that building (which had 92 apartments on 28 storeys) incurred a cost – all waste – of $25,000, the equivalent of 16 per cent of the total labour cost for building the partitions.

The builders were enthusiastic supporters of the change. Gone were the stepladders, stools and work platforms, and the ergonomic problems that accompanied them. After working this way for some time, one of the builders remarked "Don't ask me to go back to working that way. If you pile pallets one on the other, you'd better bring someone else to take them down." What was once just an accepted part of the work became an unthinkable nuisance.

The next major change was increasing the "level of detail" of the wall in the design phase. Rather than just specifying the overall dimensions of the wall, the team modelled the blocks in the BIM model, fitting them together like virtual Lego blocks. The goal was to minimize the number of cuts and the amount of wasted material. By detailing the entire apartment this way ahead of time, they could also identify exactly which off-cuts could be used for other locations that needed partial blocks. Investing the time to create a detailed plan before the work was to be done in the field had the potential to greatly reduce the waste. The outcome of this planning was a "shop drawing" that was handed to the builder, showing exactly which blocks would be used where in constructing the walls.

Some cuts were inevitable – each course of blocks must be offset from that below it by half a block, so that the seams do not line up vertically. Thus, even if the wall length is an exact multiple of the block width, there will still have to be cuts. But it was

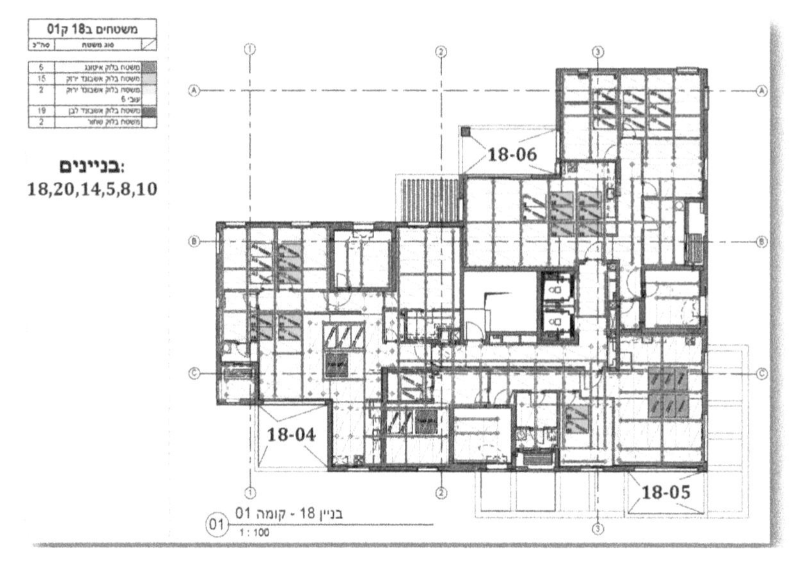

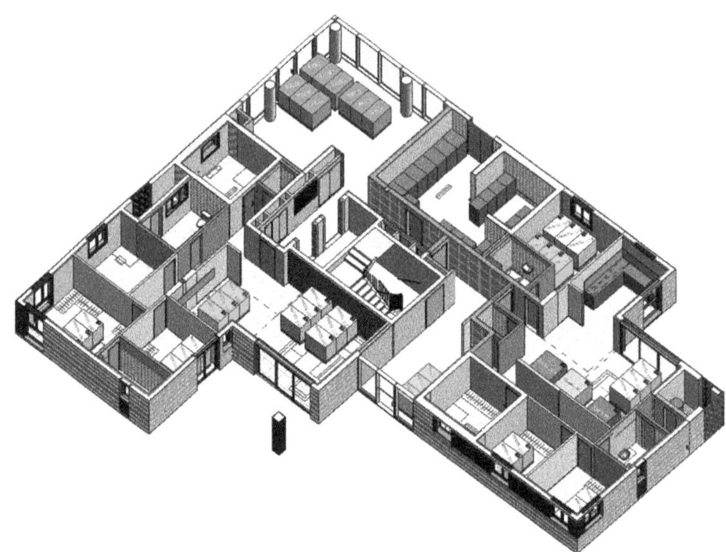

Figure 8.7 (top) Tidhar layout plan for placing block pallets on a storey, showing the type of block in each pallet and their placement between the planned locations of shoring poles that support the slab formwork; (bottom) an isometric view of the BIM model provided to the crane operator and the signaller

theoretically possible to configure the blocks in such a way that there would be no cuts to finish the wall vertically and close the gap between the last full-size course and the ceiling. Doing this, however, meant that the height between the floor slab and the ceiling slab had to be designed to accommodate the block sizes.

The team achieved this with an innovation that also solved a different problem. The first row of blocks was laid directly on the concrete slab, before installation of the flooring. Since the ceramic floor tiles required a base layer of gravel, the lower part of

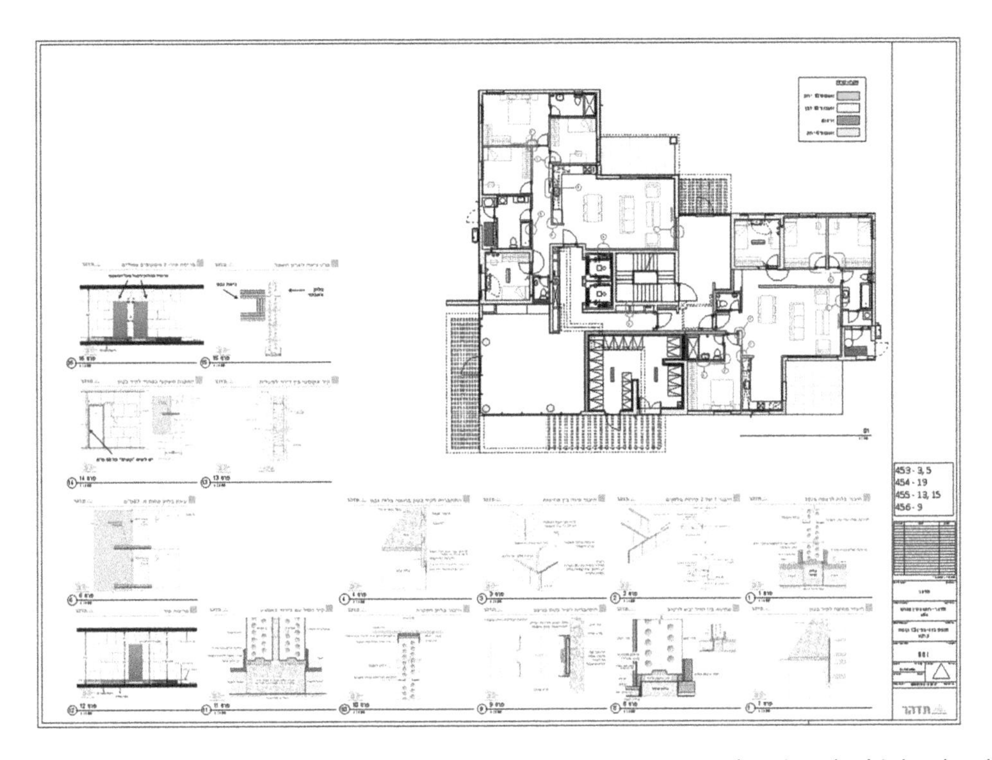

Figure 8.8 Tidhar shop drawing for the partitions in an apartment, showing the higher level of detail that was planned to reduce waste from non-standardized work.

the blocks was potentially exposed to moisture. Previously, therefore, the first row was built with special moisture-resistant blocks (tinged green to be easily distinguishable from the regular blocks). These are more expensive than the standard white blocks, and much of their 50 cm height was unnecessary as only the lower 15 cm or so was needed. The team changed the partition design to use a concrete block at the base row, and Tidhar contracted with a concrete block supplier to manufacture blocks with a height of 18 cm. This was a special dimension block, manufactured only for Tidhar. But it was the exact height needed for the gypsum block rows to reach the ceiling with no cutting. The changes made upstream in the supply chain – in the architect's product design and the component fabrication dimensions – removed significant waste at the work face.

With all of these changes, it was possible to be more accurate in the quantity of blocks brought to each floor. This meant they could avoid situations of either having too many blocks on a floor (which would then be thrown out) or too few blocks, which required manually hauling blocks up or down the stairs to cover the shortfall. This level of detailed planning was only possible through fully utilizing the BIM system. It allowed the team to create a bill of materials requesting exactly the number of blocks of each type required for each apartment. In fact, the numbers were so accurate that they decided to shoot for zero waste of blocks on each floor. The precise number of blocks would be supplied – the only safety stock would be in "rounding up" the number of pallets required to exactly outfit the floor to a whole number (i.e., they would not require a pallet to be divided before being brought to the floor).

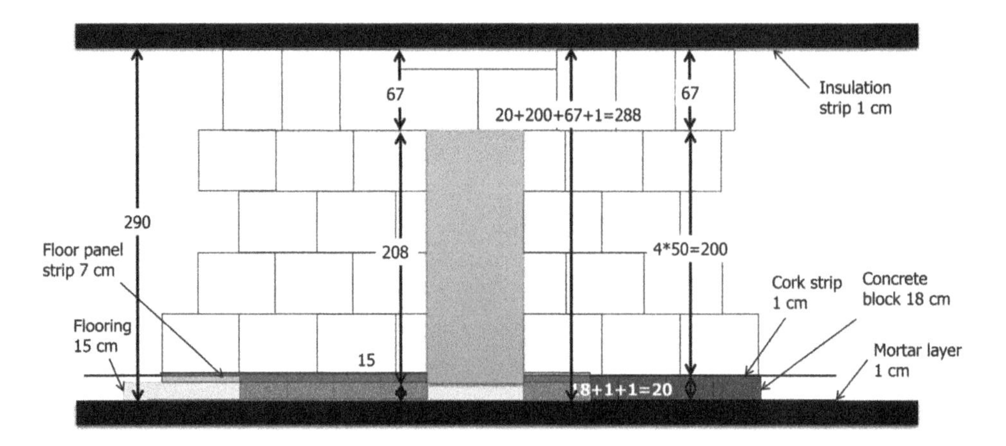

Figure 8.9 Production detail of a block partition designed to minimize cutting of blocks

Another change made the physical waste of blocks visibly tangible on site. The team decided to equip the builders on each floor with one polypropylene "big bag", which has a volume of about 1 to 1.5 cubic metres. The builders were instructed to put all of the scrap pieces of blocks into only this volume, and no more. Also, they were required to bring the scraps with them from apartment to apartment as they moved around the storey constructing the partition walls in each. If they needed a spare piece, they were to first check through the bags before cutting a fresh block. This process significantly reduced the amount of discarded scrap.

More than just the material cost of the wasted blocks that were saved, the real savings was in no longer having to transport so many additional wasted blocks. As Gil lamented at the beginning of this chapter, scrap blocks had to be transported to the construction site from the factory, up by crane to the storey under construction, back down the garbage chute after they were deemed scrap, and then hauled away from the site to be properly disposed of in a landfill.

Ultimately, all of the changes proved themselves. In the project in Yavne, what would have typically been 40 big bags of wasted material was reduced to four – a saving of 90 per cent wasted material and associated transportation, to say nothing of the ecological value of that much less landfill. Also, the saving was apparent in the amount of time that was required to build a wall – less waste meant less time required per storey, which means the overall duration of the work was reduced as well.

Learning to see waste

Identifying all of these cases of waste does not mean that the company must or is even capable of immediately revising the process to remove the waste. There will always be some element of transportation, unless – *ad absurdum* – the gypsum block factory was located inside each apartment being built. Likewise, every other category of waste will inevitably exist in some form or another, no matter how many improvements are made. Toyota themselves often comment on how much waste there is in their processes, and they have been continuously improving their processes for over 50 years.

The point of identifying the waste is to begin to change the way the people in the organization see and how they think about the work they are doing. In particular, the language of waste allows people to start thinking of ways to reduce or eliminate the wastes in the process, even if this is done in the smallest little chunks. This is the very essence of continuous improvement.

The quote from Gil that opened this chapter is a great example of the change that takes place in the ways that people think. The wastes were always there in the process, but until he started learning to see through Lean lenses, they never occurred to him as such. Everyone, including senior managers, just accepted them as part of the normal process of building buildings, and never thought to question the wastes or think of ways to minimize them.

Returning to the example of transportation, once the organization recognizes that all the movements of the blocks in and around the site do not add value, they can begin asking questions that are geared to reducing the amount of movement and thus the amount of waste. Rather than the delivery truck arriving whenever is convenient for the supplier and unloading the pallets wherever the driver can find a free corner on the construction site, why not schedule the arrival of the pallets for a given storey to be JIT, right when the crane will be available to lift them directly from the truck to the building? This sort of synchronization requires that the rhythm of the structural work be sufficiently stable (i.e. unvarying) to make the required delivery times predictable (to be able to give a precise delivery time to the supplier, the planners must be able to reliably predict when it will be needed). This assumes of course that the truck travel time is itself relatively stable; the less predictable it is, the more buffer time the truck driver will have to allow in order to raise the probability of arriving on time. Given these challenges, a better solution would use an automated pull signal from the site to the supplier. The blocks improvement team did not consider this option, and we will discuss it in the next section.

Another pertinent question is: can the deliveries be arranged in such a way that the pallets for each apartment can be placed directly into those apartments by the crane? Questions like these are the engine of continuous improvement, since they point people towards the issues that need to be dealt with in the system in order to remove waste incrementally.

As the stories above illustrate, many of the wastes were not caused by builders' negligence or incompetence, but by a lack of proper design of the production system. This may have been another case of local optimization: it is much easier for a designer to specify "partition wall" and let the builder try to piece it together than it is for the designer to take the time to assemble the wall virtually, ahead of time and optimize the dimensions and constructability. But when the project has hundreds or even thousands of apartments, the investment of time to detail the partitions and prepare the necessary shop drawings pays dividends.

And indeed, after recognizing the inherent wastes in the process, the work team spent most of their efforts on the design and planning phases of the project. It was here that they could have greatest impact.

What was not done – Future opportunities

While there was a great deal to celebrate in the improvements that were made, there is, naturally, more that could still be improved. We will consider three possible directions:

improving the material supply process; preparing and delivering the blocks in pre-prepared kits; and moving to prefabricated partitions that reduce the site work to installation alone.

Tighter control of the process, resulting from all of the changes described above but particularly from more detailed planning, could move the company in a direction where they could institute a supplier-replenished pull system ("kanban") for material delivery. If the pull signal can be issued with a time buffer greater than the response time that the supplier can reasonably commit to meet, then this will be clearly preferable. It would relieve Tidhar from having to constantly monitor the inventory levels and juggle multiple order quantities. However, such arrangements depend on the stability of the production cycle time, the stability of the delivery time and the delivery capacity of the supplier. There may be economic trade-offs in meeting these conditions that can be negotiated or arranged between the contractor and the supplier.

Using an approach similar to that of the Heathrow Terminal 5 project,[4] a staging location could be used to decouple the supplier from the structural works process into which blocks must be delivered. This could simply be a convenient predetermined spot on the site within the coverage area of the tower crane. Spaces would be clearly marked for the number of pallets needed for an appropriate buffer size (equivalent to a little more than the amount needed for the apartments on a single storey). Each day, the supplier would perform a "milk run" in which he would note how many pallets had been "consumed" by the project (pulled from the kanban into the building), and resupply exactly that quantity. This mechanism leads to reduced waste since it reduces the batch size of the resupply and eliminates much of the communication back and forth between the purchasing department and the supplier. Smaller batch sizes mean less unevenness in the supply chain (less Mura), less waiting and less overproduction. But this sort of far-reaching improvement, which would necessarily need to be carried out in cooperation with the supplier, is only possible after the company gets its own affairs in order and has taken the time to properly plan and improve its internal processes. Once demand has been stabilized, then supply can be adapted accordingly.

Yet having learned to "see" the waste of inventory, there is still something glaringly wrong with the delivery of the blocks weeks ahead of the start of the partition-building step. In addition to the waste of inventory, Tidhar failed to address an additional problem with this process. The superintendents for the reinforced concrete process did not always ensure that the full complement of pallets was delivered before proceeding to pour the reinforced concrete slabs. As discussed in Chapter 6, the reason was fairly simple – their performance was measured based on the volume of concrete poured and the skyward progression of the building according to schedule, and not at all by their success or failure in getting the blocks in place for the following crews. While they are responsible for placing the pallets of blocks on each floor before the ceiling slab is poured, no mechanisms of accountability were created to ensure this responsibility is fulfilled. This led to situations where delays in block deliveries resulted in the need to deliver blocks after the floors were completed, a slow and laborious process.

One could imagine numerous relatively simple ways to correct this problem of compliance, and we leave it to the imagination of the reader to propose them. But the fact remains that this delivery is premature and inherently wasteful. It would be better in principle to deliver the blocks in much smaller batches, just before they were needed and with a pull signal. In fact, a Lean Construction implementation common in some construction companies in Brazil for just this block delivery problem appears to fulfil all three of these requirements.

As a first step, one might consider delivering pallets of blocks JIT to storeys using the temporary steel cantilevered staging platforms that are commonly used with tower cranes for delivery of finishing materials (floor tiles, plumbing and HVAC equipment, etc.) to tall buildings. However, as mentioned above, this is a slower process that is inefficient in its use of the tower cranes. It requires additional effort and equipment to distribute the pallets within the floors, and the steel platforms are inherently unsafe.

The solution developed in tall-building construction in Brazil solves this problem by using hoists or elevators for transporting the blocks rather than tower cranes. But its real innovation lies in the idea of preparing "kits" of blocks for each partition and delivering them to the builders with clear assembly instructions, just the way IKEA does for the home consumer. The first step is to prepare a detailed BIM model of the partition walls with complete details of the electrical and plumbing installations, and then to prepare shop drawings with part lists for each kit. The shop drawings guide teams of builders located in a ground-floor staging and prefabrication area on the site in preparing and assembling the kits. Figure 8.10 shows an apartment in the BIM model, a shop-drawing elevation, and ceramic blocks prefabricated with electrical outlet boxes. The kits are loaded onto small pallets that a single worker can move with a trolley to the works elevator, and delivered to the assembly crews at the destination storey. The result is minimal waste of material and reduced non-value-adding work for the builders. In the local business context, these savings outweigh the addition of the fabrication steps and the double handling of the blocks.

Given the extent of automation of both design and fabrication possible with BIM, there is yet another way to further improve the process. The basic idea is to prefabricate whole partition panels in a factory setting, deliver and install them before the ceiling slabs are poured, and thus eliminate entirely the process of block-on-block partition building. Interestingly, the same autoclaved concrete 50 × 60 × 10 cm blocks used by Tidhar are in fact manufactured in large panels and then cut to size by a cutting machine at the factory. If a technical solution could be found to transport the panels without damage, then they could be cut to shape by the existing machine in the factory (using computer-aided manufacturing directly from the BIM model), labelled, delivered and installed on site with no on-site cutting, less waste and less non-value-adding work.

Lessons learned

Lessons learned are:

- Any activity that does not add value to the product as defined by the customer is waste (non-value-adding work). There are seven kinds of process waste (Muda): transportation, inventory, motion, waiting, overproduction, over-processing and defects. In addition, there are the wastes of unevenness (Mura) and unreasonableness (Muri).
- Once you identify one example of non-value-adding work, you begin to see many others. One way to identify non-value-adding work is to observe one of the workers, either on the site or in the office, and ask whether the customer would be willing to pay for the work that is being done at each point in time.

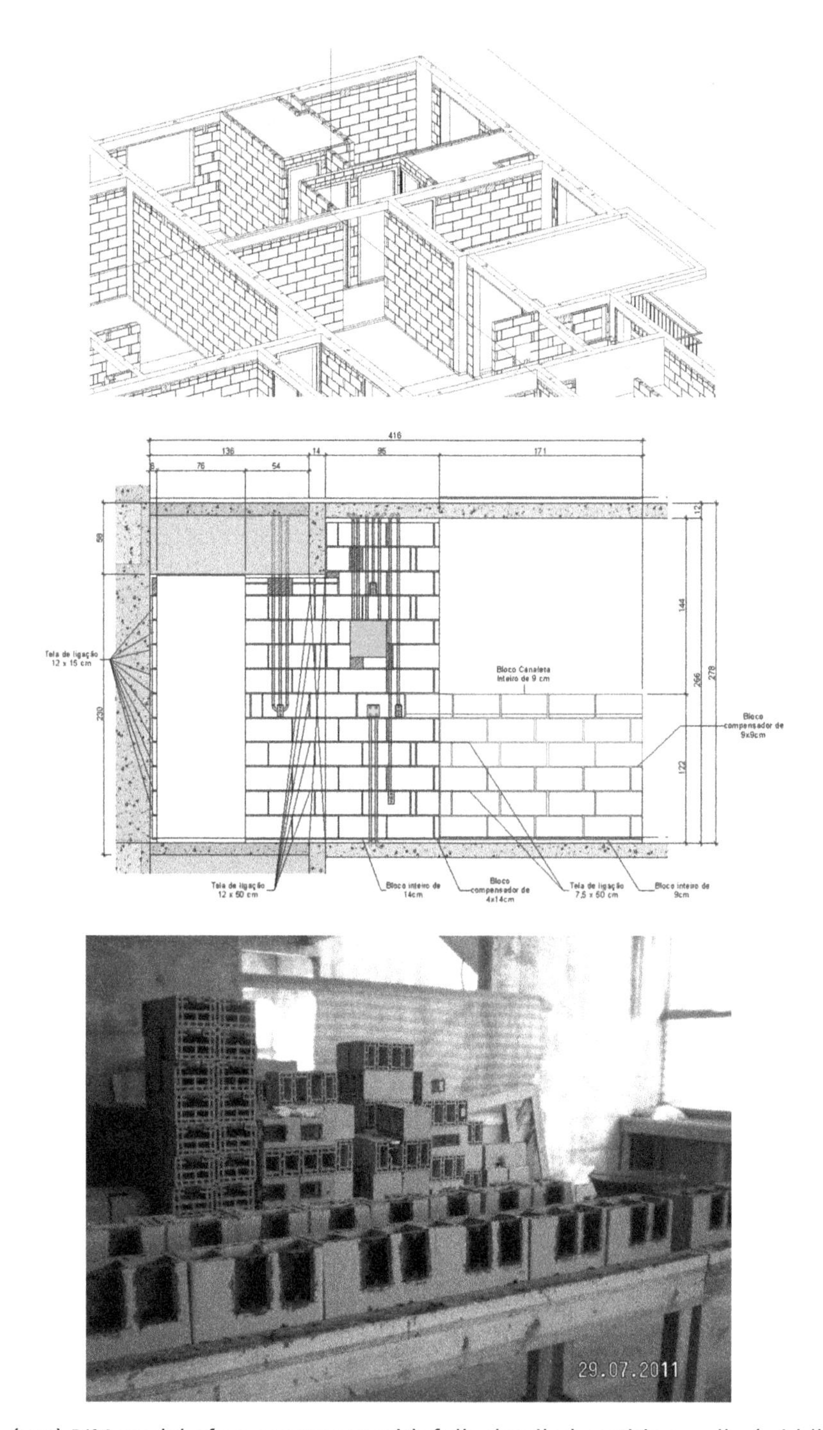

Figure 8.10 (top) BIM model of an apartment with fully detailed partition walls, (middle) elevation view of a partition kit, (bottom) fabrication of ceramic blocks with embedded electrical outlets ready for fitting

Source: Images courtesy Ferreira Construtora Cozman and DWG Arquitetura

- Often, investment in detailed engineering and preparation of shop drawings and enhancing the resolution of design can help avoid waste on site because it eliminates improvization by the workers concerning decisions about the best way to work. But this requires engineering and/or planning functions within the general contractor's staff, which means that the improvements can only be made if the requisite infrastructure is available. This is one of the reasons why many construction companies find changes like these difficult to make. There is a fundamental strategic divide between companies whose leadership see site management as an "evil" overhead whose cost should be cut as far as possible, versus a company like Tidhar, whose leadership see the site management as an essential support service to the production itself that easily generates benefits well beyond its cost.
- BIM allows preparation of fabrication details quickly and effectively. A general contractor can use BIM for a wide variety of production design and planning functions, thus spreading the overhead cost and making it a powerful tool for improving a wide range of processes on and off site.
- Accurate estimation of material quantities can help minimize material waste; BIM makes this very easy to do as a by-product of detailed fabrication design.
- Standardization of work methods is an essential basis for continuous improvement – there must be a common baseline that can be measured.
- BIM also enables general contractor staff to communicate standard procedures quickly and effectively to workers in the field, whether through simple 3D instruction sheets like those used by Tidhar and shown in this chapter, or through more sophisticated field computing devices such as tablets or smartphones.
- Think global optimization of the process, not just local optimization of the operations.

Notes

1 Frank and Lillian Gilbreth were pioneers in the field of industrial engineering, made famous for their time and motion studies. See Sheldrake (2003).

2 Customer changes and their effect on workflow are discussed in Chapter 11, "Virtual design and construction". Detailed analysis of the associated problems can be found in Goldin (2007) and in Sacks and Goldin (2007).

3 The role of "project production engineer" was in itself a new addition to the site management team, an outcome of the Lean and BIM processes. The role is described in Chapter 20, "Pulling it all together".

4 In complex projects, such as the Heathrow Terminal 5 project, an intermediate solution to this supply chain problem is to set-up a warehouse near the site to provide a decoupling point for material deliveries. In this way deliveries of material can be made to the site itself from the local warehouse with very short response times and in small quantities. Of course, this adds the non-value-adding waste of the warehousing. The key is to seek the overall system optimum in any given context, rather than to isolate and optimize the individual parts. For a discussion of the Heathrow Terminal 5 project see Lane and Woodman (2000).

9 Learning to see

"Learning to see" means learning to see production as a process and to see the wastes in the process. It requires developing the sensitivity to look beyond the blinkered view that sees only the work of transformation, to see the flow of products, the bottlenecks and the problems in between.

Cheni Kerem was a consultant who worked with Lean Israel Ltd., a management consulting company. Tidhar engaged Lean Israel to guide and support their adoption of Lean Construction as a part of their "Growing the Margin" initiative. Cheni had years of experience with Lean, but this was his first experience in the construction industry. Before joining Lean Israel, Cheni served as the CEO of an automotive battery manufacturing company, where he led a successful Lean transformation.

"Learning to see" as a Lean principle

The concept of "learning to see" (LTS) is a critical part of instilling organizational change, since it allows people to experience first-hand how much potential there is for improvement.[1] Most people see just a short section of a business or production process, and fill in the gaps with their own preconceptions about what other people do and what and how things flow. As most Lean consultants do, Lean Israel trains workers from all levels to observe the processes in their company with a critical eye, to look for the inherent wastes. One of the goals is to encourage people to "go to the Gemba" (the place where value for the customer is actually created), not just once, but to make it a part of their routine. These "Gemba walks" give managers an opportunity to see the challenges and opportunities for improvement first-hand, not filtered through indirect reporting or second-hand hearsay. They also give front-line workers and the field team an opportunity to bring up issues they're facing and to propose ideas for improvement. Direct access to managers encourages workers to keep identifying wastes and developing countermeasures, a key part of continuous improvement.

Among many other tools, Lean Israel teaches their clients the concept of "two-second improvements", learned from Paul Akers, the CEO of FastCap.[2] The idea is that each day, workers in a company should try to find opportunities for improvements that save just two seconds from their processes. This makes the challenge seem achievable and takes the pressure off looking only for the "home run" improvements. Over time, as people learn to constantly be looking for the small opportunities and work every day at improving, they develop their problem identification and solving skills. They learn to accept change and improvement as a part of their experience at work.

First encounter with the construction industry

Cheni began working with Tidhar from the start of the "Growing the Margin" initiative:

> With my background in industry and having seen what can be achieved when Lean becomes an integral part of an organization's culture, I was amazed when I began work in the less orderly world of construction. I began to see that in a project-focused world like construction, there are major barriers to managers being able to think in terms of process flow, value creation, and incremental improvement. Despite being so close to the project, project managers and construction engineers spend very little time focused on improvements when they come to the Gemba. They seek to home in on the bottom line, on the results: the productive activities that they can see. They spend very little time focused on the process, examining it to understand what parts add value and which are waste. Lean thinking teaches that focusing on the wastes reveals opportunities to make lasting improvement.
>
> A number of other things about construction also stood out. I saw a very different environment from the manufacturing industry I had experience with. For starters, in industry, the workstations and the workers are always in more or less the same locations, with the products moving from one to another. **In construction, the opposite is the case; the "product" stays in the same location, and it is the workers and the work tasks that circulate between the spaces of a building**. Simply locating where work is being done and where value is being added can sometimes be a challenge, since there is no fixed place where it reliably occurs. In construction, the Lean directive "Go to Gemba" can require time just to find the work.
>
> In manufacturing, most of the front-line workers are employed directly by the company, and the company "owns" the value-adding processes. But **in construction, most of the workers who are actually adding value to the product are not employed directly**. Instead, subcontractors do almost all of the work. The ability to influence behaviour is limited, since workers naturally answer first and foremost to whoever pays their wages. There are conflicting interests within the project and among the group of people who are ostensibly trying to work together to build the building. The subcontractors "own" the value-adding processes, but each of them is pursuing its own local optimization and profit, which often conflicts with project goals and the overall project optimum.
>
> In financially stable manufacturing companies, turnover in the workforce is commonly low. It's not uncommon for people to work together for decades. Contrast that to the situation **in construction, where each project employs a completely new set of people**. Building stable processes (one of the foundations of Lean) will be difficult at best, since people are constantly shifting in and out.
>
> Another difference is the makeup of the workforce. In most industries, the workers are drawn from the community, residents of the local area. **In the construction industry in many countries many of the labourers are foreign workers who speak a foreign language**. Communication between managers and workers is often

> indirect, via a foreman who translates. Cultural differences make creating a shared sense of common purpose or direction in the project very difficult, to say nothing of trying to make a change.
>
> **Yet another big difference between a factory floor and a building project is the sheer amount of space and the number of distinct, separate locations that the construction management team is responsible for**. In a traditional factory, the product is small enough that a good amount of raw materials, partially assembled products, and finished goods can all be stored under one roof, in addition to the tools of production, workers, offices, support staff, etc. Each of the production and production support departments will typically have a manager who is in charge of the entire area. The factory manager oversees all of these together, but there are usually not more than ten or so locations in total. Contrast that to the situation in a construction site. In a residential project, there might be two 25-storey buildings. On each floor, there could be four apartments and a shared lobby. This means that there are at least 250 work areas (more when including the public spaces, exteriors, landscaping, parking, etc.) that the project manager is responsible for. Each of the project manager's subordinates (the superintendents and foremen) may be working across a wide swath of those, since the tendency is to have lots of parts of the project under work at the same time (lots of WIP).
>
> And finally, in most cases, **site construction engineers have little or no freedom to select the construction methods**. Methods are determined by the head office and the building's designers, who have already chosen the product components, often with little regard for the specific constraints of construction in the field. If the company's head office is distanced from the Gemba, designers are even more so, in both space and time. Toyota succeeded in part by bringing production people into product development. Traditional construction does not do this.

When considered together, the differences Cheni identified between construction and other types of industry explain why the "contractual" mindset is so pervasive in the construction industry. General contractors find it very difficult to manage the actual construction processes. It is easier to buy a product than it is to manage a process and seemingly far less risky; production management (and with it, the risks of low productivity) appears to become someone else's problem.

Beginning work with Tidhar

Cheni observed that Tidhar's site managers often competed with one another to mobilize resources to their own needs. But, given all the wastes he could readily identify on site, in his opinion the site managers actually had *too many* resources, not too few. Most of the resources were not creating value, due to the wastes inherent in the way that work was managed. These included much of the time workers spent waiting or moving around, lack of tools or materials, attempting to start work before the prerequisites were met (task readiness), defects and rework, high inventories, damaged/discarded materials and others:

> At the end of every "waste walk" that I conducted with Tidhar managers on site, I would ask: "Out of all the people we just met, how many were actually creating

value?" Most workers were busy moving materials and equipment around, preparing work spaces, moving themselves, waiting, checking on work that had been done or other non-value-creating activities. The amount of value creation we would see was minimal.

With an increasingly clear view of the need to help project managers, construction engineers and even site superintendents to see the process flow, to see work as it was done from beginning to end, Cheni proposed a "Lean boot camp". The Lean boot camp was described in detail in Chapter 3, "Education and motivation". The goal of the boot camp was to educate managers in Lean concepts and to train them in methodical problem solving, so that they could become change agents. The theoretical component highlighted the differences between traditional management and Lean management, as outlined in Table 9.1.

Two practical components of the boot camp were designed to strengthen people's ability to see waste in processes: (1) a site tour in small groups, with each group assigned

Table 9.1 Differences between traditional and lean management

Traditional management	*Lean management*
Produce product as quickly as possible and whenever there is material (push)	Produce only when there is pull from the customer (usually the next step in the production process)
Attempt to reach local optimization for every point along the process and hope that this will translate into optimization of the whole (reductionist view – the whole is the sum of its parts)	Attempt to reach global optimization of the entire process (holistic view – the whole is something greater than the sum of its parts)
Working according to standards limits the ability to be creative	Standards are the basis for continuous improvement, and creativity is applied in improving the standards
Hide or gloss over problems, to avoid punishment	Value the exposure of problems, since they represent opportunities to improve
Never stop the production: the schedule must be adhered to	Stop and solve problems, so that they don't reoccur in the future and cause more problems
Workers are a liability	Workers are an asset that, if properly developed, can appreciate over time and pay dividends
The manager is the boss, responsible for giving orders	The manager is a teacher and leader, responsible for asking questions
Make decisions based on reports and dashboards in the office	Learn and decide according to what is experienced first hand in the Gemba
Plan quickly (but end up getting delayed in execution)	Plan slowly, arrive at consensus, so that the plan can be carried out swiftly
Experts (managers, engineers) are responsible for solving problems, workers for working	Everyone is capable of solving problems

to identify different types of wastes in the work processes they encountered, and (2) two days in which each participant was assigned to work on site as a member of a construction crew, doing jobs that labourers do every day:

> The goal of the site tour was to give participants experience learning to see and identify wastes in the Gemba. We split into small groups and sent everyone to observe the work underway in different parts of the site. The instructions were to first identify the value in the activity they saw, and then identify which wastes they could see in the process. Here, for the first time, the lights went on in people's eyes, as they started seeing an environment, which they thought they knew intimately from years of experience, in a totally different way. At first, they found it difficult to let go of their preconceptions about what activities create value for the customer. For example, one group observed a worker repairing broken flooring tiles. They identified excess movement or waiting times as wastes. But I challenged them to identify not just the obvious waste, but to look for the value in the process. After some discussion, they realized that the entire process was waste. Customers don't want to pay for mistakes.

Figure 9.1 "I feel like this has been a formative experience, one unlike any other I remember for a long time" (Tal Hershkovitz, Tidhar Construction CEO, during the Lean boot camp)

The second component – spending two days working on site with the crews – was a particularly informative and motivating experience for most of the participants. Typical comments before the work day included statements like "What, you don't think I know about construction? I'm there every day, I see how the work goes, I know exactly what they're doing." But in reality, a manager's point of view is very different

> from that of a worker. The work itself has a different level of resolution, one that cannot be seen unless you do it yourself. For example, one manager worked with a plumbing crew. His job was to install covers on the inspection traps for a variety of drainage pipes. To get the parts, he had to walk back and forth across the site ten times. Experiencing it first-hand, actually walking those laps back and forth, was the only way to understand what that felt like, and how it affected his productivity.

The underlying goal of the work experience was to show the managers what workers in the field experience every day: much of the production waste is systemic, due to constraints beyond their control. If Tidhar could improve the systems and remove some of those constraints, workers could do their jobs more efficiently.

Challenges to learning to see in construction

After the Gemba work, participants returned to the classroom energized and full of ideas for addressing wastes they had observed, both big and small. Unfortunately, this was a one-time gain. Construction people in general work under severe time pressure, which usually precludes zooming out to take a wider view and seek system improvements. Project managers need to learn to see, but too often, they don't even get out of the trailers and to the Gemba. When they do, they are focused on the completed work, the results, not on the process through which it is created. For example, site engineers are required to perform quality checks on reinforcement work prior to casting. They usually arrive just before a scheduled pour, once the rebar work is complete and leave soon after. They see the finished work, but not how it is done.

Cheni encouraged site engineers to visit earlier in the process, to interact with the workers and explore whether the work was being done correctly. In this way site engineers can provide added value, because only once they see and understand the process can they identify and tackle bottlenecks and waste. It's not just about creating an ideal sequence of trades; it's about observing how the work of each trade is done and asking how the work can flow more effectively. This is the difference between discovering errors after they have happened and seeing them as they occur.

Unlike the situation in a factory, where the environment is fairly stable, in construction the work space itself is constantly changing. This makes it hard to nail down the processes, because the location of the work changes as crews move. Further complicating matters, since so much of the work is subcontracted and workers come and go, production processes are variable and thus difficult to observe and measure.

Another part of the problem is that most construction professionals have difficulty seeing the full cost of a process. Focusing on unit prices for products or labour, as traditional project management in construction does, obscures the site and management overheads, the cost of rework, materials wasted, warranty claims and so on. These indirect costs cannot be negotiated with a supplier, but they are no less real and they represent opportunity for improvement. Project managers get immediate positive feedback when they negotiate a price reduction with a subcontractor, but no one accounts for the damage to production flow that may accompany the discount. Toyota and other excellent companies did not succeed by buying the cheapest possible parts.

With a clear view of the challenge and of the limitations inherent in his work as a consultant, Cheni summed up his views:

The question is: how much does the organization and its leadership want the change? How willing are they to commit the resources to making the Lean transformation? There are no secret shortcuts, just hard work. Ultimately, it's up to the company executives to make that commitment and put in the effort; no external consultant can make the change by themselves. A consultant can teach the principles, can demonstrate and guide them, but the change must come from the leadership of the company; they must make it a priority. And in order to have the drive to keep persisting over time, you need a visionary leader who can take the long view and keep driving him- or herself and the organization toward the future state.

Lessons learned

Learning to See can be a paradigm changer, since seeing reveals the contents of the opaque black boxes that traditional production planning and control sees (for a more detailed explanation, see Chapter 16 "Production planning and control in construction"). The story of the "four-day floor" (see Chapter 18, "Raising planning resolution") is a good example. Before the team learned the work process step-by-step, they did not really understand it nor could they identify its bottlenecks. Only once they broke the process down into hourly chunks could they have real impact and improve the production rate. Lessons learned are:

- Learning to See is one of the crucial shifts of mindset required for a Lean transformation to succeed.
- Experiencing the Gemba first-hand through the eyes of the workers, by spending protracted time on the Gemba, is one of the most effective ways of teaching executives to "see".
- Consultants can teach and guide, but they cannot lead. Lean improvements demand persistence, and persistence depends on leadership.

Notes

1 See Rother and Shook (2009).
2 Paul Akers and the two-second improvements are also explained in Chapter 6, "Continuous improvement and respect for people". See Akers (2011).

10 CU Fiat project

Project profile

Figure 10.1 A general view of the ground floor commercial spaces in the CU Fiat project. Inset: an architect's rendering of the same space

Project description

CU Fiat is a campus of four office buildings with four underground parking levels, commercial space on the ground floor and four office towers, in Tel Aviv, Israel. Buildings B, C and D have eight floors each, while Building A has five floors but is planned to expand to 15 floors in the future. Above-ground offices and retail space amount to 32,930 m^2, while the four underground parking levels are 39,580 m^2 (total built area 72,510 m^2).

The project was developed by a private investment group whose members are the individual owners of the offices and commercial spaces in the building. The group was formed to collectively purchase the property and was directly involved in the whole life cycle of project development.

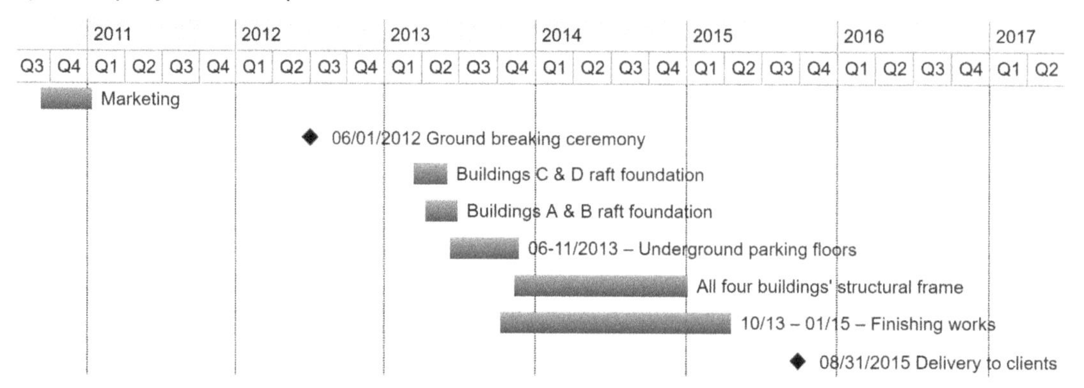

Figure 10.2 CU Fiat project timeline

Tidhar's project team consisted of a project manager, a design and planning coordinator, two project engineers, two site superintendents, a part-time building systems engineer and a BIM modeller on site.

Construction methods

All the buildings are cast-in-place reinforced concrete, with aluminium and glass curtain walls.

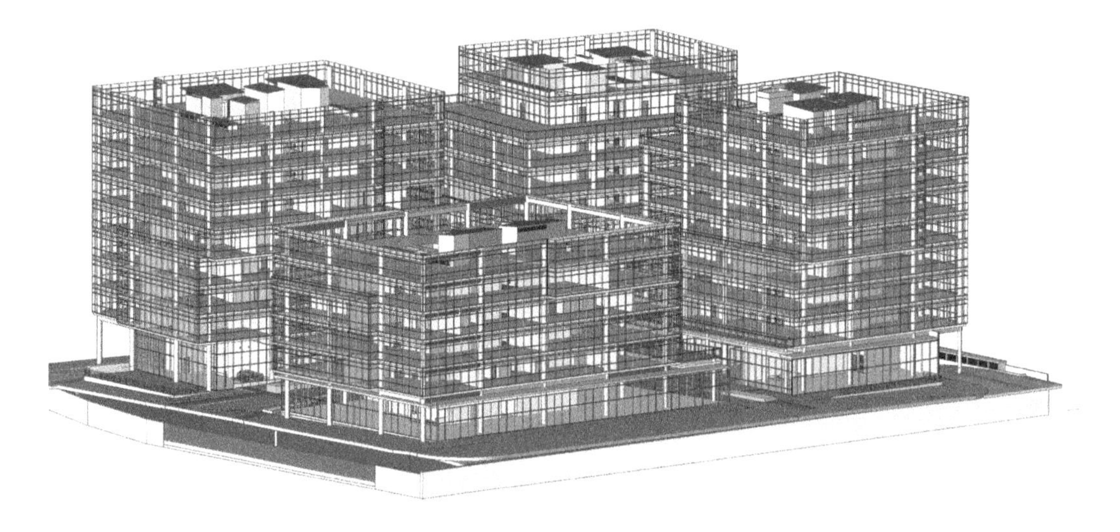

Figure 10.3 The BIM model of the above-ground parts of the CU Fiat project

Application of Lean and BIM

The project was modelled in two separate files, one each for the underground and above-ground models. The models were used for site organization, MEP coordination in the underground parking and building cores, area planning and management, representation of design options and as a tool for decision-making, material take-off for the construction teams and quantity estimation of items such as doors, windows, railings and so on. The architectural model was prepared by the project architect. Tidhar modellers integrated the MEP and construction data using Revit and Navisworks.

11 Virtual design and construction

> The basic principle is that we "build the model" before we build the building.

What is virtual design and construction?

The term "*virtual design and construction*", or VDC, is used to emphasize the difference between design and construction using BIM on the one hand, and design and construction using traditional tools, primarily 2D drawings on the other hand. The word "virtual" distinguishes VDC from "real" construction, in the same way that virtual reality is a digital expression of reality. Where fully implemented, VDC entails:

1. generating a digital design model using BIM tools;
2. simulating and analysing the performance of the digital building model using software;
3. planning the construction process using BIM tools, by rebuilding or refining the digital design model to generate a digital construction model, and by compiling an integrated digital construction plan;
4. simulating and analysing the digital construction plan using software;
5. building the building according to the digital model by executing the construction plan in reality, including digital monitoring and recording of the construction process;
6. recording the "as-built" state of the building by updating the digital construction model.

Steps 1 and 2 are "*virtual design*"; steps 3 and 4 are "*virtual construction*"; step 5 is "*construction*"; and step 6 is measuring, monitoring and reporting. Naturally, the VDC process is cyclical, with each consecutive step providing feedback for the previous step, which may then need to be repeated in part as more is learned about the project and the building.

The VDC steps can be compared with the traditional approach:

- design the building;
- plan construction;
- build the building;
- record the as-built state.

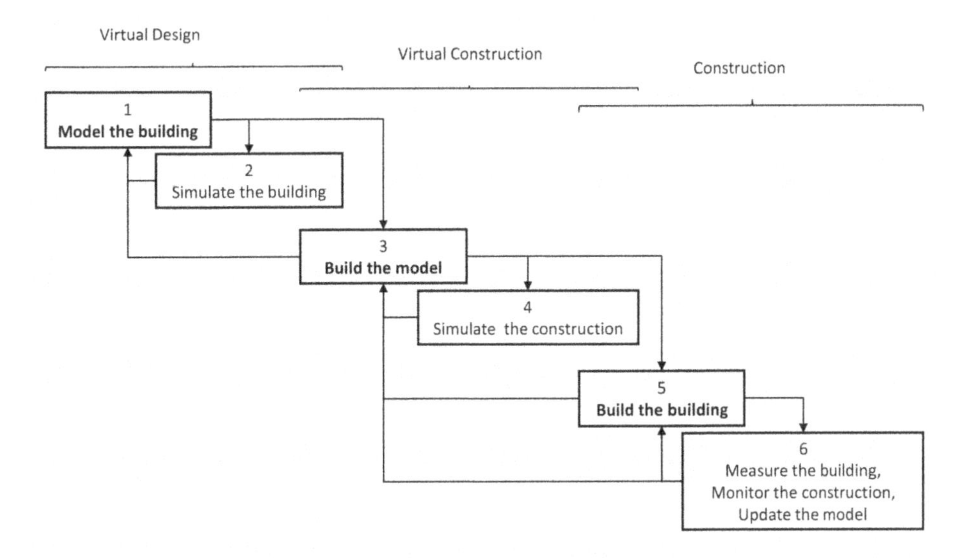

Figure 11.1 Modelling the building, building the model, building the building

The two main differences are (1) the use of BIM for modelling the building design, planning, construction and reporting, and (2) the use of model-based simulations to support design and planning, i.e. compiling and testing a design prototype (step 2) and compiling and testing a process plan prototype (step 4). Design of a building in the traditional process includes analysis and review, as does construction planning. The reason for listing simulation and analysis in the VDC approach as distinct steps (2 and 4) is to emphasize that BIM makes prototyping far more accessible, accurate and cost effective than it is using traditional mock-up methods. Simulation and analysis are integral and essential to the VDC process.

Using this framework, one can see that a construction project that claims that it is using BIM may in fact be anywhere along a relatively wide spectrum that runs from using BIM solely for digital design (step 1 in the VDC process above) all the way to a full VDC implementation (incorporating steps 1–6). A construction company that builds a digital model in order to prepare a cost and schedule estimate in response to an owner's call for proposals in a construction tendering process can also be said to be using BIM, even if they do not implement steps 4–6. For the remainder of the book, we will assume that VDC means an ideal, holistic implementation of the full process, including all steps from 1 to 6, and we will measure the extent of BIM use in practice using this scale.

VDC implies collaboration

Just as designing and constructing a building is a collaborative endeavour that needs many people with a wide variety of skills, virtual design and construction also requires many professionals to collaborate. The UK government BIM mandate, which calls for BIM adoption by all suppliers to government departments and agencies, specifically defines levels of collaboration as the main measure for the depth of BIM adoption. The PAS 1192-2:2013[1] document defines three distinct levels of BIM implementation (CAD is level 0): 3D modelling and drawing production (level 1); simulation and analysis of profession-specific (domain-specific) models (level 2); and intelligent collaboration and full sharing of models (level 3).[2]

Figure 11.2, which uses a ladder as a metaphor for the level of sophistication of BIM usage, makes a finer distinction at level 3 between the use of federated multidisciplinary models to coordinate design and production among different construction trades, at the lower rung, and full collaboration on an integrated "live" model, at the highest rung. A federated model is a compilation of static models, exported from each of the "live" models from the distinct disciplines, which can be used for coordination reviews and for clash detection, but which cannot be edited directly. This also means that some of the simulations and analyses have longer and more cumbersome cycles.

In any event, no matter what the level of BIM use in a project, VDC implies extensive collaboration. The benefits of collaboration with BIM are significant. Some of those benefits are the core motivation for the "big room" approach described in Chapter 14, "BIM in the big room".

A Lean view of VDC

Design is an iterative process, progressing from synthesis to analysis and back again, in loops in which designers learn and improve their understanding of the problem and of the solutions in each successive cycle. With a few esoteric exceptions,[3] all building design is virtual. Designs may be expressed, stored or analysed in people's minds, on paper or in digital form, all of which are virtual representations of the intended physical artefact. Construction, by contrast, is a linear process in which components are fabricated and assembled to produce the whole building. Construction is physical, and it is only iterative where faults or defects are identified in the design or in the execution.

Iteration in construction is rework; rework is waste. Removal of waste is a key goal of Lean. Iteration in design appears to add value, as it enhances understanding. But as Ballard pointed out,[4] some of the iteration in design can also be considered wasteful, such as when iteration results from design errors or from poor sequencing of design decisions within multidisciplinary design teams.

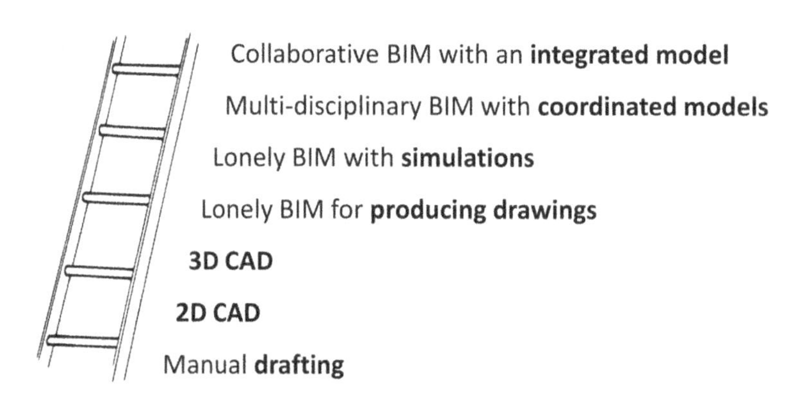

Figure 11.2 Ladder of sophistication of BIM use.

Note: "Lonely BIM" refers to the isolated use of BIM tools, without sharing models among the project team.

From a Lean perspective, VDC adds value and contributes to removal of waste in design and construction. The use of BIM for virtual design (steps 1 and 2 of the VDC process) can enhance value where visualizations and simulations enable designers to empower clients and end users to define their requirements and to test whether the designed building will function as intended. This is the value of *virtual prototyping*. Virtual design can also reduce the waste of waiting by reducing the cycle times for design iterations, and it can reduce negative iteration by enabling designers to identify and remove errors.

The second major contribution of VDC is the introduction of the virtual construction phase (steps 3 and 4). The ability to perform what is essentially a series of *virtual first run studies* for various assemblies of the construction process means that construction teams can benefit from the practice of test production runs that is almost taken for granted in manufacturing. Where errors are identified, they can be removed by changing the design or the construction plan without incurring the costs of physical rework or demolition. Furthermore, processes and methods can be optimized by trial and error, again at minimal cost. VDC brings positive iteration to construction.

Tidhar's first steps with VDC

The "Weather Vane" building, part of a complex of 32 residential buildings in a town called Yavne, was one of Tidhar's first projects to benefit from a VDC approach (for details of this project, see Chapter 7, "Tidhar on the Park, Yavne"). The building and all its systems were modelled by one of Tidhar's own modellers from the consultants' design intent drawings, under the supervision of Maia Davidson, the project's design manager at the time. This was the project architect's first experience with BIM, but none of the engineers and other consultants used BIM: all coordination and communication during the design phase still relied on 2D CAD drawings. The MEP subcontractors were all small firms that did not have the capacity to model their systems themselves.

Construction detailing and coordination review meetings were conducted at the site offices during the weeks leading up to the start of construction. The integrated model of the building, compiled from distinct discipline-specific models and displayed using Navisworks, was the focal point of each meeting. Decisions taken at the meeting were implemented by the modeller from week to week.

The participants in the review meetings included:

- Tidhar's construction engineer, construction superintendents (for structural works and for interior systems and finishes), project architect and BIM modeller;
- project managers from each of the relevant subcontractors: plumbing and sprinkler systems, electrical and communications, air-conditioning and ventilation, aluminium windows, kitchens and drywall;
- representatives of the architectural, structural and MEP design firms.

Although the architectural, structural and MEP designers sent representatives to the meetings, Tidhar had extensive leeway to adapt the construction details because Tidhar was also the project owner. At one particular meeting, early in 2013, two issues came up that clearly underscored the value of the VDC process. The modeller walked the participants through the basement floor, highlighting the locations of drainage pipes, water supply pipes, sprinklers, lighting units, ventilation ducts, cable trays and various other systems. After reviewing and resolving a number of physical clashes between air ducts and sprinkler lines, attention was drawn to the emergency power generator to be installed in the parking basement. The designer had routed the exhaust pipe for the diesel-powered generator out of the generator room, along the ceiling of the parking floor and finally through the exterior wall alongside the parking entrance ramp. From the inside of the parking floor, everything seemed to be in order. The exhaust pipe was properly routed, there were no clashes with other systems and the pipe would be high enough off the floor to prevent damage to the insulation that protected the hot pipe. However, the subcontractor tasked with installing the generator and its exhaust pipe raised an objection: "Can you show us where the exhaust pipe comes out on the other side? We have to install a vertical riser and a protective cap for rain – will we have access?" As soon as the modeller moved the walk-through position to view the relevant wall from the ramp, the MEP designer's face turned red:

> Oops – we'll have to change that. The pipe comes out at about 150 cm above the ramp, which means that people will be able to reach it, which is not allowed because when the generator is in use it is a hazard. We hadn't noticed that in the 2D plan view. We thought that the ramp surface at that point was much lower than it appears in the model!

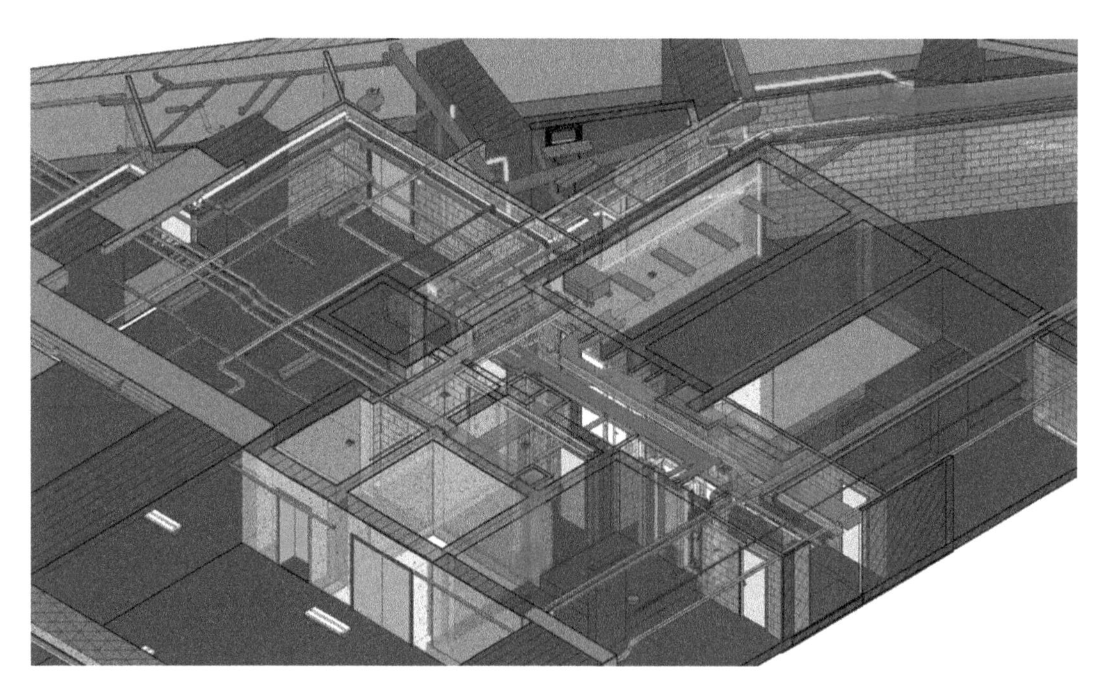

Figure 11.3 A view of the model of the basement floor of the "Weather Vane" building in Yavne, highlighting the locations of drainage pipes, water supply pipes, sprinklers, lighting units, ventilation ducts, cable trays and various other systems

This example is fairly typical of the issues that arise in model review sessions, but it accurately reveals how the virtual construction activity helped to avoid an error that would certainly have resulted in extensive redesign and rework had it only been identified when the installer reached the end of the pipeline installation. The flexibility to change other aspects of the design to accommodate such a correction at that point in construction would naturally have been far less than it was before any real work had been done. Yet the following example, raised by the structural works superintendent Dudi Nasi in the same meeting, had much more far-reaching consequences for this building and for most future Tidhar projects.

Dudi said:

> If only my father could see this! He was a construction superintendent for 30 years, and he complained about these kinds of problems every day in construction. It's useful for us to see what we are about to build, so that we can solve our problems before they happen. This is "virtual construction in action". Here's a practical example: when we build the reinforced concrete structural frame, we never quite know what vertical pipes and ducts there will be, so we leave big rectangular holes where the shafts are in each slab for pipes and ducts of various kinds. In our typical residential buildings, these are about 2 × 0.6 m in size. They are usually located within the core area of the building, in the lobby area. The holes in the slabs are left open until the MEP trades install their risers; after that, we cast the slab infill around the pipes. This is difficult because it is fine work, and it is difficult to install and to remove the wooden formwork around the pipes. It is also disruptive to the RC [reinforced concrete] crew, because they need to assign workers to these tasks when they are really needed (and planned) to be working with the main RC crew at the top of the building. The holes are also a safety hazard for falls throughout the construction period.
>
> But I can see clearly that if we can design the piping thoroughly, and "build it virtually" in advance, then we can have the sleeves for the pipes themselves in the production model. If they can be shown on the fabrication drawings for the RC slab, instead of leaving a large hole in the slab (the full size of the shaft), we can block out only the sleeves needed for the pipes from the concrete pour. No danger, no rework, no problem. And I can see that since we're all here with the MEP designers and detailers, there's a good chance we can make it happen.

He turned to the key MEP people in the room – plumbing, electrical and ventilation – and they agreed enthusiastically. It turns out that their work is much easier if they have a slab to stand on at each level as they install pipes and ducts, instead of having to work while suspended upon temporary wooden platforms in a tall shaft.

Using the BIM model in this way – virtually building the building by compiling a digital construction model and then using it for multidisciplinary review in collaboration with the relevant construction people – represents steps 3 and 4 of the VDC process. Despite the inability of the subcontractors in the meeting to use BIM technology directly, the collaboration was full in the sense that they could view and interrogate the model as if it were a real prototype, solving problems and removing potential conflicts and errors that would otherwise have been discovered only during construction.

From a Lean point of view, the change to construction practice that Dudi describes is a good example of the process of continuous improvement. Thanks to the BIM construction detail model, the pipes and ducts in the vertical shafts could be detailed collaboratively, which had direct benefits in the construction process: less rework, removal of one instance of re-entrant flow, safer work and greater productivity for reinforced concrete crews and building systems crews as a consequence of all three. The investment in BIM modelling and coordination meetings pays off in the production itself. This has become standard practice in all Tidhar projects.

From the point of view of exploitation of BIM technology, one may ask if a single modeller working for the construction contractor, modelling all of the systems, is a particularly sophisticated use of BIM? Surely the designers, or the subcontractors, should be providing the models? On large and complex projects, it is certainly preferable to have the trades model their own systems and coordinate them in collaboration with one another, and we will discuss how later in this chapter. However, for a relatively small residential building, use of BIM in this way is in fact "multidisciplinary BIM with a coordinated model", and it empowers workers to collaborate and to improve their construction methods. In any situation where subcontractors and fabricators do not yet have the capacity or ability to work with BIM, or the scale of the project does not allow them to dedicate their resources, modelling by a construction company's BIM modeller can still have a strong impact.

In fact, there is a corollary benefit to this practice – the modeller virtually builds the building, becoming intimately familiar with all its nooks and crannies, and thus is in a unique position not only to flag problems and issues, but also to identify opportunities for improvement of construction methods. This capability is magnified when the model is used to empower the teams that will actually build the building to join in this virtual construction exercise. This is the heart of collaborative VDC.

Formalizing Tidhar's VDC methods

Given the clear value gained by the process piloted at the "Weather Vane" building in Yavne, Tidhar decided to implement it across the company, extending it to significantly bigger projects. Adopting the Lean principle of standardization, they set about preparing the groundwork for broad implementation. This included a standard job description for recruiting modellers and a "level of development" specification to define the required model contents at each stage of a project.

Project BIM modeller's job description

Box 11.1 shows the job description document from Tidhar's organizational knowledge base. By 2016, Tidhar employed nine modellers under this job description. They each worked on a particular project, spending much of their time at the job site, providing model information directly to the site staff and to the subcontractors. As their competence grew, they became the focal point for design information on the site. Their ability to provide printouts of detailed views of specific aspects of the work, on demand, and complete with 3D views, proved to be of major value to almost all of the trades.

Box 11.1 Job description: BIM modeller

1 Position in the organization:

Division/Organizational unit:	Project development
Department:	Engineering
Person responsible for salary, employment, dismissal:	CEO
Direct manager:	Entrepreneurial engineering manager
Professional manager:	BIM Manager
Direct manager of:	None
Professional manager of:	None

2 Core role (Mission)

- Creating and managing BIM models from consultants' and designers' plans.
- Assisting planning and implementation of the project in light of the design programme, budget and project schedule.

3 Day-to-day work (Framework)

- Work in accordance with Tidhar's method and management tools.
- Maintain and uphold the company's charter, fostering its corporate values and culture.

4 Responsibilities (Objectives)

- Receive design drawings from the project's design manager and model the project using BIM software (architecture, structure and systems).
- System coordination, superposition.
- Help evaluate and optimize building designs (in coordination with the design manager).
- Ongoing communication with designers to obtain clarifications and additional information about the design.
- Provision of production quantities (take-off) from the model.
- Help in establishing a detailed project budget (in coordination with the Tender Department) and maintenance of the project budget over the life of the project.
- Preparation of a construction schedule in a format suitable for 4D BIM (in coordination with the project manager) and maintenance of the 4D schedule over the life of the project.
- Preparation of a construction cash flow plan suitable for 5D BIM (in coordination with the Tender Department) and maintenance of the cash flow plan over the life of the project.

5 Objectives and performance indicators

Item	Target	Weight
A	**Meeting schedule** – according to the timetable set.	
B	**Construction within budget** – keeping the construction expenses as determined at the project budget planning meeting.	
C	**Meeting quality targets**:	
	1 Quality of plans/drawings (quality grade assessed by the direct manager):	50%
	2 Stable schedule performance:	10%
	3 Budgetary stability over the life of the project:	40%
	A. From the beginning of the project until six months before the end:	10%
	B. The final six months of the project:	90%
D	**Labour relations management** – 360 feedback score:	
	1 Co-workers assessment: (design managers, construction managers, project managers and relevant construction engineers):	30%
	2 Direct manager:	70%
E	**Achieving customer satisfaction for internal customers** (design managers, construction managers, project managers and construction engineers): departmental unit feedback score (departmental performance, not worker performance)	

6 Evaluation cycle:

Annual review

7 Characteristics, personality traits and essential skills required for the job

Topic	Description
Personality traits	**(EQ) Emotional intelligence:** capable of working with people, a team player. Energetic (those who work opposite the person or with the person feel that he or she contributes energy) **(V) Values:** integrity, reliability, excellence and a service orientation
Education and professional skills	Civil engineer with experience in BIM software (Revit preferred), strong skills in reading design drawings
Language skills	Hebrew – mother tongue English – at least at the level of conversation

Levels of development (LOD)

With the growing investment in BIM, more modellers and a growing dependence on the information the models provide, it became necessary to standardize the level of detail that Tidhar would require for each trade at each stage of the virtual design and construction process. Five distinct stages were defined and labelled as BIM 0 (where no model information was required other than the site boundaries and town plan setback lines) to BIM 4 (the as-built model). The levels of detail that would be required in the model were defined for each of the major building systems and for management functions such as cost estimating, as shown in Table 11.1.

Problems with VDC implementation

As we have seen thus far, the first stage of Tidhar's BIM adoption process focused on the "low-hanging fruit" of building system coordination through virtual construction by the general contractor's own modeller, supporting model review and coordination meetings with the designers and trade crew managers. One of the goals was to stabilize the production work on site directly by reducing the degree of uncertainty about what was to be built and how it was to be built. This works well overall, but it does have limitations. The impact of late design changes can be very disruptive, as the following example shows.

In June 2013, after a six-week hiatus, the "Growing the Margin" steering committee met in Tidhar's seventh floor boardroom at the Ra'anana head office. Frustrations had grown over the period, and tensions were bottled up. Gil Geva's late arrival for the start of the meeting did not improve design VP Zohar Raz's mood.[5] Zohar opened the meeting with a review of progress on the BIM front:

> We have reached a point where we are in full control of the BIM technology. All of the construction detail drawings for Tidhar on the Park in Yavne, CU Fiat and Dawn Tower are being produced directly from our models, our modellers are doing superb work. But it's all worthless. It's worthless because the project managers, along with the construction engineers, are making detailed design changes on the fly without any regard for the models or the drawings. They're doing whatever they feel like, exactly as they have always done. Our culture is just as messed up as when we began, despite all our investments and all our efforts.
>
> Take for example what happened last week at Yavne. The project manager for the "Weather Vane" building sat down with the superintendent, and on the typical apartment floor partition plans, they started marking up with a highlighter which walls would be built from blocks and which would be drywall. The work was due to be done the next week, and here they were changing the designs! And this after our design manager had spent countless hours with the design consultants, going through the combined models using Navisworks, tweaking each and every wall and partition to get it right, considering all the piping and electrical systems, and here they were undoing all that hard work …

Tal Hershkovitz felt that, as general manager of the construction division, he had to stand up for his loyal and well-intentioned construction leadership team at the "Weather Vane" project:

Table 11.1 BIM LOD definitions for in Tidhar's construction projects

	BIM 0	*BIM 1*	*BIM 2*	*BIM 3*	*BIM 4*
General description	Business development, design brief	Initial programme, technical requirements, vertical shafts scheme	Final design, coordinated architecture and structure, no building system coordination	Construction detail, building systems fully coordinated by the subcontractors	As-made
Architectural detail	Zone areas and volumes defined for each function	Generic walls, generic floors, approximate levels	Final geometry and material assignments	Association of elements to work zones, unit costs (for subcontracting), catalogue ID numbers for purchasing and delivery	Red-line
Interior design detail	None	None	Model generic finish materials	Modelling of working details, attribute values for purchasing and delivery control	
Structural detail	None	Walls, beams, columns and generic slabs	Final coordinated geometry, estimated reinforcement content ratios	Finalized geometry and dimensions, concrete pour definitions, association of elements to work zones	Red-line
Aluminium (windows) detail	Modelled according to brief	Modelling according to architectural design, areas of window per wall	Modelled according to detailed façades, coordinated with architectural model, structural model and checked by safety consultant	Windows and door schedule, coordinated with the subcontractor, association to work zones	Shop drawings
Metalwork detail	None	Metalwork schedule using Tidhar detail library	Metalwork schedule using Tidhar detail library, coordinated with architecture	Metalwork schedule for construction, catalogue numbers, unit costs	Shop drawings

Table 11.1 Continued

	BIM 0	*BIM 1*	*BIM 2*	*BIM 3*	*BIM 4*
Electrical detail	None	Vertical shafts, technical rooms	Modelled according to programme, lengths and types of ducts	Attributes for equipment purchasing, association to work zones	Manufacturer, model, reference to relevant data in facility file, maintenance instructions
HVAC detail	None	Estimated longitudinal ductwork, types of mechanical equipment	Longitudinal geometry, approximate requirements for mechanical equipment	Attributes for equipment purchasing, association to work zones	Red-line for longitudinal ducts. Mechanical equipment: manufacturer, model, warranty, service provider, reference to relevant data in facility file, maintenance instructions
Plumbing detail	None	Estimated longitudinal ductwork, types of mechanical equipment	Longitudinal geometry, approximate requirements for mechanical equipment	Attributes for equipment purchasing, association to work zones	Construction budget
Site detail	Gross site development areas			Construction budget	
Costing detail	Budget for real estate deal	Design estimate (gross unit costs)	Design baseline budget		Actual cost

> What do you expect! There was an earlier decision to reduce the sale prices for the apartments in that building, and so there were no ceiling drops designed for the air-conditioning units. Now all the buyers want to install air-conditioning, they're only getting that information now, and they have to build next week. How else can they deal with it – we should be grateful that they can respond to change so quickly! And in any case, maybe the drawings your modellers gave them didn't have enough information on them!

This was too much for Zohar to take:

> What?! Have you seen the drawings? They are the most detailed drawings anyone on a construction site like that has ever got. They show every wall, every layer of every wall. All the tiling, all the piping openings, all the heights – everything is shown! Your people just don't know how to read the drawings.

Ronen Barak joined the defence: "Here, look for yourself". He had brought up a typical floor drawing, generated directly from the REVIT file of the building, and projected the PDF on the meeting room screen:

> At this intersection here, you can see how there are tiles shown on the bathroom side, and the hatching clearly shows water-resistant drywall boards, and here are the dimensions for the height of the tiling from the floor. The designs for the parking basement floors were thoroughly coordinated, and the systems went through all three phases of coordination, LOD 1 internal check, LOD 2 internal check, and LOD 3 models with all the engineers and the architects present. The site superintendent was also present at the last session, in fact he has been involved from the start. They just have a huge discipline problem – they're not willing to accept the change to building from the model, not making things up on the fly like they have always done.

"But do the site people know how to read all that detail off the drawing?", someone asked. "The project manager is new to the project, maybe he didn't know about all the earlier coordination work?" Tal joined in again:

> Yes, and even if he did, the only way for him to deal with the unexpected need for the dropped ceilings for the air-conditioning ducts was to mark it up on the drawing and give it to the drywall crew the next day. Your BIM modeller for the project, Gal Salomon, is only at the site office two days a week, maybe they just couldn't wait for him. In any case, the fact is that the site people are still living in the "old world" where they are praised for ingenuity and they get satisfaction from last-minute fire-fighting.

Gil cut off the discussion, as he often did in these meetings when they became too heated or risked becoming repetitive. He tried to shift the focus to the implications of what was being said and what needed to be done about it:

> These are good problems, guys. It shows that we're reaching the point where our emphasis on thorough design and planning before construction has a chance of

> becoming fulfilled. But clearly, we still have some problems here, with the people and with the processes. The good side is that we seem to have gotten full control of the technology – and that's no small achievement, considering that when we saw it being implemented last year in Denmark, we thought it was beyond our peoples' skill levels. The bad side is that looking at the Yavne case from the Lean perspective, we cannot expect to achieve stability in the construction work itself if we let them make changes like that! It messes up the whole flow for the subcontractors, and it reduces the value of the model, it brings us back to where we started. So here are the key issues:
>
> 1 The basic principle is that we need to be "building the model".
> 2 We need to make sure that our site people can read the new drawings and understand all the information they contain.
> 3 We need to figure out a way to route these late design changes through the model. We cannot have a bypass route for information flowing to the work crews.
> 4 The response time for the modelling support must be shorter. The long cycle time for the modeller to get the changes into the model is a major problem.

He concluded with a simple but challenging directive: "Solve these problems, and things will improve".

Transitioning from traditional practice to VDC

When a construction company adopts BIM and Lean practices, the transition needs to be carefully considered and planned. The change must be paced to match both the employees' growing technical capabilities as well as the gradual change in their mindset. This can be managed by introducing change in specific steps, moving from one step to the next, from project to project.

Figure 11.4 displays the traditional design and construction practice. While specifics vary across different regions and countries, the process is fairly standard: from design, to construction detailing, to construction itself. The designers' drawings convey the design intent, and the builders (contractors and suppliers) are responsible for detailing the design, setting construction methods and performing the construction work itself. Where shop drawings are prepared, they are submitted to the designers to confirm that they conform to the design intent.

The first step toward implementing VDC is to introduce building modelling as a coordination tool, as shown in Figure 11.5. This can be done for design and construction separately, depending on the tools and skills available, but the best results will be achieved when both design and construction coordination make full use of BIM. Figure 11.5 shows building modelling applied to both phases.

The design itself may be done using BIM, making the design coordination task more efficient. But as long as the design models are not provided to the builders, the picture remains the same from their perspective. Similarly, where one or more of the detailers uses BIM to model their systems, efficiency and accuracy may be improved, but the nature of the information flows remains the same as in the traditional methods.

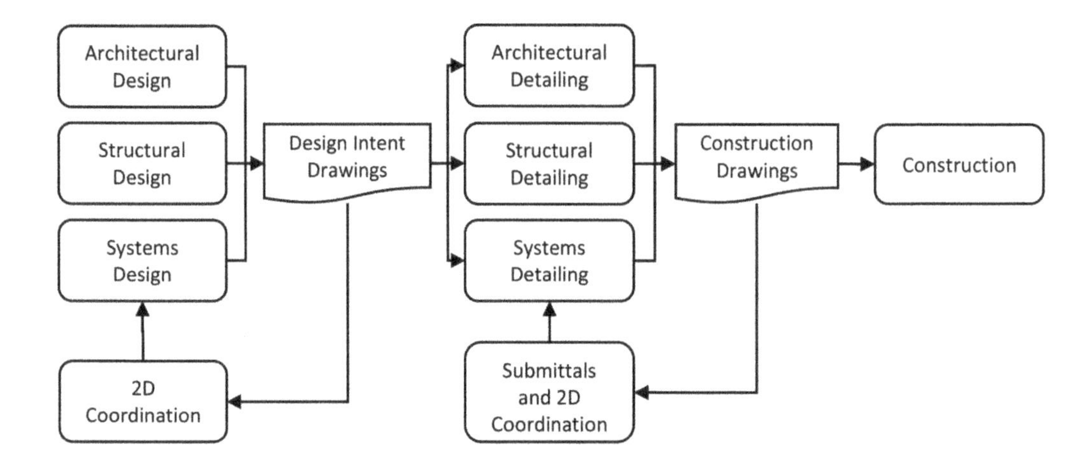

Figure 11.4 Traditional practice: design and construction with 2D drawings

The next step, which Tidhar attempted to introduce right from the start of its BIM implementation, is to disconnect the flow of information to the site using drawings and replace it with a flow that ensures that all information delivered to the construction site comes directly from the coordinated model. In this mode, shown in Figure 11.6, the major difference is that the general contractor assumes central control of the information flow, placing the modelling activity between detailing and construction, thus ensuring that all parts of the building must be built virtually before they are built in reality.

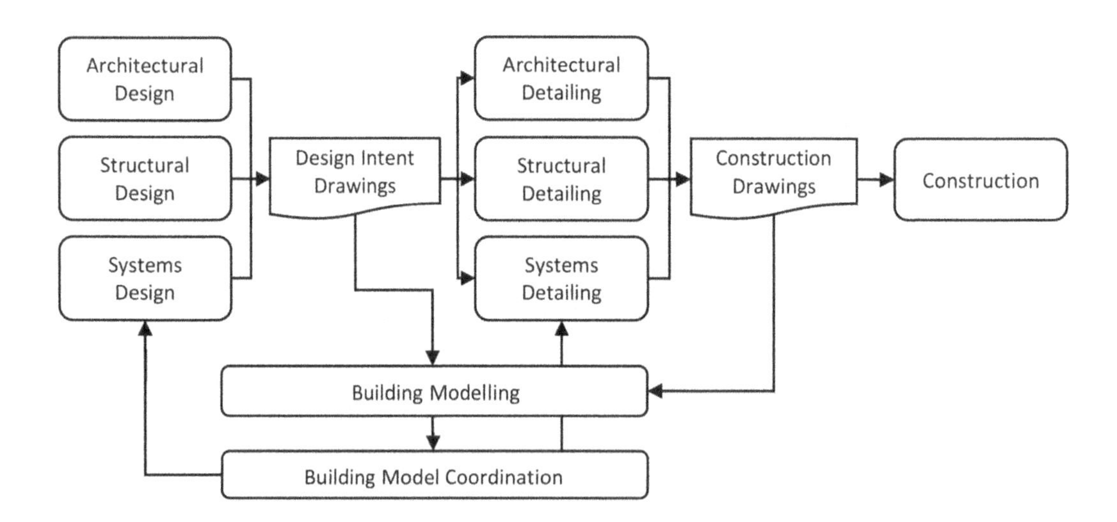

Figure 11.5 First steps in VDC: BIM as a coordination tool

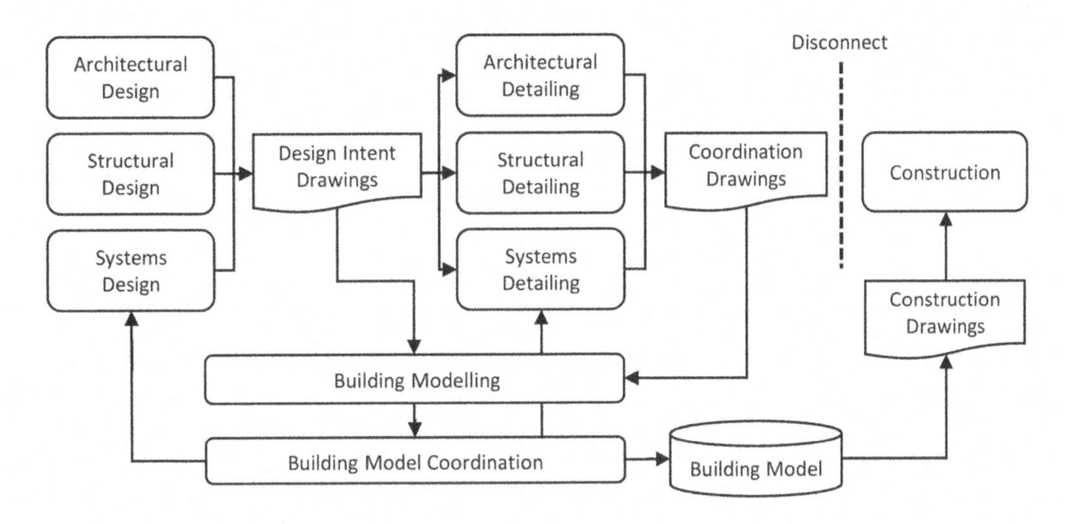

Figure 11.6 Next step in VDC: building from the model rather than from the drawings

The problem that Tidhar encountered, as many others have in this situation, was that although the construction drawings were derived directly from the model, managers had not paid sufficient attention to the change required in the mindset of the engineers and superintendents on the site, nor to the business processes that determined the information flows. The agility with which the site personnel were accustomed to responding to late design changes, such as the demands from individual apartment buyers to adapt their apartments to accommodate central air-conditioning systems with dropped ceilings, turned out to be a challenge to the company's attempts to move to a VDC process. The fact that the communication from the model to the construction site was still dependent on paper drawings enabled the site personnel to intervene at the last moment, making changes to the drawings and disrupting the consistency of the system.

Only by routing the information through the modeller could they ensure that the information would remain internally consistent, i.e. that the building "as-built" would accurately reflect the model (and vice versa, of course). If that consistency is lost, then the value of the model information is reduced because the accuracy and utility of any further work instructions that flow from it are dubious. If this happens, builders' confidence in the system degrades and "virtual construction" can become a futile exercise.

Ideally, builders at the work face should use the information from the model directly. This is becoming increasingly feasible and common with the introduction of handheld devices (tablets, smartphones, etc.), but wherever paper remains an important medium of communication, management must ensure that construction engineers, superintendents, work crews and all others on site recognize the importance of the model and behave accordingly.

Box 11.2 Printed model views displayed on site

One way of working around the limitation imposed by the perceived inconvenience or expense of using handheld computing devices to view a model on a construction site is to provide printed views of the model and to make them available as widely as possible on site. Tidhar adopted this approach for all of the complex MEP work in the Dawn Tower project, a 56-storey mixed-use commercial, residential and hotel tower in Tel Aviv (details of the project can be found in Chapter 4, "The Dawn Tower"). The same approach was used by Shimzu Construction Ltd and its subcontractors in the Mapletree Business Park II commercial development in Singapore. Figure 11.7 shows one of many large noticeboards used at the site to display model views. The images were used by superintendents and trade crew managers in daily "huddle meetings" with crews to make sure workers understood what was to be done and the construction methods to be used. Copies were also distributed to be taken to the work face.

Figure 11.7 One of many large noticeboards used at the site to display model views at the Mapletree Business Park II commercial development in Singapore

Source: Image courtesy of DCA Architects and Shimzu Contractors.

The next step in VDC implementation is to place the model at the centre of the design and construction activities. Figure 11.8 shows how this works in principle. In addition to coordination, the BIM model is the centre of the information flows. It serves for simulation and analysis of the building's functional behaviour, providing feedback for the designers. It also serves for simulation and analysis of the construction process, providing feedback for the planners. The former helps designers optimize the product; the latter helps planners optimize the process. Building model coordination functions to maintain consistency and integrity within the model.

Ideally, the building model at the centre of the figure is hosted in the cloud and is accessible to all project participants at all times, albeit with user-specific locking mechanisms to protect the model's integrity. In practice, given the variety of software in the market and the as-yet unresolved technical challenges of central model management, in the majority of implementations the building model at the centre will be a federated model. Each professional discipline and each construction trade uses the BIM software with which they are most productive and efficient, but uploads their local revisions to the central federated model at short, regular, predetermined intervals.

This set-up represents implementation of the first five elements of VDC (as defined at the beginning of this chapter). To close the loop and complete the VDC picture, information about the as-built state of the building must be fed back into the model as the building is built, as shown in Figure 11.9. The relevant information can include the

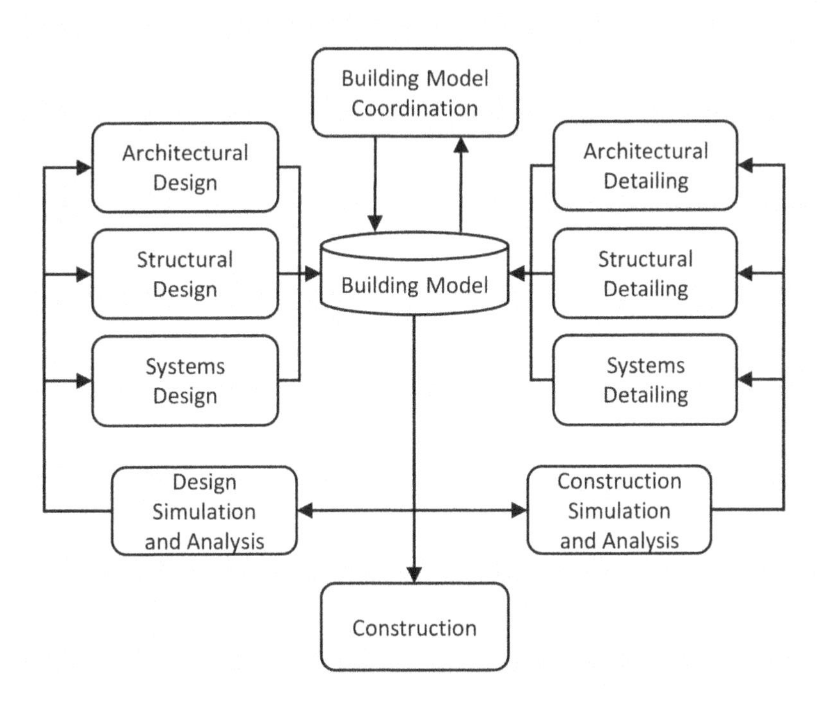

Figure 11.8 VDC with the BIM model at the centre of the design and construction process

results of quality control checks, measurements taken in the field, equipment serial numbers and other data for installed components, etc. All of these are alphanumeric data, so collecting them and associating them with building model objects as properties is relatively straightforward. For some applications – such as measuring an elevator shaft prior to installing the machinery, or assessing the planar tolerance of a large slab – it may be necessary to measure and record geometry. Addition of this information to a building model is more difficult to automate, and despite the availability of laser-scanning and other tools, it is not common.

Finally, VDC can be further improved by applying Lean thinking to the process. Smooth flow is characterized by short cycle times, small batch sizes and low levels of inventory, and this applies also to the flow of detailed product design information from the detailers to the builders. Pull control is a well-understood production control paradigm that helps achieve smooth flow, and it can be applied to detailed design information as well as to production in construction itself and to manufacturing. Figure 11.9 shows dashed arrows for the flow of pull signals – called kanbans – from the construction site

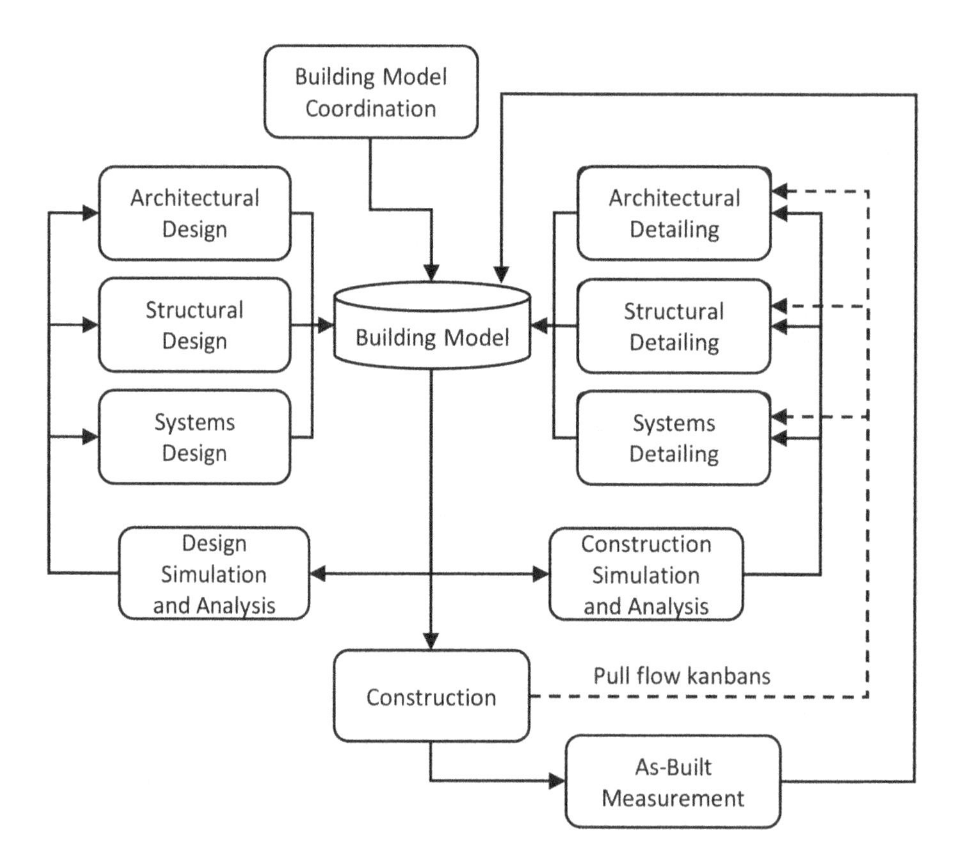

Figure 11.9 Lean implementation of a thorough VDC process: pulling information as needed for construction

back up the chain to the detailed design function. The idea is that as construction proceeds, the look-ahead plan flags locations that are due for construction and signals to the detailers that they should detail the systems in those locations in order to ensure that the information can be developed and reach maturity in time for the work to be done right the first time. In this way, detailed information is compiled only when needed, reducing the waste of premature design decision-making that could otherwise lead to rework.

Box 11.3 Building the model, not the drawings

The next chapter tells the story of BIM and Lean adoption and integration at Lease Crutcher Lewis, a large design build company active in the states of Washington and Oregon, USA. Their story highlights the many ways in which they are literally "building the model". In cases where Lewis has developed a specific way of detailing and constructing some part of a building, they will ask the owners' architects not to detail or dimension those parts of their design, preferring to model the areas themselves to ensure that they are constructible. Then they build the model, not the architects' drawings.

The company's VDC process is a standardized practice that begins with BIM modelling at the earliest possible stage (preferably in the framework of preconstruction services), continues with Lewis compiling and maintaining the model of record as the design progresses, and ends with delivery of a comprehensive as-built model ready for facility maintenance that encompasses all of a building's systems. Lewis considers VDC an essential component of the service they provide to their clients, and over time, they have changed the design and construction culture in their market in subtle ways, with long-term impact on the designers and subcontractors they work with. Their clients appear to appreciate the quality of service they receive – fully 90 per cent of Lewis's construction work is done for repeat clients.

Growing BIM capability

Figure 11.10 shows the steady growth of the use of BIM within Tidhar over the first three years after its introduction. It took one full year from the time that Ronen Barak was first employed as the Lean and BIM lead – with part of his time devoted to the tasks of a BIM manager – until the decision was made to employ the first full-time modellers as part of the "Growing the Margin" initiative. The meetings with BIM modellers on construction sites during the study tour to Finland and Denmark, particularly at Skanska Finland and at E. Pihl and Sons in Copenhagen, were a major factor in building confidence within Tidhar's management that this was a good move. Once the first pilot of system coordination proved so successful at the "Weather Vane" building in Yavne, two full-time modellers were hired and went to work on Tidhar's bigger projects in Tel Aviv, the 56-storey Dawn Tower (see Chapter 4) and a complex of four high-end office buildings named CU Fiat (see Chapter 10).

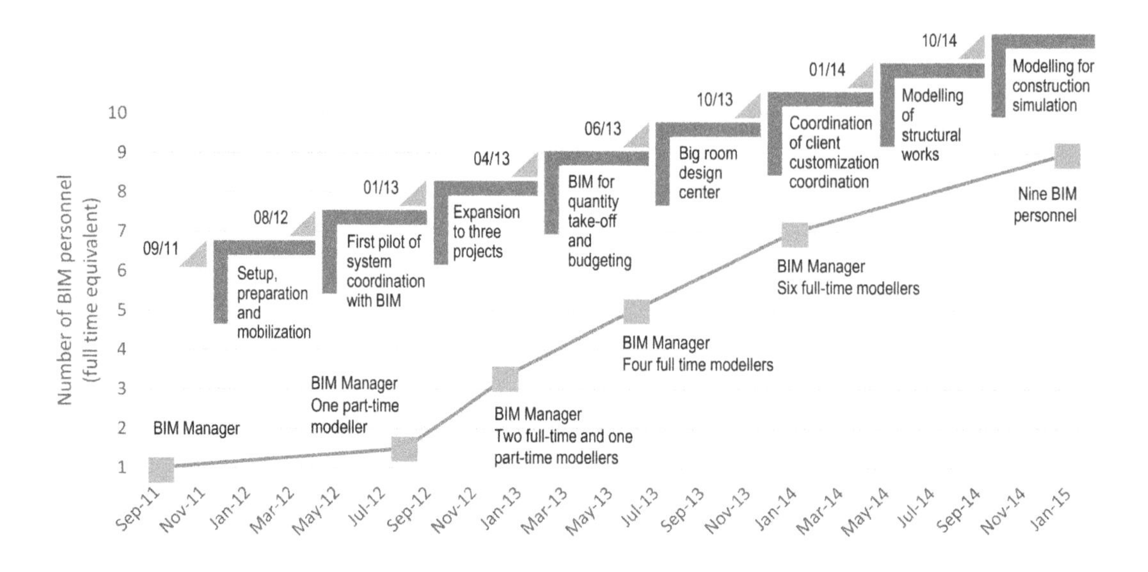

Figure 11.10 Growth in the number of employees in Tidhar's BIM team

Note: The steps in the figure represent the Lean idea of *kaizen*, in which processes are continuously improved in small steps. Like a ball pushed up a staircase, each change must be standardized and secured before starting to climb the next stair.

Box 11.4 Virtually constructing the curtain wall façades of the CU building: the value of modelling

Tanya Pankratov is an architect with a Master's degree from the Technion. Her MSc thesis explored the possibility of building responsive building façades. The idea was to use a distributed approach to control the dynamic components of cladding systems. Each section of a façade would function as an independent unit, capable of self-evaluation and individual adaptation to current climatic and lighting conditions. Neighbouring façade units would influence each other's behaviour, so that the façade as a whole could function optimally. Naturally, such a system was most easily designed and simulated using sophisticated parametric BIM tools, and she became an expert.

When Tanya graduated, she opted for experience in construction over a design job. Tidhar was growing its modelling capacity, and she quickly found herself hired to be one of the team and responsible for all of the BIM modelling for the CU Fiat project. This project was uniquely suited to her skill set – all four buildings were entirely enclosed by a sophisticated glass and aluminium façade. She explained:

> When we tried to visualize the façades to figure out how to build them, to try to understand what the geometry would be, and how they could be detailed, I realized that the architect's design drawings of the façade left us with many uncertainties. We did not have accurate quantities of the

various curtain wall types, nor did we have any idea about the connections to the structural frame ... The first quantity take-off and cost estimate for the project was performed for an initial architectural design that had changed over time, and the modifications to the architectural design in the detailed design stage would clearly result in significant deviation of the prices we would get from façade subcontractors from the initial estimate. So, we set about modelling the curtain walls to production level detail.

The curtain wall system was modelled late in the design process during the final stages of cost estimation of the curtain wall construction. The goal was to support the project manager's decision-making by providing precise quantities and cost estimates for any changes that he or the architects might propose as we finalized the design and moved toward tendering. The façades were modelled at a very high level of detail. The model included all of the objects and their parameters that were needed to extract quantities for the various types of curtain wall systems.

Figure 11.11 Detailed curtain wall façades in the BIM model of the four CU Fiat buildings

Tanya continued:

The detailed façade in the building model supported the team in a number of specific ways:

1 Setting the optimum height of the curtain wall extension above the roof level. The top of the curtain wall had to be high enough to conceal all of the air-conditioning vents and other mechanical equipment on the roofs of the four buildings, but given the high unit costs of the system, Tidhar naturally wanted them to be no higher than necessary. The architectural design called for a height of 3.8 m for all four buildings. Testing different heights and visualizing the model from the vantage points that people would have from the top floor of each of the buildings led to significant reductions in the total area of curtain walls above the roof level. For building D, we reduced the height from 3.8 m to 3.5 m, which dropped the area from 432 m^2 to 400 m^2. In buildings A, B and C, which were taller, a height of 2.0 m proved sufficient once we lowered the mechanical equipment itself. The total area for these three buildings was reduced from 1,772 m^2 to 1,044 m^2.

2 The curtain wall was designed to have two types of panelling systems: a flat constructive glass system and a system with circumferential decorative beams and "bullnose" details. The depth of the decorative details as proposed by the architect (60 cm deep) was both too expensive and

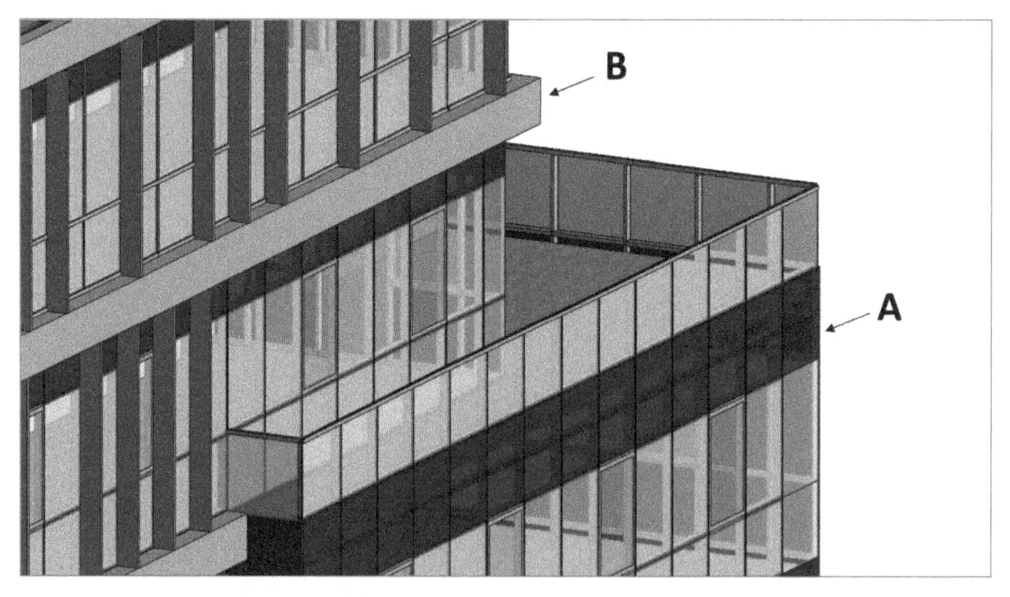

Figure 11.12 Two type of panelling systems: (A) flat constructive panels, (B) decorative beams and "bullnose" details

too complicated to construct. Various alternatives were developed in the model to find a design that would be both visually satisfying and financially acceptable. The final version of the decorative elements was just 19 cm deep. We developed it in collaboration with the aluminium curtain wall contractor, and we used a steel mould that he had available from a previous project. This dramatically reduced the cost of manufacturing – the final costs for these sections of the curtain wall were one third less than they would have been for the 60 cm deep option.

3. The initial architectural design had horizontal mullions on all the panels at a height of 110 cm above the floor. During walkthrough simulations within the model, we noticed that the horizontal division would obstruct the view of any occupant seated at a desk. With the architect, we decided to use the horizontal mullions only where windows could be opened. All the other panels had clear glass from floor to ceiling. This really increased the value of the building for the occupants.
4. Once we finalized the design, the model played a crucial role in the negotiations with the subcontractors tendering for the work. We provided the models and the quantities to the subs – and the prices per unit area that we got were significantly lower than in other projects, thanks to removing uncertainty for the bidders.

Tanya sums up the benefits they gained by modelling the façades along with the missed opportunity of late introduction of BIM in the process:

> The building envelope model played a key role in the decision-making process and directly reduced the cost of the curtain wall construction. And yet, it was more detailed than it needed to be for the purpose it served. Part of the reason was that we were trying to develop in-house knowledge in the BIM team about parametric modelling, to improve our mastery of the BIM tools at the same time as serving the project. I think it's essential for the team to provide the project manager with as much "live" information about the project as they can with reasonable resources. Having said that, the BIM team must evaluate the time and the effort needed and consider each step carefully with respect to the required level of detail and the best way to achieve it. As my boss, Zohar Raz, often says: "We don't model for the sake of modelling!"

As one can see in Figure 11.10, the number of BIM modellers in the company grew steadily, as more projects began using BIM and as the level of sophistication of BIM grew, too. The initial growth, over the first half of 2013, enabled expansion of the initial VDC approach of modelling the buildings and their systems to identify clashes and other design issues prior to construction. After that period of consolidation began a period during which the use of BIM was extended into design support, including modelling of projects at the conceptual stage to evaluate the financial prospects of different design alternatives. This included quantity take-off and estimating, preparation of construction budgets and development of pre-construction documents for sales, permitting and construction planning. The most ambitious move in this area was the "big room" that was set-up for the Rosh Ha'ayin project – that story is told in Chapter 14, "BIM in the big room".

The next step in Tidhar's BIM journey, starting in early 2014, was to use BIM for three activities that had a stronger and more direct impact on the construction sites. Introduced in parallel, they were:

- modelling of individual apartment buyers' design customizations using BIM, and feeding those models directly to the trade crews using pull;
- modelling the structural works in detail to support off-site fabrication;
- use of construction simulations with models (4D CAD) to evaluate and refine the production planning of reinforced concrete work.

The latter innovation is explained in detail in Chapter 18, "Raising planning resolution – The four-day floors". The first two innovations are the subject of the next two sections.

VDC for apartment buyers' design customization

Changes to the design of a building are always disruptive to flow, moreso if they are introduced late in the process. They are particularly disruptive if construction is already under way. VDC helps reduce the scope and frequency of changes that result from lack of coordination among design disciplines by catching the inconsistencies between them in the virtual prototype.

Yet design changes may still arise very late in the construction process from the desire of apartment buyers to customize their homes. This could occur very close to (or even after) the time that work crews are due to perform the work in the relevant apartments. Common examples of design changes in the high-rise apartment buildings (that form the major bulk of Tidhar's residential projects), include selection of flooring and wall tiles, changes to partitions to enlarge or reduce the size of rooms, addition of air-conditioning systems that require drains, electricity supply and dropped ceilings for ducts, and extensive customization of kitchens and bathrooms. Most construction companies employ staff whose job it is to meet with apartment clients, record their requests, coordinate the customization changes they desire with the relevant design consultants, budget the work, provide quotes, elicit clients' final decisions and compile drawings that reflect the required changes.

This process has three significant drawbacks. First, the process cycle time can be quite long, because the flow of information from the client to the builder has multiple stops along the way. The client changes representative prepares the drawings and the bill of quantities, and sends them to the architect and the various design consultants for approval (mostly to check that they don't violate any aspects of the building codes). Where issues arise, the information loops back to the client changes representative and to the client for correction. Once approved, changes are communicated through the project's purchasing department to the material suppliers and via the site superintendent to the appropriate subcontractor trade crews, and hence to the builder, as shown in Figure 11.13.

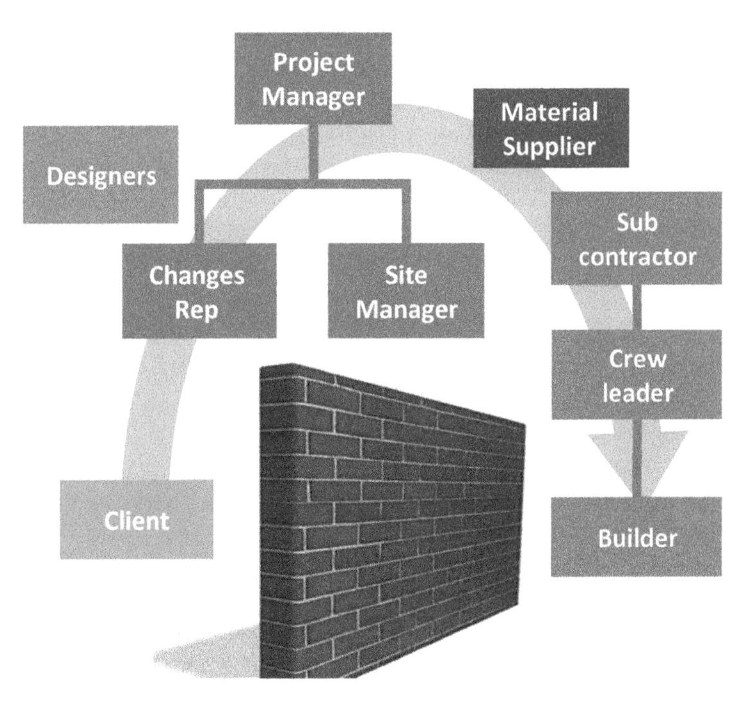

Figure 11.13 In the traditional process, client's apartment customization instructions flow through many stations on their way to the builder

Second, given the nature of communication using 2D drawings and bill of quantity documents, the process is prone to miscommunication, misunderstandings, and human error.

Third, the sequence in which clients reach their decisions does not necessarily match the sequence in which the work is performed on site. Traditional construction planning for residential high-rise buildings of this kind calls for steady progress of the trade crews performing interior finishes and installing building systems from the first floor upwards, executing the work in the order of the floors. Apartment sales, however, often follow a different pattern (or lack thereof); although apartments higher up in the building are considered more desirable and tend to sell sooner than lower apartments, the sales sequence is mostly random. Once clients sign a purchase contract, they begin the slow process of considering their options for customizing their apartment, which can take some time. Thus, the sequence of delivery of their change requests usually does not match the planned work process, and, generally speaking, construction companies find it quite difficult to entice or to coerce all of the clients to deliver their design change information in time for trade crews to begin work on their apartments. This was exactly the problem that Alpha Construction ran into back in Chapter 2, "False starts".

Apart from the wasteful information management process itself, the result of these two issues it that in some cases, design changes are delivered after an apartment has largely been completed, which results in extensive waste because the original work has no value for the customer and thus must be demolished to make way for what the customer actually wants. This is the waste of *overproduction*, where unnecessary or unwanted products are produced before there is a demand for them. The need for intensive coordination on the part of the design consultants, project superintendents and the subcontractors, together with the resulting waste of overproduction, leads general contractors to demand unit prices for the changes that are significantly higher than the price for the same work if it were part of the original design.

Tidhar's success in applying VDC was thus initially limited to the shared, or public, areas of their buildings. The challenge it faced in introducing VDC to the interior finishing and systems work in customized apartments was threefold:

- reduce the cycle time for information flow from client to builder;
- ensure that the model would remain accurate and reliable despite late changes;
- ensure that work crews would not perform work in locations in which designs had not been finalized.

The solution was to empower the client design coordinators to model the clients' design changes directly in a Revit model of the individual apartment, using the BIM model as the focal point for communication of the design to all those who needed to approve it or supply materials, as well as feeding the information directly from that model to the subcontractor trade work crews as needed. This created a single-source model for all the work in an apartment together with a pull system in which crews received information when and where they needed it, but not before it was mature.

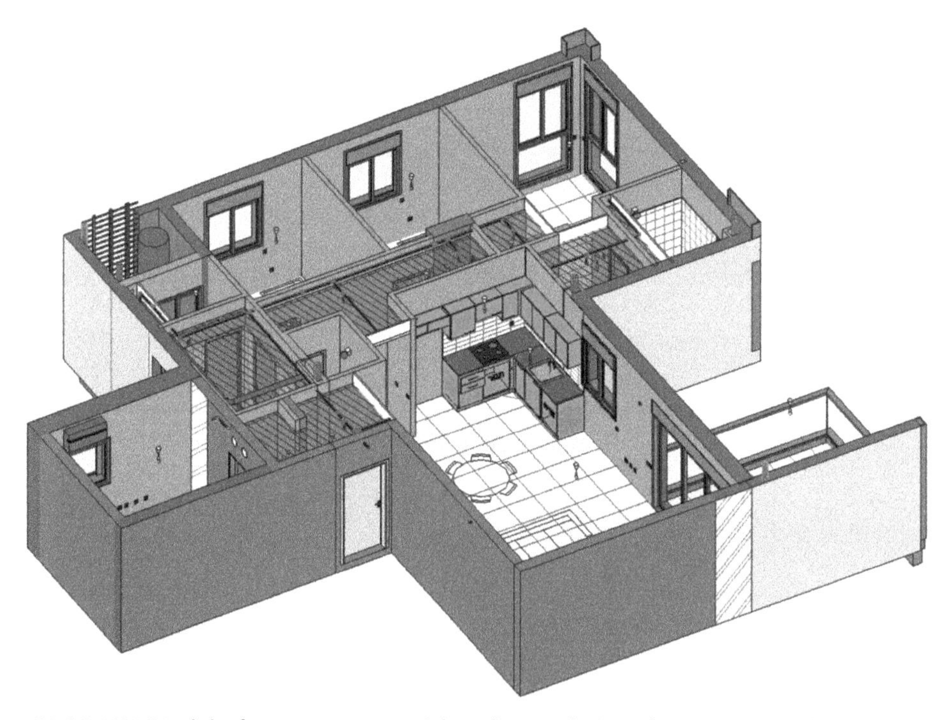

Figure 11.14 BIM Model of an apartment with a client's design changes applied

Note: The false ceilings for the air-conditioning ducts, the sprinkler system, kitchen cabinets, floor and wall tiles, plumbing and electrical systems and interior doors are all modelled in detail.

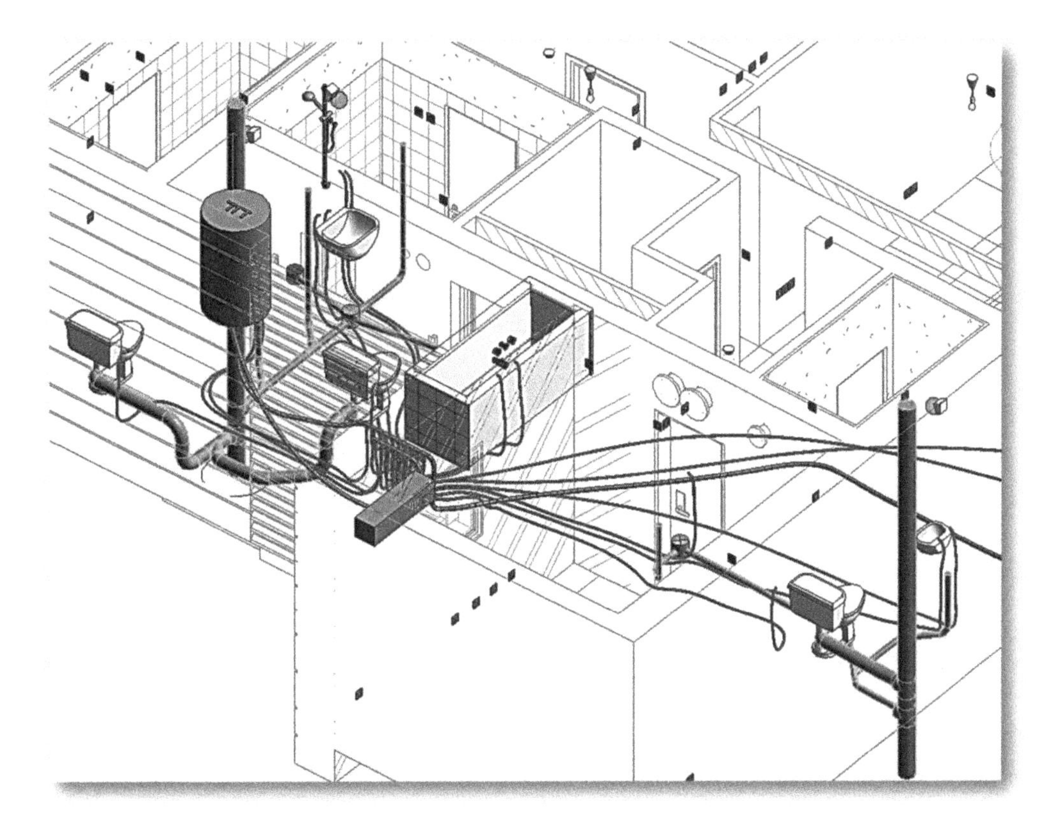

Figure 11.15 An axonometric view of the apartment model highlighting the plumbing system and electrical outlets

The new process was first applied at the Rosh Ha'ayin project (see Chapter 13, "Tidhar on the Park, Rosh Ha'ayin"). By the end of the project, the design change database showed some 65,000(!) individual line item changes, over the total 1,036 apartments. The client change representatives occupied a suite of site offices, where they met with clients, and – significantly – where superintendents and trade crew leaders could come directly to them to view the models and receive the final work package instructions for each apartment just before proceeding to begin the work.

Box 11.5 Modelling procedures for customer changes

Tidhar's procedures for the use of BIM models to compile and manage customization of the apartments detail all of the steps that the designers, modellers and client coordinators are to follow. The documents include flowcharts, BIM model templates, spreadsheets and guides. The following description outlines the process stages and highlights the solutions that were standardized. Readers may find the description helpful in defining their own detailed process for managing apartment customization.

Stage 1: Prepare the models

The BIM model manager begins by preparing template models of the various apartment types:

- a One model of each apartment type is extracted from the overall BIM model of the project.
- b The fixed parts of the apartment which the customer cannot change (concrete slabs and walls, windows, entrance door, etc.) are extracted from the apartment type model and saved in a separate file, called the "structural core model".
- c The remaining objects, which are those that can be customized, are saved as an "apartment type template".
- d The structural core model is linked into the apartment type template as reference geometry.
- e A set of views are prepared in the apartment template model for each aspect that can be customized: kitchen, HVAC, electrical and plumbing.
- f The apartment type template models are provided to the design coordinators.

At this point, the design coordinators must verify that the models accurately reflect the contents of the apartment exactly as they were stipulated in the developers' design brief and in the standard sales contract for each apartment type. The steps they perform are:

- g Produce bills of quantity and drawings for each of the trades.
- h Work with the purchasing department, the project manager and the various suppliers to confirm which sanitary fittings, kitchen cabinets, etc. will be offered and what electrical, water supply and drainage requirements these impose on the design.
- i With the help of a BIM modeller, apply all of the agreed design decisions to the model – sanitary fittings, water and drainage pipes, electrical fittings, etc.
- j Produce updated bills of quantity and drawings to confirm the validity of the updated model with the suppliers and the project manager.

The result of this procedure is a set of Revit templates, one for each apartment type in the project. Extracting the structural core from the template of items that can be changed by the customer prevents inadvertent changes, and it also has the benefit of making the individual apartment models much smaller in file size and therefore much easier to manipulate in the software during a meeting with a customer. Extracting the apartment models from the main model has the drawback of loss of information of parts that fall between apartments, but this is generally not a problem; the larger problem is that if any changes are made to the structure, the responsibility for maintaining the structural core models (i.e.

updating them according to the changes to the main model) falls to the model manager. Changes to the structure at this stage are rare, but they do occur – for example, the location of a window may be adjusted due to an error in the original model or due to conditions in the field. Maintaining the model integrity by propagating the change to the structural core models for all the apartment type templates is a manual operation that is not supported by BIM software and requires strict process controls. Given the technological constraints of the software, this drawback was felt to be a small price to pay for the agility and flexibility of small model files.

Note that up to this point, apartment models have not yet been produced for any specific customer. They are produced "just in time" for the meeting with the customer. The customer meeting is the pull signal for creation of an individual model from the apartment type template.

Finally, there are two further steps that the design coordination team must complete before meetings with customers can begin:

k Set up the cost estimation file for each apartment type in the company's legacy change management database system.

l Extract the model quantities, apply corrections for gross quantities for flooring and other line items whose model quantities are net amounts, and fill in the line item quantities in the database system.

The benefit of this procedure is that the quantities are extracted from the model and measured automatically. However, the downside of this way of working is that information is duplicated, which introduces the opportunity for errors or discrepancies. Tidhar decided not to invest in developing a live update link between the apartment models and the line item quantities in the database, and so subsequent updates must be managed manually. Note also that the quantities extracted directly from the BIM model cannot be used for purchasing of materials such as floor tiles, because they provide the net floor area without accounting for the need to cut tiles. A direct model-to-database data link would need to compute the waste factor and apply it when updating quantities across the systems.

Stage 2: Customer coordination meetings

The first meeting with each customer is devoted to making the customer familiar with the apartment model. Tidhar's design coordinator explains the 3D model to the customer, showing them how they can visualize different aspects of their apartment, and explaining the scope of the design customization changes that they are allowed to make. The model is particularly useful for explaining options that require spatial understanding, such as the extents of dropped ceilings for the different air-conditioning system options.

In the second meeting, the customer delivers their choices for customization. There are two possible processes. If only minor changes are requested, the steps for the design coordinator are:

a Create a model file for the customer's apartment by opening the apartment type template and saving with the apartment number.
b Model the changes directly in the apartment model.
c Produce the necessary drawings and quantities from the model.
d Update any changed quantities in the change management database.
e Produce a proforma invoice from the database system and obtain the client's consent to the change package.

This process is effective because it ensures that the model reflects the customer's requirements reliably. It gives the customer the opportunity to view the changes directly in the model, in 3D, and within the full context of the apartment. It also carries costs because the design coordinators must be proficient in the use of the BIM model. Most of Tidhar's design coordinators are architectural technicians, and it proved possible to train them to use Revit software effectively for this purpose.

In cases where the customer requires extensive changes, such as changes to partitions and doors, relocation of kitchen cabinets, or special features, an interior designer or architect is employed. Tidhar considered providing the apartment core template to the designers so that they could work directly on it and return a corrected Revit model, but decided that the risk of receiving information that did not conform to their data standards was too great. In theory, this would have had the advantage of feeding the customer's requirements "as-is" to the construction crews, through the models. Instead, the policy chosen was to receive the customer's design as 2D CAD drawings and have the design coordinator implement them in the apartment model, similar to steps (a) to (e) as above. Where necessary, the model manager supports the design coordinator in applying more complex changes.

In the third and last meeting with the customer, the design coordinator records the customer's choice of finishing materials – floor tile colours and sizes, door finishes, paint colours, etc. These are entered in the model and in the change management database.

Stage 3: Preparation for construction

Once the customization process itself is complete, the individual apartment model is now ready for construction. The model manager merges the apartment model into the overall building model, replacing a placeholder apartment model. Replacing the template model with customized one is therefore relatively straightforward. Where the overall building model is large, it can be split into single-storey models, with the individual apartments inserted directly.

Replacing the apartment template with the specific customized apartment models has the great benefit of ensuring that the general contractor has only one model for each floor. The original "standard design" model is replaced entirely, and cannot be used inadvertently for construction. A single model is maintained for each building or floor, ensuring that trade crews receive only up-to-date information from the customization process. Provided that the building or floor models are updated before the finishing works commence on a floor, working drawings produced "on-demand" from the model reflect the current status of the customers' requirements.

Modelling structural work

Modelling of the reinforced concrete structure – piles, foundations, beams, walls, columns, slabs, etc. – has clear benefits. The structural model guides the preparation and installation of formwork and of shoring, it provides essential and accurate quantities of concrete, and it is central to construction planning using 4D and other tools. It is also a platform for moving production off site, which may range from fabrication of repetitive yet complex structural components, such as stair sections, in moulds in an on-site workshop, to full precast concrete construction.

However, in some cases, prefabrication of components can work against the flexibility needed to achieve other efficiencies, and so one cannot simply assume that, at the micro level, everything should be modelled for fabrication. The case of rebar prefabrication at Tidhar provides a good example of this dilemma.

One of the problems that Tidhar faced in their existing process was a prevalence of discrepancies between the architectural and the structural drawings. As was common in local practice, the structural engineers would detail rebar by drafting the shapes onto a structural 2D plan or detail section that was extracted at some earlier stage from the architectural BIM model.[6] This resulted in differences between the architectural geometry and the structural geometry design because the structural engineers often did not inform the modellers of the changes they made to the geometry during design. Despite attempts at coordination, conflicts persisted into construction itself.

After some discussion in the steering group, two options were proposed: (1) require contractually that the structural engineer model using Revit or Tekla, and then coordinate their models with the architectural model using Navisworks; or (2) if the structural engineers proved unable or unwilling to model, then Tidhar would assume the responsibility for modelling the structure, and produce concrete geometry and rebar detail drawings directly from the model. In this case, the structural engineer would no longer be required to produce a set of drawings. Instead, as is common in US construction practice, the engineer would be asked to approve the drawings submitted by the general contractor.

Gil insisted that the drawings that reach the workers on site must contain all of the necessary information, and no more, and must never show the same information in more than one form (such as architectural and structural representations of the same wall, column or beam):

> Our own drawings, produced from the model, are the best quality, coordinated and reliable. Therefore, we must take the step and decide to build only from our model, not from consultants' drawings. I am not willing to see two sets of drawings on the job site that describe the same geometry, with conflicts between them.

Despite this proclamation from the CEO, there was still intense discussion, with some participants claiming that a separate structural BIM model is essential and others claiming that it could lead to discrepancies among the models because the engineer would continue to make changes to the geometry until late in the project, even when rebar detailing was in progress. Surprisingly, given the proven modelling capabilities that the company had developed by the time of this debate, strong leadership was required to push the team to commit to the second process in the event that the first could not be achieved.

Some 14 months after returning from the study tour in Finland and Denmark, the company as a whole had made great strides in developing in-house BIM capabilities. A number of projects had Revit modellers leading the information management, and the company was adopting BIM for management of the client changes process described above. And yet, the notion of VDC as a process that is fundamentally different from traditional practice, in that it interposes a prototyping phase between design and construction, was still not clear to many of the senior managers. They tended to fall back into an understanding of the contractor's goal in using BIM as limited to quality control of design information, as a last step in the design process; they had not made the "leap of faith" to see it as the first step of the construction process. Their attitude was reflected in questions like "do we really need to model all the building systems?" The answer was either "yes", if you want to prototype the actual construction; or "not necessarily", if you are only checking information quality (some designers provide relatively well-coordinated drawings). The mode of modelling as a process for checking constructability was not yet understood nor accepted.

To model, or not to model? That is the question!

For those that saw the need for thorough modelling, the need to model the concrete reinforcement bars was clear, while for others it appeared to be superfluous. The former pushed for adoption of Tekla Structures for rebar modelling; the latter were willing to suffice with CAD drawings of rebar detailed on top of geometry provided directly from the model.

In a follow-up meeting two weeks later at the Rosh Ha'ayin site office, the discussion focused specifically on the question of the need to model the rebar with the goal of producing electronic rebar bending schedules automatically for computer numerically controlled (CNC) rebar bending off site. Those in the meeting from the design/BIM side (Ronen Barak, Zohar Raz) saw clear value in this automation, so they pushed strongly to adopt modelling using Tekla Structures for rebar detailing. However, Tal Hershkovitz disagreed. He questioned the assumption that off-site fabrication is always better. He explained that in many sites, the project manager decides to purchase rebar in raw lengths and cut and bend it on site. The reasons were twofold: (1) some structural reinforced concrete work crews include the bending in the services they quoted or offer to do it at a low rate (having this extra work on site is a way for them to address operator balance issues introduced by the varying work volumes that typify cast-in-place concrete works, like closing formwork or pouring concrete); and (2) the price for raw steel is about

40 per cent less than for bent rebar. If there is space on site for a bending station, and the weather is good, it can be profitable to do the bending on-site.

Tal claimed that the freedom for the project manager to decide should be maintained: those PMs who find modelling and off-site fabrication of rebar economically justifiable should select that process, while other PMs could still select rebar fabrication on site.

This approach may seem anathema to readers in countries where the wages of construction site workers are relatively high, yet it makes intuitive sense for those where lower labour costs may make automation (modelling and CNC fabrication) appear more costly. However, Tal's approach contrasts with a central idea of Lean – that of standardization. At any given time, there should be one stable and acceptable way to perform a task for any process, which represents the best way that the organization currently knows how to carry out that process. This is not the absolutely most effective or economical way to perform the task (because we assume that there is always room for improvement), but it is the best way that is currently known. Potential improvements must be introduced in a rational way using small-scale experiments, based on clear and demonstrable comparisons between the current state and the proposed future state of any given process. But when processes are not standardized, it is difficult to evaluate the merits of any specific improvement.

The failure of the steering group to reach a consensus on the issue meant that, by default, BIM modelling and CNC fabrication of rebar was not introduced at any of the company's projects. It was not until a second project at Rosh Ha'ayin was begun, in late 2015, that one of Tidhar's modellers was given the job of modelling the reinforced concrete and rebar thoroughly, with the structural engineer fulfilling a role similar to the "engineer of record" in the US paradigm.

Lessons learned

- **VDC implies intensive and extensive collaboration**. The benefits of using BIM as the tool of collaboration are significant because coordination among disciplines is essential for accurate and effective design of a building. BIM enables VDC because it allows everyone in the team to understand the full scope of the design and to work concurrently, rather than working on large batches of information in separate silos. For a complex and highly-interdependent product like a building and its systems, this is key.
- VDC adds value and contributes to removal of waste in design and construction through **virtual prototyping** and **virtual first-run studies** of production.
- To the fullest extent possible, **build from the model**. Information for construction can be extracted from the model directly in the form of working drawings with local views, or viewed directly in the field using mobile computing devices.
- The timing of information extraction from the model for construction – such as producing working drawings for trade crews – is critical. If done too soon, drawings may be out of date. Thus, **maintaining the discipline of just-in-time pull of information is essential**. Working drawings should be generated as late as possible, only when needed for the start of construction on any given building section.
- **Avoid splitting the model**. In Tidhar's apartment customization process, splitting the building model into individual apartment models for client customization creates an

anomaly in which information temporarily exists in more than one place. Given the constraints of the BIM tools used, this was considered necessary in order to provide small files for the design coordinators to manipulate in meetings with customers. However, it created an opening for coordination errors to creep into the process.

Notes

1 A PAS is a publicly available specification document of the British Standards Institution (BSI 2013). A PAS is usually commissioned and prepared by groups of representatives of industry-leading companies. It does not have the formal and binding status of a standard.
2 BSI (2013).
3 Where people begin with materials and build directly, designing as they go, the design may be said to be physical. Even in this case, people compose and manipulate mental models of the structure they are building, which may be considered virtual expressions of the design.
4 See Ballard (2000a).
5 Gil Geva, Tidhar Group CEO, was first introduced in Chapter 1, "Introduction: Growing the Margin"; Zohar Raz, the design manager, was first introduced in Chapter 3, "Education and motivation".
6 With respect to the responsibilities of structural engineering design consultants, construction practice in Israel is similar to that in Europe, in which the structural engineer not only performs structural design but is also responsible for detailing the reinforcement.

12 The Lease Crutcher Lewis way

> "We'll ask for your approval for the construction model, and then we'll build from the model, not the design drawings."

Lewis virtual design and construction

Lease Crutcher Lewis, a Seattle-based general contractor with projects in Washington State and Oregon, was established in 1886 and has a strong tradition of construction expertise. Lewis, as most people call it, is not a new company, yet it has embraced the technology of tomorrow through a thorough, well-documented and conscientiously implemented VDC process. Lewis VDC coordinators tell clients' designers:

> Please don't model in more detail than is needed. There are elements in the model that we won't be able to dimension until we have selected a manufacturer. During our clash detection process, we will verify that the manufacturer's modelled elements work with your design. We will encourage the supplier to produce shop drawings directly from that model. We'll ask for your approval for the construction model, and then we'll build from the model, not the design drawings.

Box 12.1 Lowrey's story

Lowrey Pugh is Lewis's senior VDC coordinator. Here is the first instalment of his story:

> My father was a union plumber. I got out of school, got married young and had to get a job, so it's not surprising that I became a union plumber too. I advanced quickly on site. During my apprenticeship, I took a CAD detailing class and I liked it, but I went back to work in the field because I was good at it. But after many years the work felt repetitive, I felt trapped and I hated all of the rework we had to do when designs changed or systems didn't fit together.
>
> After a back injury, I had the opportunity to take a different professional path. I studied a two-year degree in architectural and engineering graphics, thinking I would become a mechanical detailer. Just one month after

graduating, a professor connected me to Lewis, and I started the next day, replacing Lewis's only BIM detailer at the time. That was six years ago.

In the degree programme, Revit was very new, but I liked it. I could see how effective it was, and I used it to do all of the homework projects in the other classes. My predecessor at Lewis did everything: modelling, BIM management, providing information to projects, etc., so the work was lacking in depth at the time. But when I joined, Lewis had a new residential project with two 27-storey towers and had sold the owner on an intensive VDC service role. I was the first VDC coordinator to be dedicated to a project. I began to put my own stamp on the self-performed work: I compiled integrated drawings (of multiple trades) from the models, used a lot of colour, and placed an emphasis on readability. We saw a great return on investment in terms of reduced construction costs. At one point in the project, we realized that one of the subs had much more rework and quality issues than the others. Upon investigating, we found that their crew was using the contract drawings rather than the drawings I had been producing from the model; all the others were building off the model. That was an "Aha!" moment for all of the crews, and it really increased everyone's respect for the model.

Lana [Lewis's VDC manager] arrived soon after that project, and over the following five years we grew from 2 to 12 people in the VDC department. We have learned mainly by experimenting. For example, what size of job justifies a dedicated VDC coordinator? How scalable is the service? Is it worth having a dedicated VDC coordinator and BIM detailer on a smaller $15 million job or only on larger $150 million jobs? We initially had a single BIM detailer servicing multiple projects, but as our VDC procedures matured, our department grew from a BIM detailing role to a dedicated VDC coordinator role.

Three years in, Lewis built a university science building and research laboratory in a GC/CM (general contractor/construction manager) capacity. We realized that with the traditional RFI (request for information) process there was no way we were were going to meet the schedule, so we started the BIM construction process (which we developed into our current CDR [collaborative design resolution] process). We shared models better with the design team, we got stakeholders to the table and we enabled them to make decisions quickly. We took the detailing weight off the designers, moving it to the detailers at Lewis. We spent about $1 million on modelling before construction – it was a very complex STEM (science, technology, engineering and mathematics) academic building. That enabled good solutions within short cycle times. For example, we were able to coordinate inserts and fittings for mechanical trades during concrete casting in record time. Bringing the trades together not only helped solve the technical problems, it accelerated team building and collaboration among trades on the job.

Lewis and VDC

Lewis has developed highly sophisticated VDC capabilities thanks to four factors: strong leadership from management, careful selection of VDC coordinators with the right skill sets, application of BIM and cloud technology, and methodical Lean practice. The company's VDC process is a standardized practice that begins with BIM modelling at the earliest possible stage (Lewis's preference is to do so in the framework of preconstruction services), continues with Lewis compiling all of the professional and trade models into a model of record and maintaining it as the design progresses, and ends with delivery of a comprehensive, as-built model ready for facility maintenance that encompasses all of the building's systems. Lewis considers VDC an essential component of the service they provide to their clients. Over time, Lewis's influence has helped shape the design and construction culture in their market in subtle ways, with long-term impact on the designers and subcontractors they work with. Their clients appear to appreciate the quality of service they receive – fully 90 per cent of Lewis's construction work is done for repeat clients.

Lewis's practical goals in using BIM are clearly stated in the company's VDC policy document as follows:

> The purpose of this Building Information Modelling Execution Plan (BIMxP) is to orchestrate the use of BIM by the collective project team and to continuously improve the delivery of the building as a collaborative effort. Lease Crutcher Lewis is dedicated to the principles of Lean and the use of BIM as a tool to maximize efficiency in the field, reduce waste, and guide our project delivery processes. Our ultimate goal is to have a safe, productive, collaborative project through the mitigation of RFI's and work stoppage due to rework. BIM is a method of communication used to identify issues, concerns, discrepancies, and omissions, allowing the team to respond thoughtfully and not be forced to react in a hurried environment.
>
> **We are looking to build respect and trust not only in the people but in the process, including the model.**

The VDC mission statement calls for "All Lewis Employees Accessing the Model to **Leverage** its Information", for five purposes:

- estimating, preplanning and safety;
- coordination and RFI mitigation;
- sequencing and prefabrication;
- shop-drawing review and clash detection;
- building from the model – robotic total stations and work packages.

Standard VDC procedures

The Lewis VDC procedures are laid out in a detailed flowchart that defines each and every step required of the company's staff and of the designers and subcontractors. Further details are provided in a standard table of VDC responsibilities, a modelling standards document and a BIM project execution plan template document, and the process is embedded in the company's project management communication software as a set of tables.

The VDC procedures have been refined over time. The company's VDC manager, project managers and VDC coordinators have the authority to conduct controlled experiments, be it with new software or new processes, to measure the benefits and to update the standard procedure when innovations are found to be worthwhile. No specific budget is allocated for BIM development, but ample resources are provided as needed. This is a great example of the principles of continuous improvement and respect for people. Lewis employees call a small improvement step a "Plus One". Having their own language is a sign of the company developing a Lean culture, described further in the section on Lewis Lean below.

Wherever possible, Lewis prefers to get involved in a construction project before the design is complete, preferably no later than 60 per cent through the detailed design (DD) phase. Every project begins with a BIM/VDC preplanning phase, in which the project team is formed and an internal Lewis BIM preconstruction meeting is convened. For any new employees in the management team, software training, VDC process training and observations of VDC meetings in other projects are mandatory. The preconstruction meeting reviews Lewis's standard "preconstruction BIM agenda" and sets the process in motion.

The BIM/VDC preplanning phase is usually initiated once a contract is won, but wherever possible, Lewis will begin working on a project before the contracts are in place, in the interest of providing excellent service through early contractor involvement, using BIM. This is naturally more common on projects with repeat clients, where strong relationships built on trust can develop and success in winning the contract is more certain. Although Lewis does not participate in IPD projects, it aims for alignment of interests with its clients and subcontractors as a core company value.

The goal of the next phases for the BIM team, modelling start and budgeting, is to compile a "single source of truth" BIM model at level of development (LOD) 200 (and no more) by the end of the design team's schematic design (SD) and concept design (CD) phases. A Lewis BIM modeller is assigned to compile the model. This is done from the designers' documents even if they have models, to ensure that the model is fit for construction-specific needs and to ensure that the level of definition is appropriate – i.e. nothing is modelled in any more detail than the current status of decision-making warrants. For example, the edge conditions of a concrete slab are detailed in a sequence of handoffs in which each member of the team brings the model to the appropriate level of detail, as Lana Gochenauer explains:

> We ask the architect to define the edge of building and place a slab edge but to understand that may change when the exterior wall system is procured. At that time, the final slab edge will be selected and modelled in detail. The architect will model the slab edge at LOD 200 or 300. The structural engineer takes the slab to LOD 350 because we are missing exact dimensions for the enclosure. The exterior enclosure sub-contractor will take the model to LOD 400. This is one reason we created our own matrix for Model Element Authorship (MEA). None of the matrices that we found from the American Institute of Architects or other sources captured the handoffs in the way we required.

Architectural Model (200)

Lewis Federated Model (300)

As-Built / Operations (500)

Figure 12.1 Levels of development

As Lewis's modeller complies the construction model, the designers' models are linked in for verification. The result is a federated design model that includes elements that are to be self-performed by Lewis.

Why does Lewis insist on compiling and maintaining their own models even when the architects and/or the engineers provide models? The reason is that estimation and construction models are different from design models in scope and content. For example, an architect may model a single column rising 23 floors. But for estimating, for scheduling and for work packaging (adding all of the detail needed for construction in the field), Lewis must break the column down into sections for each floor level. Quantity take-off and estimating is done at this stage using Innovaya and Timberline software, and with their standard process, concrete quantity take-off is now usually within 0.1 per cent of the traditionally-generated take-off quantities. Lewis project managers and engineers have very high confidence in the accuracy of their models.

What happens if there are significant changes in the design at this stage? When the design changes involve anything more than adjustments to dimensions or simple addition, deletion or substitution of specific objects, Lewis prefers to remodel major parts of the building, or even the whole building from scratch, as a matter of policy. In such cases, the potential cost of the risk embedded in a model that is edited to reflect design changes, which may turn out later in the construction process to be inaccurate or inconsistent due to human error, is considered to be greater than the marginal cost of remodelling from scratch. A reliable model is key, because unreliable project information can undermine confidence in the virtual construction process, leading to far more waste downstream.

As a matter of policy, Lewis employs construction people trained in BIM, rather than BIM people trained in construction. Experience has shown that the transfer from the field to a BIM-related role is more effective than attempting to teach people with little construction experience how to model in a way that incorporates the construction knowledge that is essential for an effective virtual construction process. Lowrey Pugh, Lewis' senior VDC coordinator, enjoyed a long career as a plumber and a site supervisor before learning BIM.

While all Lewis engineers and managers are expected to be competent in using BIM models, there are two BIM-specific roles: VDC coordinator and BIM detailer (modeller). To support projects with an annual average turnover of over $500 million, Lewis currently employs three BIM detailers, seven coordinators and a VDC department manager. The VDC coordinators function as "project information managers". Their role is to manage the federated construction model; they ensure that the models are updated according to the protocol and schedule set for each project, they monitor the quality of models received from the subcontractors, and they coordinate Lewis's own BIM detailers to complete aspects of the model that will be self-performed by Lewis personnel.

The next major phase in the process is coordination. A VDC coordinator is assigned to the project and a project-specific BIM schedule is prepared based on the company's standard template. BIM/VDC coordination meetings are usually held weekly. As the coordination effort progresses, clashes are recorded using images and parameter values. The clashes are assigned ID numbers, and they are logged in the CDR table on Lewis's Bluebeam Studio project collaboration platform. The responsibility for resolving each clash is assigned to a specific person – the architect, the Lewis modeller or some other designer or subcontractor. Clashes are resolved through negotiation of the technical issues with all those involved. Once resolved, the responsible person updates the models to reflect the decisions made.

Note that coordination is quite different from clash detection, the phase that follows. Coordination is done at LOD 300, and it aims to allocate spaces, or volumes, in the building to the different building systems in such a way that physical conflicts or clashes are avoided by design. Lewis's BIM execution plan document defines the spatial modelling method using relatively simple rules, such as this:

> This method will be applied to all projects. Modelling will be defined by space. For example; it will be decided that Mechanical sheet metal will go tight to floor, electrical will take the first 8" above the ceiling and the space in between will be reserved for plumbing. Not only vertical space, but also in the corridor: it will be decided if mechanical will stay tight to right side and electrical cable tray will stay tight to the left, etc.

The coordination effort is sequenced to align chronologically with the planned construction schedule. The same subdivision of zones or locations in the building that is used for construction planning is used to define the coordination work packages in the information development schedule. For any given zone, coordination is followed by detailing, clash checking, information delivery, fabrication and construction. The modelling tasks are integrated within the project schedule as a whole, and they are generally driven by using Lean principles and tools that account for the lead times needed for procurement of the construction materials and components in each zone. Reverse phase scheduling is applied from the zone construction milestones back through procurement lead times and finally to the modelling process. Modelling is then pulled to satisfy the preconditions for procurement and construction. The Last Planner® System is applied for production control in three distinct phases: during preconstruction, during construction detailing and coordination, and during construction.

The "clash detection" phase begins only once trade subcontractors have been brought on board and begin to prepare their own models. Once the subcontractors have been hired, a subcontractor kick-off meeting is held and Lewis's model is shared with them. The subcontractors are required to model the work to be done at LOD 350. Subcontractors add their trade models to Lewis's model using Navisworks, creating a federated model that is updated weekly. Lewis uses Autodesk software for all its projects, but they will hire subcontractors who use other BIM software packages on the condition that the subs can successfully merge their models with the Lewis model in Navisworks (using IFC file exchange). The standard procedure document details a "first-run" study that must be performed to ensure this basic level of interoperability.

Clash detection continues even after construction has begun, but it stays ahead of the construction zone by zone to make sure the clash detection is complete prior to beginning work in any given area. Clash detection in a zone is designed to be a three-week process, with clashes identified in the first week, resolved in the second week and signed off on by the third week. A clash-detection schedule is prepared based on the zone construction schedule and distributed to the trades. Subcontractors are required to do their homework before each clash detection meeting:

- The system subcontractors must check for clashes between their system and those of the other subcontractors in the zone before the meeting. For example, the HVAC duct supplier and the sprinkler system supplier are required to coordinate with one another before they come to check their systems against the building as a whole. The number of clashes is generally small, thanks to the coordination process that precedes detailing. Making the subcontractors responsible for clash checking before the meetings tends to motivate them to invest in system coordination with the other trades before proceeding with detailing.
- They must upload their models to the online project management system at least two days before the meeting.
- The VDC coordinator checks each subcontractor's model one at a time using a predetermined set of rules, groups the clashes found according to the trades that are involved, and distributes a Navisworks NWD format report file to all trades.
- The report is used as the agenda for the meetings and the solutions agreed upon are recorded.
- Clashes are tracked through the defined three-week cycle; any that are not resolved and signed off on within that time frame are flagged. These are then resolved in dedicated meetings with the project manager, the VDC coordinator and the relevant trades.

A key principle of the VDC procedure is that subcontractor trades must prepare all fabrication details and shop drawings solely and directly from their model, once it has been approved through the coordination and the clash-check process. This ensures a single-track process, and it is essential to maintain the integrity of the virtual model's reflection of the physical construction.

Once a trade's fabrication details and shop drawings have been prepared, reviewed and approved by Lewis's engineers, they are the basis for the next step of the VDC procedure: work package definition. The goal of this phase is to compile work packages for the field work and prepare the documentation needed to support the workers in the field. Using the zone plan, zone sheets are prepared in Revit for each trade, reflecting only the work to be performed in that work package by that trade. The work packages are reviewed by the project engineers and the foreman or supervisor of each trade.

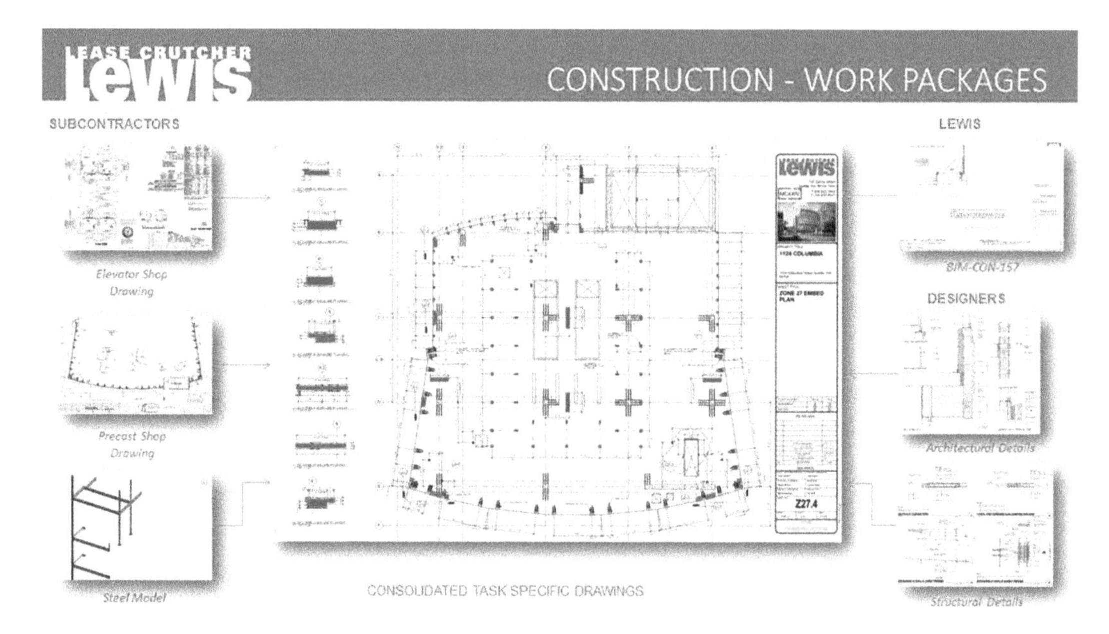

Figure 12.2 Construction work packages provide workers in the field with a single representation from all sources. Unlike traditional drawings, this drawing displays the information from multiple design disciplines, alleviating the need for workers on site to correlate instructions from multiple drawings

From this point forward, the work packages become the units of production that flow through the process – through the make-ready process, the weekly work planning meeting, the actual construction and the per cent plan complete monitoring steps of LPS. In Lewis's procedure, definition of the work packages using BIM is a key enabler of the Lean Construction process on site. This is a good example of the synergy described in the BIM/Lean interaction matrix discussed in the Introduction, where the BIM functionalities "automated generation of construction tasks", "visualizations of process status" and "online communication of product and process" help achieve the Lean goals of reducing variability and cycle times and reducing batch sizes.

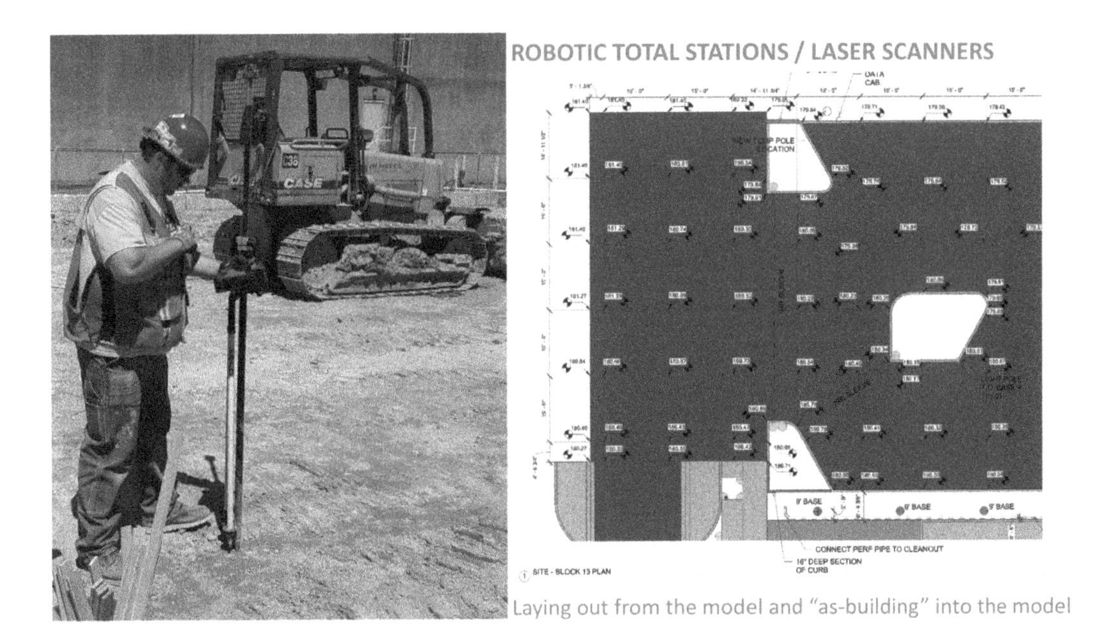

Figure 12.3 Robotic total stations and laser-scanners: laying out work directly from the model

Figure 12.4 BIM in the field: virtual plan table and BIM box

During construction, the construction crews access the model information using several technologies. Lewis staff use robotic total stations and lasers to lay out points from the model directly onto work surfaces. The model and the work packages are directly accessible on iPads, on "virtual plans tables", and in booths with large-format screens called "BIM Boxes".

Box 12.2 Detailed modelling supports off site prefabrication

Lewis self-performed the wood framing on the Oregon Department of Veteran's Affairs Lebanon Veterans Home project, which was the winner of numerous awards. To take advantage of the repetitious design and meet the budget requirements, the team chose a panelized wall system. The team modelled the entire MEP and fire sprinkler systems in BIM prior to the completion of the final wall panel design. This helped direct the wall panel factory, which framed walls with mechanical openings, moved studs for plumbing and installed backing for fixtures. The additional shop-drawing coordination and wall-framing modifications were done at virtually no cost while saving thousands in potential reframing and mechanical conflicts.

The final step in the Lewis VDC procedure deals with the preparation and recording of addenda. An addendum is needed whenever additional information is provided in a design change or a change order. The procedure defines how changes or updates are coordinated among the trades for execution in the field and how they are recorded within the BIM model. It includes steps for obtaining approval from the designers if required, for issuing a "for construction" order and for updating the models and the VDC work packages rapidly. Short response times are achieved by routing the definition and coordination process through the online BlueBeam Studio application, distributing information to modellers and finally by ensuring the changes are properly reflected in the current BIM model.

At the end of a building's construction, Lewis hands over an as-built model to the owner and conducts a BIM debriefing session for the project. The debriefings are part of the organizational learning process, itself a form of continuous improvement. The lessons learned in these sessions shape future methods, and they are recorded in the VDC procedure documents so that future projects can build on the wisdom gained through practice.

Box 12.3 Lease Crutcher Lewis's LOD definitions based on AIA G202 document

(LOD – 100) The model content includes overall building masses indicative of area, height, volume, location and orientation and will be modelled in 3 dimensions or represented by other data. The model will be analysed based on volume, area and orientation by application of generalized performance criteria assigned to the representative model elements.

(LOD – 200) Elements are modelled as generalized systems or assemblies with an approximate quantity, size, shape, location and orientation. The model will be analysed for performance of selected systems by application of generalized performance criteria assigned to the representative model elements.

(LOD – 300) Elements are modelled as specific assemblies, accurate (within acceptable tolerance) in terms of quantities, size, shape, location and orientation suitable for generation of traditional construction documents.

(LOD – 350) Elements are modelled as specific assemblies, accurate size (within acceptable tolerance) in terms of quantities, size, shape, location and orientation suitable for generation of traditional construction documents. The model will be analysed for location of selected systems by application of clash avoidance for the representative model elements. All required clearance zones are modelled on a separate layer.

(LOD – 400) Elements are modelled as specific assemblies, accurate in terms of quantity, size, shape, location and orientation with complete fabrication, assembly, and detailing information. Model elements are virtual representations of the proposed elements and are suitable for construction.

(LOD – 500) Elements are modelled as constructed assemblies, actual and accurate in terms of quantity, size, shape, location and orientation. Elements reflect as-built information and include facility management (FM) information.

Lowrey's story continues

Lowrey emphasizes the dependence of trade crews on design information to get the work right:

> I challenge the traditional assumption that people working on the site generally know what they should be doing. In fact, crews usually don't know the project; they are very dependent on the info they get from the superintendent, the foreman, etc. Our process got the trades directly involved, not only in the modelling but also in the decision-making. They were exposed to the designers' constraints, and they became more committed to the design and the process. At the time, I was printing a lot of A4 pages in colour of model views from Navisworks, and posting them on the site. But that was unsustainable, so we built a BIM kiosk with a tempered glass screen, to make the model available on site. It became very popular; we were pleased to see all of the field crew using it. Our concrete foreman realized that it was better to have the junior apprentice read the model himself instead of coming to the foreman to ask about every little detail. It was very productive, and he did not find that it challenged his authority.
>
> When trades found a conflict between systems in the drawings while in the field, they could go to the model and see how things should be. It really helped resolve questions quickly. If decisions were made on site to change something, say an electrical routing, then the changes were made back into the model: the field guys would mark up a printout that was provided especially for this purpose, so that our BIM detailer could make the necessary changes to the model.

Coordination between design and construction

Lowrey continued:

> On another project, we began working with the steel detailer who was working in Tekla Structures. At first the project engineers were given 2D printout sheets to review and mark up. I asked for the Tekla model, so that the engineers could review the model directly. They didn't have Tekla licenses, nor were they familiar with Tekla, so we converted the model to Sketchup. Then the engineers could edit the model and mark up the changes they made with flagged call-outs. This was easier for the steel detailer – he imported the Sketchup models back into Tekla to use as a 3D overlay, but he still had to fix his Tekla model to match the changes in Sketchup. Now the detailers use BIMSight, but they still can't edit the model during review.[1]
>
> At Lewis, when project managers consider whether or not to pay for VDC, it's usually the superintendents who push for it – "Remember our previous project? All the things that went wrong were the parts of the building that were not in the model." They grow dependent on the VDC guys and build strong personal connections, partly due to a "foxhole" community ethos – they've been in the trenches with the VDC coordinator on a previous project. On the first site I worked on as a VDC coordinator, they didn't understand what it was all about, so they wanted to put me in a back room of the site office. Now, the VDC coordinators' workstations are placed in the centre of the site offices – they are literally the info hub.
>
> At this point in our journey, we push for subs to submit models and have the project engineers review the 3D models directly. We prefer to have the engineers work directly on the model. One way of encouraging them to use Revit, for example, is to build parametric parts that make it quicker and easier for them to use, such as stud rails, exterior embeds, etc. These are scopes of work that they can cope with, with low risk. Once project engineers do that, it means less work for our BIM detailer.
>
> We've deepened the collaboration with designers over time, because we have learned how to exploit the win-win opportunities collaboration offers us. One of those is using the same model. For example, if I provide an architect with a set of Revit families of walls the way we build them, it's not really any extra work for them to choose my 7 5/8" CMU (concrete masonry unit) wall than the standard 8" wall that comes with the software, so that it will already be modelled the way Lewis needs it. We add penetrations of our own directly to the architects' models. With a bit of give and take it's been easy to define modelling methods so that more of the models could be shared. We set up a "Managed Area of Responsibility" matrix and a "Model Element Authorship" matrix, which essentially say that whoever can do it best should do it, but in a coordinated way. In that way, we can maintain the model as the "single source of truth".
>
> Traditionally, designers have only been concerned with getting permits, since after that they are not making money. At Lewis, we have a very different objective – to build from the model. One thing that helps us get over designers' reluctance to engaging with our VDC process is when owner's representatives get on board with our VDC mission. We want everyone to use the model and everyone to get the most value out of it. So, when an owner's rep says, "How about getting on board early and having one model for all?", it helps motivate the collaboration.

The role of trade subcontractors

And more:

> Not all trade subs have the capability to model, and very few general contractors can support the process. So, there are some GCs out there who are ostensibly pushing their subs to model (since they write it in the contract), and thus the subs commit to modelling (by agreeing to the contract). But when the time arrives, they ask the GC, "Do you really need us to model?" and the answer generally is, "No, not really". Even when the GC does insist, the sub typically outsources the modelling so that they can fulfill the contract terms and nothing else. But at Lewis, we see it differently. The model is central to coordination. If a sub typically can't model, we don't let them outsource it, because if they do then they will be incapable of maintaining it during construction. Instead we provide a modelling service where we will do the modelling for them in-house. In return for doing the modelling, we ask the sub to provide the time of a foreman a month before the job to review and plan the work. Then we give them the model, and call them in for a constructability review. The end product of the meeting is a set of questions to be explored. In the third meeting, we compile all the information, resolve the issues, and end up with a final set of CDRs. We ask for final feedback, clear up any remaining issues, and prepare the dimensioned plans for each work package. The main idea is to get the subs to be as prepared as possible. Using BIM, we coach them to remove their constraints. Lewis is taking on the process, engaging them to resolve their own issues. We build trust by getting the subs to see the value in what we're doing. What we're trying to achieve is to get them to make money by being productive, not by doing change orders.
>
> Over time, some subs are increasingly willing to work with Lewis. Because they know our VDC process, and they know how they will be treated, they quote us lower rates for work. They know their productivity will be higher, they know their material waste will be lower than on other jobs, and they know that the schedule will be reliable because work packages are properly structured. We win some jobs partly due to better prices from subs, so there's no question that the VDC procedures are contributing to Lewis's competitive advantage. Apart from everything else, I know that when I was a tradesman, I hated demolishing stuff and rebuilding it, no matter how much I was paid. So I know that subs appreciate the fact that they are given the right information, so that they can do the work right the first time.

Perhaps one of the unique characteristics of Lewis is that they are an employee-owned company with a very flat organizational structure. This may be the key to people being flexible enough to take the initiative yet responsible enough to maintain standards when necessary. Lowrey stated:

> We're a best practices company. VDC people have a strong role in projects partly because some of our value is that people at Lewis go beyond the formal job descriptions and responsibilities in other companies in order to do what's needed to get the job done right.

Aligning subcontractors with the VDC process

VDC can only be successful if all parties are fully engaged with the process, because once any one trade's physical work diverges from its digital representation in the BIM model, all of the other trades and designers begin to question whether their own work can be executed in accordance with the model. Recognizing this, Lewis places strong emphasis on aligning their subcontractors with the VDC process through the subcontracts, the published procedure documents, the education and support they provide, all of which focus on generating a high degree of collaboration. Each trade is required to sign the BIM execution plan as a part of their contractual obligations. The document states that:

> By signing this document, you are certifying that you are willing to fully participate in the Lewis BIM/VDC process, that you have read Addendum A, and that you agree to the underlying concept. This includes submitting resumes of previous work, resumes of those who will be performing this work and samples of work. You also are affirming that you will be modelling not for the sake of modelling but rather to use the digital information in the virtual built environment to complete your project scope. You are also indicating that you will not engage in a two-path process, creating shop drawings independently of the coordinated model.

The "Addendum A" that this declaration refers to is an excerpt from the British Standards Institution's document PAS 1192-2:2013, which is part of the set of documents that define the use of BIM within the context of the UK government's BIM mandate from 2011 (shown in Box 12.4). The excerpt is especially significant in Lewis's process, far more than the specific declarative sentences that follow, because it calls on subcontractors to align their attitudes and their behaviour with Lewis's expectations of the way they are to work on the project. Referencing this excerpt reveals Lewis's strong commitment to collaboration and to Lean Construction, though without the formal contractual conditions imposed by explicitly collaborative forms of contracting such as IPD.

Box 12.4 Addendum A to the Lease Crutcher Lewis BIM execution plan template document

An excerpt from the British Standards Institution's document PAS 1192-2:2013 [highlighted phrases are marked in red for emphasis by Lease Crutcher Lewis]

General information

The production of coordinated design and construction information is a task- and time-based process, independent of which procurement route or form of contract is used. Each task needs to be carried out in a particular order for the mutual benefit of all those involved, otherwise known as "collaborative working". In a collaborative working environment, teams are asked to produce information using standardized processes and agreed standards and methods, to ensure the same form and quality, enabling information to be used and reused without change or interpretation. If an individual, office or team changes the process without agreement, it will hinder collaboration – a participant insisting on "my standard" is not acceptable in a collaborative working environment.

This approach does not require more work, as this information has always been required to be produced. However, true collaborative working requires mutual understanding and trust within the team and a deeper level of standardized process than has previously been experienced, if the information is to be produced and delivered in a consistent timely manner. The benefits of working in this way can include fewer delays and disputes within the team, better management of project risk and better understanding of where costs are being incurred.

Wherever possible, the principles of Lean should also be applied to reduce the expenditure of resources for any goal other than the creation of value for the employer. For example, BS 1192:2007 promotes the avoidance of wasteful activities such as:

- waiting and searching for information;
- over production of information with no defined use;
- over-processing information, simply because the technology can; and
- defects, caused by poor coordination across the graphical and non-graphical data set which require rework.

However, for the production of information to be truly Lean, it is critical to understand its future use. This is achieved by "beginning with the end in mind" and identifying the downstream uses of information, to ensure information can be used and re-used throughout the project and life of the asset. It is to this end that PAS 1192-2 has been produced. It is anticipated that this document is of equal value to small practices as well as large multi-nationals. The impact of poor information management and waste is potentially equal on all projects.

Lewis Lean

Jeff Cleator, the president of the Lease Cruther Lewis's Washington operations, explains that from his perspective, Lean is a culture within the company:

> The guiding principle I try to convey to our people is that we must excel at providing value for our clients. To do that, we need to be willing to innovate, so we've worked hard to ensure that our people are open to change. For me, Lewis Lean implies continuous improvement. I see BIM as an instrument for continuous improvement, a tool for improving a variety of processes. BIM has helped us improve our preconstruction services such as estimating and scheduling, our safety planning, and our delivery of as-built information to clients at project completion. We're continuously finding new ways in which BIM can improve these activities.

Although Jeff is a member of the Lean Construction Institute and founded the company's continuous improvement initiative, called "Lewis Lean", he attributes the culture of continuous improvement to his former mentor, at the time he was an intern at the

LEWISLEAN

Changing The Way Jobs Get Built

Minimize Waste:	**Optimize Delivery:**
• Reduce Worker Idle Time • Prevent Accidents • Avoid Over/Under Design • Optimize Just In Time Delivery • Eliminate Unnecessary Steps • Prevent Defects • Shorten Learning Curve	• Align Goals • Foster Trust • Establish Accountability • Ensure Peak Performance

Figure 12.5 Lewis Lean principles

company, Bill Lewis: "Bill Lewis, who was CEO & President for many years, believed in empowering the company's employees. Our culture is to take care of clients, and we innovate in support of that goal."

As discussed earlier in the book, Lean thinking calls for continuous improvement through a formal process of stepwise improvements that are institutionalized rigorously through standardization after the benefits of the proposed improvement have been measured to be superior to the existing procedure. Lewis, like many construction companies, struggles with this because it is difficult to standardize procedures and work methods when many of the workers on a construction project are not Lewis employees but rather the staff of subcontractors. This makes it difficult to influence the working culture and the ways people work.

Lewis uses its custom company intranet (WebPM, a customized implementation of SharePoint) to standardize procedures for its employees. Like its VDC department, each head office department is required to embed its protocols in the IT system. Lewis managers also try to encourage Lean behaviour in their subcontractors, with varying levels of success. Hiring subcontractors who have become accustomed to working the Lewis way on earlier projects is one way of ensuring Lean behaviour and compliance with Lewis's standard procedures, and it promotes a broader and longer-lasting cultural change than always working with new subcontractors. This is straightforward where the Lewis project manager has the backing of the client, who – given the CM arrangement under which Lewis usually works – has a final say in selection of subcontractors. Some clients understand the value obtained by working with known subcontractors, and in turn favour the selection of Lewis's preferred subcontractors. Others, however, still seek the lowest price for subcontracted work packages, failing to understand that the best value isn't always reflected in the lowest upfront price.

Like many builders in the US northwest, Lewis elects to do more self-performed work than contractors in other regions who have moved further toward the Construction Manager at Risk (CM@Risk) model. Lewis prefers to self-perform work for the sake of control, reliability, protecting the critical path and company morale. Nevertheless, they find that clients increasingly want to be involved in the bidding and the award of subcontracted work. In some cases, they have had to bid their internal work against external subcontractors in order to comply with the client's demand for thorough bidding. While clients tend to prefer lowest price, the Lewis management tries to make the case for best value. They make a subcontractor's level of BIM expertise, the quality of their safety programme, their schedule reliability and other key performance indicators explicit to the clients. One way to do this is to monetize these indicators, by assessing what the cost of supporting those subcontractors to achieve the required performance will be in terms of Lewis's own resources.

There are three main ways to achieve subcontractor compliance, in this order of preference:

1 Restrict bid solicitations to subcontractors who have proven their value. This encourages subcontractors to conform to the Lewis way.
2 Write bid solicitations and contract documents to demand compliance with VDC, safety, planning and other methods.
3 Provide subcontractors thorough support if they cannot comply despite having committed. This is particularly the case for VDC, where Lewis prefers to model for subcontractors rather than having them outsource the modelling. This way helps, over time, to educate subcontractors to work the Lewis way, but it requires a resource investment on Lewis's part.

Jeff Cleator believes that owners hire Lewis to protect them from the risk of poor subcontractors, and thus clients typically respect Lewis's way of doing things, particularly clients who are more sophisticated buyers of Lewis's services and understand their value proposition. It is these clients who moved away from lump sum contracts to guaranteed maximum price (GMP) contracts and are now open to even more progressive forms such as design build (DB) and IPD.

In terms of Lean tools, Lewis people use A3 process improvement reports, spaghetti maps, and elements of LPS. They have also incorporated elements from the Six Sigma methodology into their improvement methods. However, individual employees and teams are given the discretion to use particular tools based on the situation at hand. Project teams have the flexibility to tweak the way they use pull planning, short interval schedule and morning huddles according to the specifics of the job. "I reject the idea that LPS or sticky notes are synonymous with 'Lean Construction'. The emphasis must be on culture, on continuous improvement, and on driving out waste", Jeff says.

Do Lewis clients and subcontractors understand the value of the Lewis way? Jeff believes that they do. Clients appreciate that the risks (in the form of change orders and schedule slippage) are reduced, and thus they come back for repeat business. Subcontractors have told him, informally, that they offer Lewis lower unit costs than they offer their competitors, because they know they will be able to achieve far higher productivity (and thus higher profit on unit-priced work) thanks to Lewis's VDC and production control practices, than on jobs with other general contractors. Another measure of the effectiveness of the Lewis way, relative to the performance of other construction employers in the same market, is the ease with which Lewis can mobilize union labour, thanks to their reputation for providing productive working conditions. Lewis is a union company – even when work is plentiful, Lewis has no problems recruiting people.

Lewis rejects the notion of transferring risk – from clients to GCs to subcontractors, and so on down the construction food chain. Jeff explains that "Lewis prefers to allocate risk to those best able to deal with it or to control it, and to manage it collectively, because unless the project is successful everyone's in trouble." This attitude toward risk is inherently evident in its VDC process. Undertaking the modelling and the coordination for all of the trades implies that it takes risks that are often carried by designers or subcontractors. It is this attitude, preferring collaboration to competition, cooperation to conflict, that characterizes Lewis's approach to construction and underlies its adoption of BIM and of Lean Construction.

A recent anecdote reveals the approach in practice. In discussions with an owner and a group of designers, all agreed that the project at hand could best be done using a central BIM model. The legal teams were assigned the task of preparing appropriate contracts – but despite protracted negotiations, they failed to agree. The lawyers resisted accepting the model as the "instrument of service", the authoritative expression of the designers' intent. Instead, they wanted the drawings to remain the instrument of service. Lewis proposed instead that their standard BIM execution plan should simply be appended as an exhibit to the contract. This was done, and the project moved ahead with a collaborative VDC process, ultimately deemed successful by all. Lewis accepted some risk, but favoured the more collaborative and productive way of working over the defensive risk avoidance but lose-lose approach.

Summary

Lewis carefully maintains the model truth, the integrity of the virtual model's reflection of reality. The company clearly defines for architects and design consultants where to stop, recommending they progress no further than LOD 200 and providing them with an MEA table and LOD specifications for each Uniformat-classified item. The process is tailored to avoid over-detailing (for example, where details depend on procurement) and to avoid overproduction in design, which contributes to a smoother preconstruction process and a timelier job start. They run model coordination meetings from the start of their work on the project and all the way through to handover of the as-built BIM model to the owner. They provide a detailed BIM execution plan for subcontractors, with details of the procedures to be followed through design, system coordination, clash detection, composition of work packages and construction itself. The plan requires the trades to collaborate fully with the whole project team to pursue the VDC process to its fullest.

Lewis's use of BIM is also tailored to support the Lean Construction methods and tools the company uses. Design coordination, detailing and clash detection are all pulled in reverse phase scheduling according to the zone construction schedule. Compilation of work packages from the federated model directly supports the make-ready, weekly work planning and control aspects of LPS implementation.

"Lewis Lean" is the company's dedicated programme for Lean Construction, and it is the company's culture of continuous improvement.[2] All employees are strongly encouraged to propose "Plus Ones" or incremental improvements to processes. LPS and other formalized tools are not a major part of Lewis Lean, but rather tools to be used as needed in the pursuit of waste reduction and continuous improvement.

Notes

1 Sketchup is a 3D modelling software. BIMSight is a free BIM viewing software that does not support editing of a model.
2 For more information, see http://lewisbuilds.com/.

13 Tidhar on the Park, Rosh Ha'ayin

Project profile

Figure 13.1 A rendering of the 20 buildings of the Tidhar on the Park, Rosh Ha'ayin project

Project description

Tidhar on the Park, Rosh Ha'ayin was a residential development of 1,036 apartments, located in the town of Rosh Ha'ayin, 25 km from the city of Tel Aviv, Israel. The project was part of a new neighbourhood development that would eventually comprise some 14,000 apartments. Tidhar acquired the rights to the project in March 2013 in partnership with a construction management firm called CPM. CPM held an option to buy the land but needed the capital to develop it.

The project was divided into four essentially identical quarters, each with 259 apartment units in five buildings: two towers (21 storeys, 92 apartments in each) and three low buildings (9 storeys, 25 apartments in each). The towers were identical, with the exception that their entrance-level interfaces with the underground parking areas were adapted to the local topographical and site constraints. The low buildings were also similar to one another, though some were mirror-images of the same design. The total area of all five buildings in each quarter was 25,890 m², and each quarter had an 11,500 m² underground parking garage with 355 parking spaces under its five buildings. With 259 apartments in each quarter, the average apartment size was almost exactly 100 m² (most were either four- or five-room apartments).

All together, the total built area of the project was 147,560 m², with a further 11,000 m² of landscaped public park area between the buildings and above the parking levels.

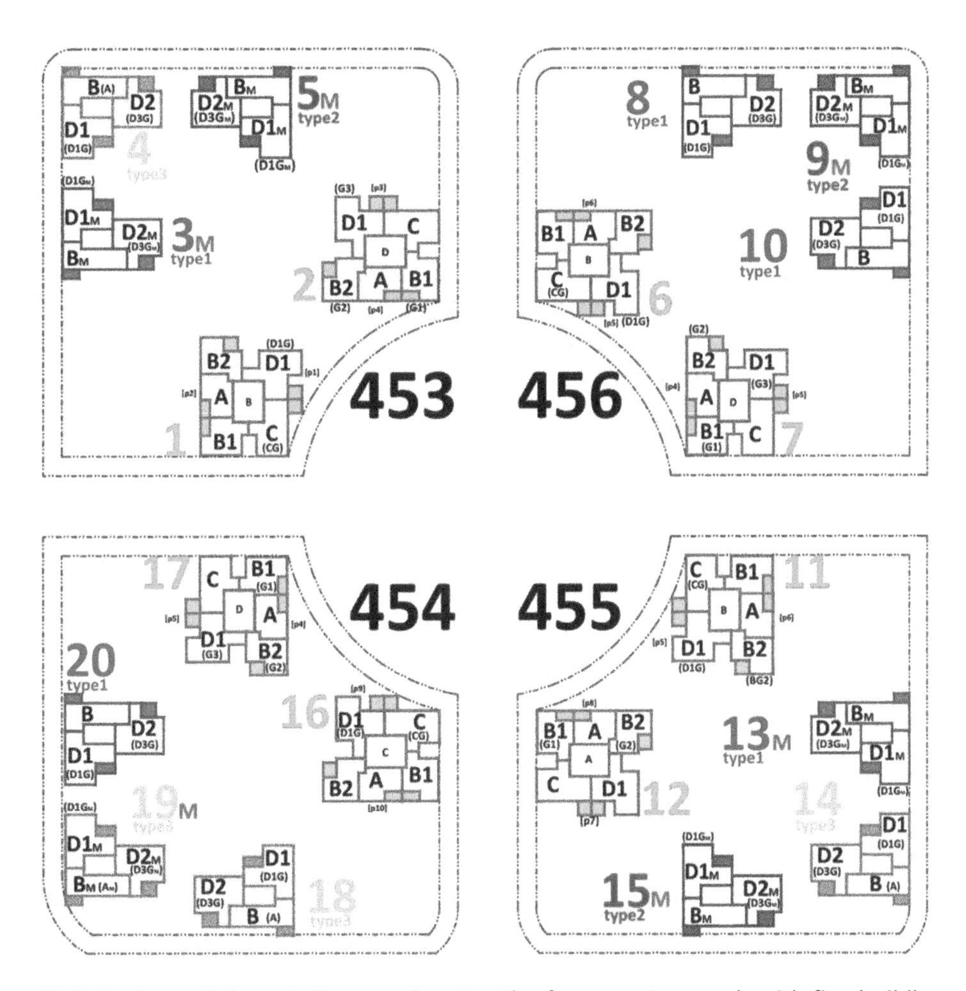

Figure 13.2 A schematic layout diagram showing the four quarters, each with five buildings

Note: The figure shows eight towers, all of the same type (labelled in green) and 12 lower buildings (coloured by type).

Construction methods

The construction methods were similar to those used in the Tidhar on the Park, Yavne project (see Chapter 7). The structures were all reinforced *in situ* concrete with natural stone cladding applied using the "Baranovich" method. The interiors had gypsum block partitions, ceramic tile floors and plastered concrete ceilings. Air-conditioning units and ductwork were hidden by gypsum board dropped ceilings.

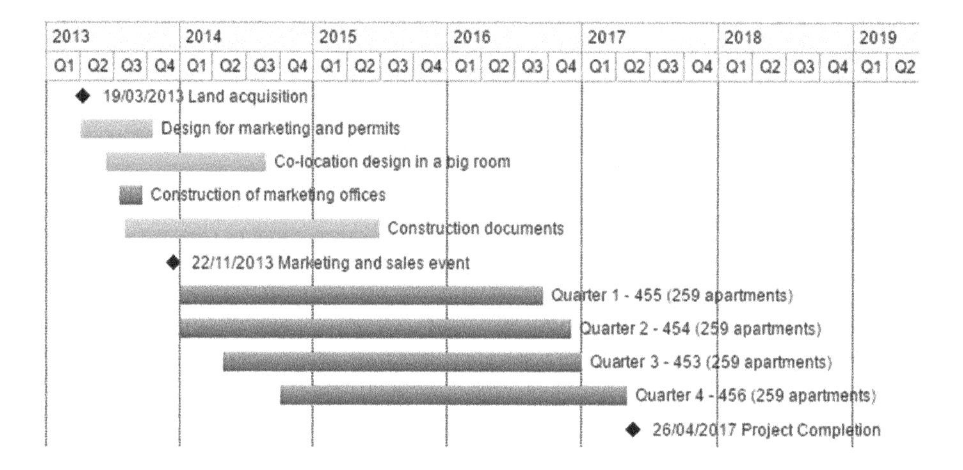

Figure 13.3 Tidhar on the Park, Rosh Ha'ayin: timeline

Tidhar's project team

The overall project was divided into five smaller projects. Each quarter was organized as a separate project with one project manager and two site engineers: one for the two high-rise buildings (182 apartments) and one for the three low-rise buildings (75 apartments). As Figure 13.4 shows, structural superintendents worked with the two site engineers, and each quarter had six finishing work superintendents (one for each of the five buildings and one for the underground structure), one installation manager (for the two high-rise buildings) and one logistics manager. The public park was the fifth project, and a project manager and three superintendents were assigned to it.

In order to manage the four quarters as a single cohesive project despite their independent organizational hierarchies, an overall project headquarters was formed. This HQ, with a regional manager, two secretaries, a technical secretary, a logistics manager, a production engineer, a BIM manager, three quality control engineers and eight client-change coordinators, provided centralized services for all four projects.

Chapter 20, "Pulling it all together", is devoted entirely to the story of Lean Construction and BIM application in this project.

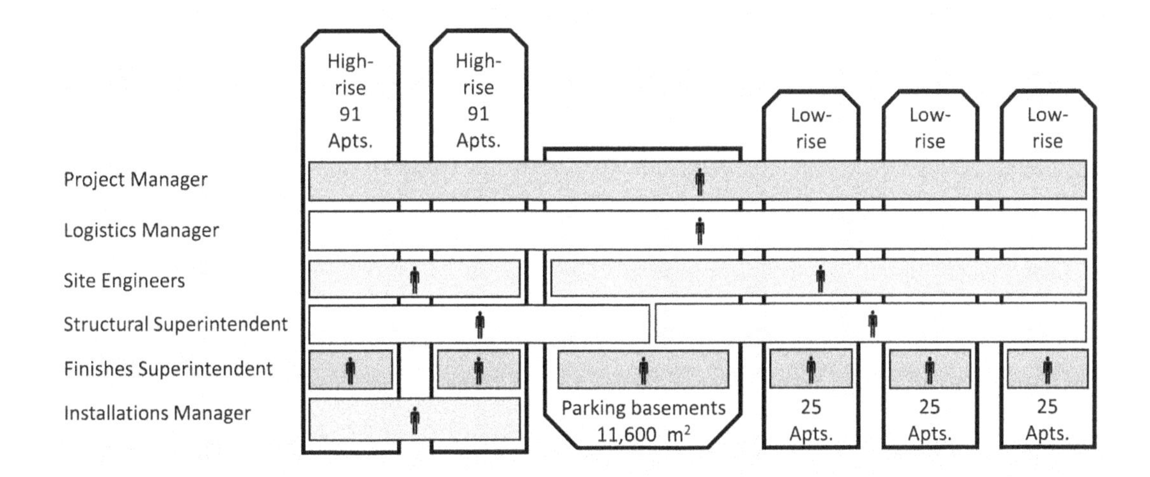

Figure 13.4 Management resources assigned to each quarter of the project in Rosh Ha'ayin

Note: At the peak, the project had over 70 Tidhar staff on site, spread over the five sub-projects.

14 BIM in the big room

Co-location of designers in a common office space enables concurrent design and full exploitation of BIM tools. Cycle times for decisions are greatly shortened and the waste of negative iterations in design is reduced. Co-location with BIM in a "big room" is perhaps the easiest way to significantly reduce the overall duration of a construction project.

The big room

The central idea of a "big room" is that all of the architects, engineers and construction consultants who need to collaborate in the design of a building work in a suite of offices very often located on or near the construction site. The name derives from the large common meeting room that is at the heart of the office suite and is usually furnished with a conference table and a set of projectors that can display multiple views of a BIM model simultaneously, allowing all participants in the meeting to share a common object for discussion.

The idea of co-locating teams for building design derives from the Lean approach to product development, in which multi-skilled teams are gathered together, breaking from the traditional ways of working in the silos of professional design departments. Co-located designers aim to achieve collaborative and concurrent design, which means that all aspects of the building are designed in parallel, with a focus on multidisciplinary problem solving.

The key mechanisms at work in big rooms are short communication cycles, efficient flow of information in small batch sizes and elimination of much of the negative iteration that is common in traditional building design processes. Negative iteration refers to all of the work invested in developing solutions to local design issues where the path developed does not ultimately converge on a solution that satisfies all of the constraints. It often arises where designers' individual views of the problem are restricted and constraints are not recognized until designers from other disciplines review the proposed solutions. Concurrent work helps eliminate, or greatly shorten the lifespan of, proposed solutions that are not feasible.

While co-locating designers from multiple design and planning disciplines seems like a "no-brainer" from the point of view of the good of the project, co-location can be very difficult to initiate and sustain due to the disparate interests of the design firms whose employees are required to co-locate. The main difficulties arise from the imbalance in effort required from different disciplines and traditional modes of managing resource allocation in design firms.

In general, architects need to invest the largest share of person-hours in designing a building, structural engineers will have a smaller workload, and services engineers even less. As long as the workload reaches at least one full-time equivalent (FTE) for any firm, having designers wholly committed to a single project is minimally feasible. Where the resource requirement dips below one FTE for any design firm, they cannot justify a full-time presence in the big room. Even if this is implemented in an organized fashion, such as specifying particular days of the week that a designer will be in the big room, it is disruptive because the principle of immediate availability for consultation and collaborative work is violated. This has a negative effect on the other participants, who begin to see declining value in their own presence.

Traditional modes of resource allocation are such that design firms tend to "overbook" their resources to avoid the inevitable downtime that results from the unstable pace of most projects. Designers shift their focus from project to project according to changing degrees of urgency as deadlines alternately approach and get delayed. Design firm managers who take this approach are often reluctant to devote full-time resources to any one single project, which is prerequisite for co-location.

Box 14.1 Toyota's Obeya ("big room") co-located Product Development Centre

Like many elements of Lean, the concept of the big room has its roots in Toyota. The Obeya ("big room" in Japanese) is a central tool of the Toyota Product Development System. The Obeya is where the product development team gathers on a regular basis (typically daily) to share information and discuss issues that have arisen. Just as in construction, the interfaces between different subsystems are critical to the overall success of the project, and it is in the big room where the different functions interact as they negotiate successful interlacing of the entire product. Big rooms are focused on visual presentations of the data, so the walls will typically be covered with charts and diagrams, metrics and objectives. These visual tools are not just decorations; they are the focal points of the team's conversations during their regular meetings. Rather than the project leader (called the "chief engineer" at Toyota) conducting a series of separate conversations with each functional department, the big room allows the communication to be multiplexed in parallel, increasing both the speed and quality of information flow which in turn enables effective simultaneous engineering.

At Toyota, not only the engineering functions are represented in the Obeya; manufacturing and supply chain, recognized as crucial to the long-term success of the product, also attend and interact. By bringing these operational functions into the design process, the groundwork for the transition to production can be laid and potential problems can be identified much earlier. And with all the relevant parties in the room, decisions can be made much more quickly. The schedule is also developed collaboratively, meaning increased buy-in from the parties represented at the meetings (this aspect is similar to LPS, discussed in Chapter 17, "The Last

Planner® System", which uses involvement in planning to create commitment to execution according to the plan).

Thus the function of the big room is twofold: information gathering and information management, while strong attention is paid to the interpersonal dynamics of the project team.[1] Use of the big room in building design and construction has grown popular particularly within the framework of IPD, a procurement model that seeks to align the interests of all participants in a project. Under an IPD contract, the owner, designers and contractors work together to design and build a project to a target cost and then share in the gain (profit) or pain (loss) that results once the final actual cost is known at the end of the project. The target cost itself is set through close collaboration among the parties.

The use of BIM is a key enabler of the IPD paradigm. In the early phases, BIM enables a common understanding of the project and computation of increasingly accurate cost estimates, as well as other simulations with very short cycle times to support evaluation and comparison of design options. This ensures convergence to the target cost while maintaining the required value. Once construction begins, the federated BIM model is the central focus of communication for collaboration. It is the central source for information for construction, making it possible to avoid many of the wastes that so commonly occur when team members do not collaborate.

There are many case study examples of the use of the big room in IPD project contexts in the literature, and these had a strong influence on Tidhar's decision to move to big-room designing. A good example, which details the use of BIM in the Sutter Medical Centre, Castro Valley, CA, can be found in Chapter 9.3 of the BIM Handbook.[2]

The 3D view and photograph in Figure 14.1 show the co-location space used for the target value design (TVD) phase of the Temecula Valley Hospital project (a \$151 million, five-storey, 140-bed, 16,445 m^2 hospital, built for Universal Health Services (UHS) under an IPD contract that included HMC architects, the Turner and DPR construction companies, and various trade partners).

Figure 14.1 A plan (left) and a general view (right) of the co-location space at the Temecula Valley Hospital Project[3]

Source: Reproduced courtesy of Doanh Do, P2SL, UC Berkeley.

The principles for those working together in the space reflect the commitment and focus on the group effort as it is expressed in everyday behaviour:

- No one person has more authority than others.
- **Speak up** – get engaged in conversation and share ideas.
- **Your opinion is important** in helping guide the team.
- **Listen to others** – focusing on what others have to say helps you understand their point of view. No side conversations.
- **Only have one meeting at a time**. Conversations should be heard and shared by all.
- Help keep the meeting and participants on track by eliminating phone disruptions.
- **No multi tasking**. This includes laptop computers and personal digital assistants (PDAs).
- **Stay on time**. This includes start time, end time, break times and agenda.

Permits within six months instead of two years

When Tidhar began the work on the Rosh Ha'ayin project discussed in the previous chapter (Chapter 13, "Tidhar on the Park, Rosh Ha'ayin"), it was clear that BIM would play a big role throughout the project life cycle.

Figure 14.2 A view of the model of one of the four quarters of the Rosh Ha'ayin project

Tidhar's management considered this an excellent opportunity to introduce all of the Lean and BIM practices that were being developed in the company, because the product type – four- and five-roomed apartments in medium- to high-rise buildings – was exactly the product type Tidhar had come to specialize in, considering it to be its "bread and butter" product. The CEO's directive to apply BIM in all new projects was to be applied here thoroughly, but this was to be a first in another sense – it would be the first time that a Tidhar project would adopt a Lean design approach. One of the most significant features of the approach was that all of the designers and consultants would be required to co-locate their design staff to a "big room".

Tidhar quickly ordered a set of mobile site office buildings, with 166 m^2 of air-conditioned space. The layout included two open-plan work areas with a total of 14 desks fitted out with BIM workstations, communal printers and plotters, a conference room with large format plasma screens, offices for the project design manager, the company's own modeller and administrative support, a kitchenette and bathrooms. At this stage, the project was still in the schematic design phase, with negotiations with the local planning council still under way. The fact that the site itself had not yet been established did not deter Tidhar's leadership: the site office buildings were set up on another Tidhar site, so that work could begin in the big room.

Maia Davidson, an architect by training and an experienced design manager, was assigned the task of managing the design effort for the project, with direct responsibility for setting up and running the big room. She realized that the big room process was not "business as usual" but something quite different. She had to integrate designers from a variety of independent design firms in a team, which required not only moving them out of their comfort zones in their separate offices, but also changing the focus of their loyalties. The project had to come first, ahead of the needs of their own firms.

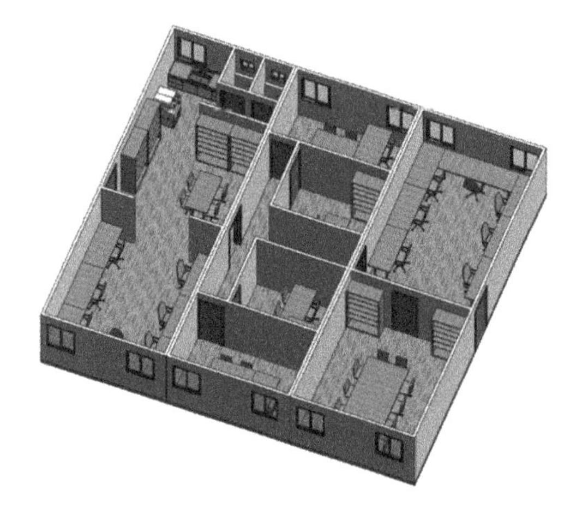

Figure 14.3 The site co-located design space for the Rosh Ha'ayin project, composed of three 12.3 × 4.5 m modular office units

Box 14.2 Maia Davidson's careful management of the big room

Maia Davidson recalls:

> I considered carefully where each designer should sit within the big room. The first thing that came to mind was the nature of the need for professional collaboration: structures near the architects, the various MEP systems in a cluster, and so on. But then I thought more about what it meant for each of them to be in a new environment, among people they did not know previously. I realized that the cohesion of the team and their commitment to the project would depend to a large extent on how comfortable they felt, particularly with regard to their new personal relationships. Given that this was a diverse team, not only in terms of professional disciplines, but also multicultural and multilingual (there were six mother tongues represented – French, Arabic, Russian, Hebrew, Spanish, English), they would need to find something in common with their neighbours that would allow the sense of belonging to develop. I assigned young mothers to nearby workstations, so that they would feel comfortable leaving at 4pm to fetch children; I considered how many hours each designer would spend in the big room through the week, and used that too to guide the plan.
>
> The staff meetings every second day were critical for coordinating work across disciplines, but no less so for the development of personal relationships. We celebrated birthdays and holidays together and we all went out to a restaurant to celebrate when the permits were eventually obtained. A sense of identity with the project and with Tidhar developed. One sign of this was that some of the designers began to check with me if they could take days off, despite the fact that they were employed by their design firms, not by Tidhar, and had no obligation at all to do so. This concern not to compromise the team's ability to meet any deadlines reflected their commitment to the team.

With the building permits still some months away, Gil Geva brought the entire project team – Tidhar's own people and all of the architects, engineers and consultants – to a kick-off meeting in the site offices. He made it abundantly clear to the team that this was not a trial or a pilot; the company was completely convinced that co-location and deep collaboration among the design team, using BIM, was the right way to achieve the quality of design information that would enable smooth production flow in construction. The target of "Increasing the Margin" would be achieved by investing intelligently in design and planning. What the company had seen in Finland and Denmark was to be achieved here, only more thoroughly. Any consultants who felt that this was not the right approach and were unwilling to assign their employees to work in the "big room" for sufficient time, or to take part in the planned coordination meetings, were free to withdraw from the project, but those who elected to stay would be expected to collaborate fully.

After outlining the company's broad goals, he set out bold targets for the project team:

- The building permits were to be obtained within five months. This was a truly aggressive target – the average time to get permits for projects of this type was 18–24 months.
- The building permit application documents would be prepared entirely from BIM models, with full automation of zone space measurement and reporting. No 2D documents would be produced.
- The BIM models were to reach the level of development equivalent to "detailed design", one level higher than would normally be the case for construction permit applications. Tidhar was determined to avoid the time required for the design stage that usually followed the granting of permits on most of its projects. Their experience was that inadequate levels of detail in permitting documents hid design errors and conflicts between disciplines, resulting in rework of the design and change submittals to the local authority before construction could actually begin.
- All designers were to work exclusively in the project big room, and were to upload their models to the big room project server at the end of every working day.
- The federated model was to be maintained at all times and conflicts were to be resolved in discussion among designers as soon as they were identified.

None of the design firm principals sitting at the table had experience of this kind of arrangement. Some of them did not even have any basic experience with BIM in their firms. Yet not one of them expressed any serious dissent, and none left. On the contrary, some expressed very favourable attitudes. If permits could be obtained in the time frame that Gil wanted, they would clearly make more profit than usual.

But could it really be done? Agreeing to do something in principle and actually doing it can be two different things. Discussion was lively, and many obstacles and problems were identified and raised. Gil's commitment and firm leadership was an essential ingredient, because motivating the design team was not trivial. The stick of discipline wielded by the CEO, in the dual role of owner of the project and general contractor, was accompanied with the carrot of increased productivity. He would not increase their fees, but he would set the target of completing the entire design within nine months, and the project as a whole within three years, which would result in significantly higher earnings per hour for all designers.

There was also a subtler implication in the company's call for their collaboration, which Gil had not voiced, but that was expressed explicitly in the talk by Rafael Sacks that followed Gil's opening presentation. Rafael emphasized that designers' behaviour in traditional design practice was governed not only by creativity, professional pride, ego and monetary compensation but also by fear – the fear generated by legal and financial liability for design errors. That fear is deeply rooted in the building construction design community. All designers carry extensive liability insurance, all have been sued at one time or another, and all print disclaimers on their drawings to avoid responsibility and liability as far as possible.

Box 14.3 "The contractor is responsible for all dimensions"

In the Israeli construction industry, standard practice is that the building's designers – architects, engineers – prepare most of the construction detail drawings themselves. Rebar detailing, for example, is not done by the reinforcing subcontractor, but by the engineer of record. Yet most designers traditionally include this caveat in the fine print of their drawing title blocks: "The contractor is responsible for all dimensions." This is absurd, given that the contractor has no control over the work of the designer; it is the designer who prepares the drawings and annotates the dimensions, not the contractor. This highly risk-averse mindset is a barrier to the kind of collaboration needed for Lean and BIM implementations.

Fear retards collaboration because it leads to information hoarding, a natural defence mechanism. Collaboration among a Lean design team requires designers to share information openly. Design information must often be shared before it is mature, because the iteration cycles are far shorter in the Lean design approach than they are in the traditional approach. Designers need to change the habit of concealing their work in progress until it is ready to be released, and instead encourage review of their work by other disciplines to get feedback as early as possible in the process. This change of mindset is essential for realizing the benefits of co-location and for avoiding the waste of negative iteration.

The method

What was the difference between the traditional ways of working to which all of the team members were accustomed and the new way of working that Tidhar was trying to implement? The key points are outlined in Table 14.1.

Working as a collaborative team with BIM requires careful planning of the way in which each discipline's models are shared. The plan must consider two constraints: the pace at which information is developed by each discipline at each stage of the overall design, and the constraints on model file size imposed by the limits of the hardware and the software.

Many national and organization BIM guides envisage preparation of a "BIM execution plan" (BEP) for major projects.[4] This gives projects the freedom they need to tailor their BEP to the peculiarities of the design process appropriate for their facility and to the specific suite of BIM software applications that their teams intend to use. Where a big room is used, the BEP incorporates specifications for the team structure, its meeting schedules, who will sit near whom, and other procedural aspects as well as the usual specifications of the model structure, the information sharing requirements and the design process.

The BEP that Maia Davidson and her team prepared for the Rosh Ha'ayin project defined all of these aspects. First, they defined the design discipline roles that would be required to be present in the big room through each design phase (see Figure 14.4). Some disciplines had more than one representative, depending on the intensity of the work that was needed during each phase.

Table 14.1 The key differences between the traditional way of design and the Tidhar way using BIM in the big room

Design team practice and behaviour	*The traditional way*	*The Tidhar way*
Design coordination	Review of completed drawings by a "design coordinator" using 2D drawings and a light table, at the end of the detailed design period.	Tidhar assumed the role of information integrator with responsibility for design coordination. Weekly design reviews of the federated design models using Navisworks software, resolution of clashes in consultation with the design team.
Design development, iteration of design versions	Each discipline develops the design in turn. Drawings are released at predetermined milestones, at which the ball is passed over the wall from one discipline to the next. Iterations have long cycle times and many repetitions are needed before the design converges.	Designers work in parallel to develop the design as a whole, working through the zones and the buildings in the project. Quantities of work in progress are limited. Design diverges and converges many times, with short durations for each cycle. More alternatives can be considered than in the traditional approach.
Preparation of permit application documents	Architects prepare all the needed documents using 2D CAD tools. Zone areas are measured and reported in specially prepared tables that are prepared manually.	All documents are prepared directly and automatically from the model. Ancillary documents, such as tables of zone types and areas, accessibility evaluations, and parking requirements tables, are all derived automatically.
Progression from permit documents to construction documents, and from construction documents to production documents (shop drawings)	Designers prepare construction drawings with most of the necessary details, but not quite as they are needed for work on site either in terms of content or detail. Requests for clarification are frequent. The contractor reviews the drawings for buildability, and some of the construction drawings are updated in consultation with suppliers and fabricators. In many cases the contractor or subcontractors or fabricators prepare shop drawings.	The model is detailed beyond the level of detail that is associated with construction documents. Designers are required to flesh out the model with many of the production details. The Tidhar project modeller maintains the federated production detail level model, which is used for quantity take-off for subcontract tenders and for producing final production documents. Many of the construction details are developed in collaboration with Tidhar's construction division engineers (who act as consultants), and with suppliers and fabricators, ensuring the designers have the right information for preparing production-level details.

Design Disciplines	Design Phases		
	Preliminary Design	**Detailed Design** (from conceptual design to permit documents and sales drawings)	**Design for Construction** (from permits to construction on site)
Consultants' Staff			
Architect	Yes	Yes	Yes
Structural engineer	Yes	Yes	Yes
Hydraulic engineer	Yes*	Yes	Yes
Electrical engineer	Yes (part time)	No	No
HVAC engineer	Yes (part time)	No	No
Traffic engineer	Yes	Yes	No
Landscape architect	Yes	Yes	No
Tidhar Staff			
Design manager	Yes	Yes	Yes
Model manager	Yes	Yes	Yes
Construction engineer	No	Desirable	Yes

* Required for design of the vertical systems only, not for floor layouts.

Figure 14.4 Co-location of the members of the design team through the project phases. "Yes" indicates that the designer(s) must be present during the relevant phase

Mark Starobinsky, a senior site engineer from Tidhar's own staff, was assigned the role of "consultant construction engineer" within the big room team. The idea, naturally, was to introduce construction knowledge as early as possible into the design process, similar to Toyota involving manufacturing experts in product design. The role definition called for Mark to make sure that: (1) the construction phase would be free of design errors, dimensional errors, clashes or conflicts; (2) the design would be perfectly constructible: and (3) the construction drawings would be legible and easy to read. This was a challenging task not only due to the high expectations set for the use of BIM in the big room, but also because neither Mark nor any of the designers had experience of such close collaboration between builder and designer. To incorporate construction knowledge into the detailed design required that the architects and other designers work at a level of detail that would otherwise have been postponed to the construction detailing phase. This had both benefits and drawbacks, which we discuss in the section on lessons learned, below.

Next, they mapped the design process in detail, identifying the activities that would be needed for each step and for each building in the project. The key milestones for each quarter were delivery of the permit application documents to the local authority, preparation of apartment sales contract documents, completion of construction drawings for both the parking basements and the typical floors of each of the building types.

They also determined which meetings were needed for the big room staff, what their aims were, who should attend and how long they would require (see Figure 14.5 and Table 14.2). The full team meetings held every second day proved to be critical for success. These work coordination meetings were equivalent to the construction site weekly work planning meetings of LPS (see Chapter 17, "The Last Planner® System"). Each designer would briefly explain their current work focus, their status and their next steps. Information was pulled, in that each would declare what information they needed most urgently from others. No specific answers were given, but plans for resolution of the constraints were set and commitments to provide information were made.

The opportunity for visual management of the process was exploited fully in the coordination meetings. The process plan was posted on the wall and green or red markers were used to indicate progress. All participated in updating the schedule and so all were committed to it. Maia emphasized her role as a service provider to the designers, working to remove constraints on their work so that good flow could be achieved. The designers changed their behaviour – moving from a culture of "excuses" for not completing design stages, such as "the architect didn't send the floor plans that she promised" and "I didn't get that information on time", to a collaborative mind-set and behaviour in which people accepted responsibility for their performance or their failures.

Lack of resources was one of the main issues that arose in the coordination meetings. Milestones for delivery were fixed and could not move, so the solution to "I can't manage" was almost always a request for more staff from that designer's home office. Close time monitoring of resources and workload meant that resources for any discipline could be increased at relatively short notice. The ability to identify short-term peaks in workload and the parallel ability to mobilize "reinforcements" from a designer's home office represented important benefits for the project.

	Sunday	**Monday**	**Tuesday**	**Wednesday**	**Thursday**
1	Full Team Meeting	Structural design co-ordination	Full Team Meeting	Other disciplines design co-ordination	Full Team Meeting
2	Co-ordinate production	Architectural design co-ordination	Co-ordinate production	Tidhar internal design meeting	Co-ordinate production
3		Consultants meeting – called as needed			Architectural design co-ordination
4					Consultants meeting – called as needed

Figure 14.5 Big room weekly meeting schedule

Table 14.2 Big room team meetings: participants, topics and planned durations

Meeting name	Participants	Topics	Planned duration (minutes)
Full team meeting	All big room designers, construction engineer and design manager	Review each designer's current work, discuss design issues as needed	20
Structural design co-ordination	Construction engineer, Tidhar structural engineer, head consulting structural engineer and Tidhar design manager	Identify and resolve problems in structural design, drawing production schedule	90
Coordinate production	Construction engineer and design manager	Drawing production schedule, discuss problems in drawings	60
Architectural design coordination	Construction engineer, Tidhar design manager and lead architect	Identify and resolve problems in architectural design, drawing production schedule	90
Consultants meeting	Construction engineer and design manager, consultants as needed	Identify and resolve design problems; drawing production schedule	90
Tidhar internal design meeting	Design manager and model manager		As needed

The technology imposes constraints in terms of the size of model that can be manipulated practically, which means that the model must be divided by discipline, by zone, or both. The way in which the models are to be prepared and then brought together in a federated model must be planned with procedures that stipulate the file structure, the access authorizations for each team member and the timing of joint model coordination activities.

For the Rosh Ha'ayin project, the Revit model was broken down into 12 distinct files in four groups (one for each of the four quarters). Each quarter had a parking basement model, a high-rise model (each with two towers) and a low-rise model (with three buildings each). Within each model there were also typical apartments that repeated themselves on multiple floors, but these were modelled as groups rather than as distinct families. In this way, unlimited local modifications could be made to each apartment to reflect client design changes, and the quantity take-off for each model could be easily obtained by filtering the objects according to group. The objects in each room were grouped and Revit space objects were modelled for every room. These were also associated with the apartment groups for reporting purposes.

The MEP systems in the parking garages, public areas and the vertical shafts were modelled for clash checking and to coordinate penetrations in the concrete structural frame. The MEP systems within the apartments themselves were not modelled by the big room team – they were modelled at a later stage, when interior finishing works began on site, by the client change coordination team at the construction site offices.

Box 14.4 Eli Renzer's role as information manager

Eli Renzer, Tidhar model manager for the project, prepared the model structure guidelines and helped the designers follow them. As this was Tidhar's first experience with BIM in the big room, Eli's role as information manager became pivotal as the design developed and work began on site. He explained:

> This was the first project on which we used BIM intensively from day one. We were learning as we went, and the time pressure forced us to be efficient. We subdivided the project into three model sets for each quarter of the project. In each set, we worked on two models in parallel – the architectural model and the integrated structure & systems model, which I maintained. The lead architectural modeller and I kept them coordinated continuously. More or less as changes were introduced, we discussed them, viewed the merged models, and took action where needed.
>
> At that time, the engineering consultants were not yet modelling themselves, so as they progressed in 2D, I modelled their systems. In fact, I was doing more than modelling – I was detailing their systems in the way that we, as a general contractor, intended to build them. I showed them our models so that they could check them for conformance to design intent, and they incorporated the details in their drawings. Coordination among the systems was straightforward, and there was very little clash checking needed, because I did the detailing for all of the systems (HVAC, water supply, sewage, sprinklers, electrical, etc.).
>
> It quickly became clear that we needed to manage not only the information, but the detailed design process itself. We needed to guide the consultants' engineers when to design in which area, and in which sequence. We could avoid situations of over-production by aligning their work through pull from the central model. This was much easier to do face-to-face every day; it would not have been possible if we had not shared the big room.
>
> Because this was our first major BIM project, our lawyers and those of the consulting engineers insisted that the consultants' drawings remain the document of record for the subcontractor trades – the subs were legally bound to build what the drawings said they should build. But in fact, as the construction got under way, the trade crews quickly realized that the model views that I could give them were far more accurate, detailed, and reliable than the consultants' drawings. After I moved to the site office, crew leaders would come to me before every work package to get printouts. I provided production details "on demand" – sleeve locations, pipe routings, invert levels, quantities, etc. – details that were never provided on any consultant's 2D drawings. These A3 printouts carried a legal disclaimer, defining them as "ancillary" and not "design" documents. Everyone knew that they were working contrary to the contractual terms, yet it worked much better than it had ever in the past. Everyone benefited, with no drawbacks.

In retrospect, Eli's role was one of "virtual constructor", in the sense that he was detailing the systems as the designers fed him the information. Under standard practice for residential construction in Tidhar's previous projects, the plumbing engineer would set the entry and exit points for a pipeline, check the slope, set the pipe material and diameter, and be done. The details would be set during the construction itself. But Eli laid out the details of all of these systems, concurrent with the early design and in consultation with the construction engineers. Information was generated earlier than it would have been under standard practice.[5] This created a challenge for the consulting engineers because they were not used to dealing with such a high level of detail at such an early stage. Eli became a pivotal link between the architects and the system engineers, with his contractor's BIM model as the boundary object facilitating communication and coordination. At the design intent level, the consulting engineers and the architects communicated directly, but at the detailed level, they communicated through Eli and his structure and systems model. According to Eli:

> Apart from the construction details, we also extracted data from the model for a wide range of other uses. During design, sales contract documents (definitions of the apartment assets – floor areas, storerooms, parking spaces) were produced directly. The areas of façade spaces were measured and used for cost estimating and decision-making, to evaluate the ratio of window space to wall space, etc. During construction, all manner of quantities were taken from the model for the construction engineers. We made specialized concrete casting drawings, drawings to guide placement of pallets of masonry blocks on the floors, façade drawings for the site supervisors to monitor and check the cleaning of the façades, and a host of other documents that had never been used before. Our "clients" were quite inventive in defining new sets of information. As they learned what the model could give them, their appetite grew.

Exploiting the information in BIM

One of the major benefits of BIM is the ability to exploit the information embedded in the model to support a variety of business workflows that are dependent on building-related information. This is naturally dependent on suitable preparation of the model and mechanisms for accessing the information.

The Rosh Ha'ayin project afforded a particularly salient example of this integration. As a real estate developer, Tidhar was required by law to provide detailed sales contract specifications to every customer. These specifications listed all the communal spaces in the building, any parking spaces and basement storerooms that are part of the apartment's property, and all of the details of the rooms within the apartment itself. Using traditional methods, collating all of the data for 1,036 individual contracts would have been a labour-intensive and error-prone process. Using BIM, all of the information could be generated from the model.

Similarly, as managers of the project design and permitting process, Tidhar had to submit detailed and accurate schedules of basement parking spaces to the local planning authority as part of the permit application documents. With over 2,000 parking spaces, maintaining the integrity of information between the design and the schedules was a challenge given the tight schedule and the frequent changes to the locations, types and numbers of parking spaces and their association to specific apartments.

The documentation challenge was solved by generating and maintaining the information solely within the Revit model. For example, parking spaces were represented using a "parking family" object, which in addition to the graphic representation of the parking space – white demarcation lines, parking space number and a handicap sign where needed – contained property fields for the data associated with the parking space. The data included the parking number, type (single or double space), handicap access (regular, standard handicapped vehicle, tall handicapped vehicle), roof (covered or open parking) and the apartment owner's name. When extracted to generate the full parking report, the basement level and the wing of the building where each parking spot was located was derived automatically and also listed. This enabled very quick and accurate reporting using Revit's schedule generation functions.

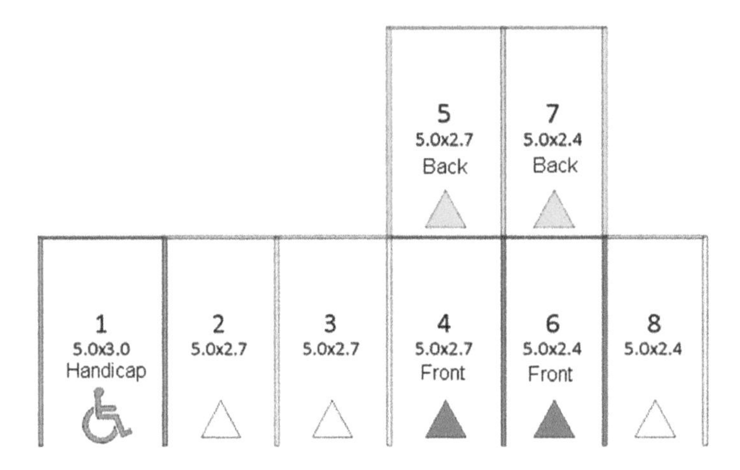

Figure 14.6 Legend for the parking space family in Revit used to manage the parking documentation for the Rosh Ha'ayin project

In the case of the sales contract specification documents, the data were stored in property fields of room and space symbols. The data were also extracted directly from the model, but in this case, they were automatically inserted using a "merge" operation into the correct fields of a sales contract specification document template in MS Word. A fully accurate specification document could be produced on demand for any apartment, with all of its details reliably and accurately reflecting the current state of the BIM model.

Future plans call for the IT department to provide sales staff with a "live" link into the Revit model data so that at the point of sale they can offer parking spaces and/or basement storerooms along with an apartment, and record the sales information directly back into the correct property fields within the model. The need arises from painful experience, where a customer was sold a parking space that had already been sold.

Conformance checking

BIM offers significant benefits in terms of the ability to use the information contained in the model directly for a variety of simulations and analyses. Clash detection to achieve thorough coordination of a building's systems is possibly the best-known example, but with some creativity one can find numerous ways to automate other design checks that are error-prone and laborious when done using traditional methods.

Checking compatibility of the architectural and structural models and checking for headroom clearance are two good examples of BIM model checking methods used by the Rosh Ha'ayin project team, largely thanks to the collaboration of designers from different disciplines finding efficient solutions to meet immediate needs. Neither of these required specific model-checking software, but they did require some ingenuity on the part of the modellers. The concrete elements in the architect's model were compared with their counterparts in the structural engineer's model using Boolean Solid operations. Subtraction of one model from the other showed the precise differences. The designers could then correct errors of dimension or omission, or delete elements that were no longer needed.

Checking for headroom clearance was necessary because the floor sloped in many places for drainage and at access ramps and because pipes, cable trays and HVAC ducts were suspended from the ceilings in many places. The team checked headroom clearance height by writing a plug-in function using the Revit application programming interface (API). The plug-in set-up a grid of points on a selected floor and projected a vector upwards from each point using the API's "*ReferenceIntersector*" class to locate the nearest building object located vertically above the point. A coloured contour map was then drawn on the floor, as shown in Figure 14.7, to visualize the headroom available.

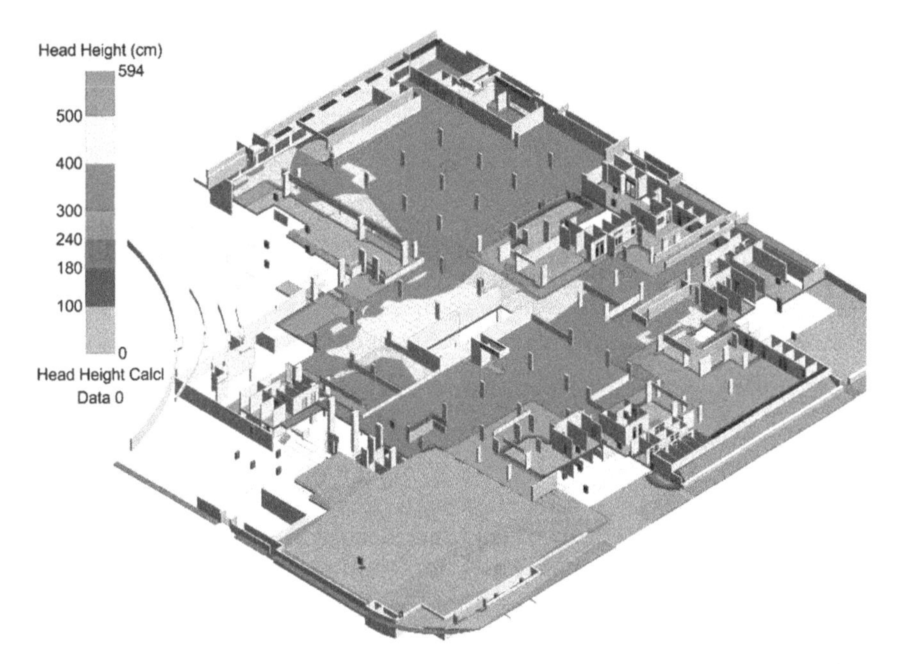

Figure 14.7 Automated checking of headroom clearance for the parking basements of one of the Rosh Ha'ayin project's four quarters

This routine was sensitive to the density of the point grid – if it was too dense, processing time would be too long; too sparse and small or narrow objects like pipes would be missed. Interestingly, one result of applying the function in the project was that while very few headroom issues were identified, the designers found that in some areas the headroom was too high! In one particular instance, above a parking garage entrance ramp, headroom was more than 5 m. Finding this space allowed the designers to extend the slab above, creating valuable parking spaces that would otherwise have been lost.

The result

The project kick-off meeting was held on 29 April 2013. The first permit applications were submitted on 30 June 2013. This in itself was a major success – in Tidhar's previous comparable project (Tidhar on the Park, Yavne – see Chapter 7), which had some 1,050 apartments in 32 buildings, it took 12 months to reach the same level of maturity of the design and to produce the permit application.

The next step in obtaining permits is to respond to the design changes that are required by the local authority after the conditional approval. This commonly requires many months for a project of this size, largely because coordination of the design and drawings is a challenge. In the case of Rosh Ha'ayin, the team enjoyed a double advantage. First, the whole set of application drawings was internally consistent, because they had all been produced from the BIM model. Second, the team had carefully defined the process for responding to the changes ahead of time and were ready for any requests. Every issue was assigned to a responsible designer, who led the consultations needed to resolve the issue and monitored implementation of any changes needed back into the model. The revised application was submitted by 23 November 2013 and the first permits were issued in December 2013 for two of the four plots. The permit for the fourth and final plot was issued in July 2014. This too was a major success. The project went from zero to issue of all building permits within eight months, which is at least four months quicker than would have been achieved without BIM and the big room.

Having secured the construction permits, the pressure was now on to issue the first construction detail drawings so that work could begin on site. Here again the power of BIM and the effectiveness of working in the big room environment proved their value. The models of the underground parking basement levels were already more detailed than was needed for the permit documents, and so the increment in design work needed at that point was significantly smaller than it would have been with 2D CAD. As a result, excavation began within four weeks of obtaining the first permits and the first reinforced concrete foundations were poured just two months later.

The work with BIM in the big room was instrumental in allowing Tidhar to achieve another record-breaking project milestone. The team had realized from day one that they could save time and effort by preparing all of the apartment sales contract drawings directly from the model. These drawings are a legally binding description of the configuration, dimensions and content of each apartment and they must be included in the sales contract. A developer cannot sign contracts (which means they effectively cannot sell any apartments), until a building permit has been issued and the contract documents, including the sales drawings, are complete. Like most real estate developers, Tidhar and their commercial partners were dependent on the cash flow that would be generated by early sales to progress with the construction phase of the project.

During the summer of 2013 the big room team prepared complete sales contract drawings for all 1,036 apartments within six weeks. At a sales fair held in Tel Aviv in November 2013, the company sold 780 apartments in a single day (see Chapter 20, "Pulling it all together"). Here too, the project set records.

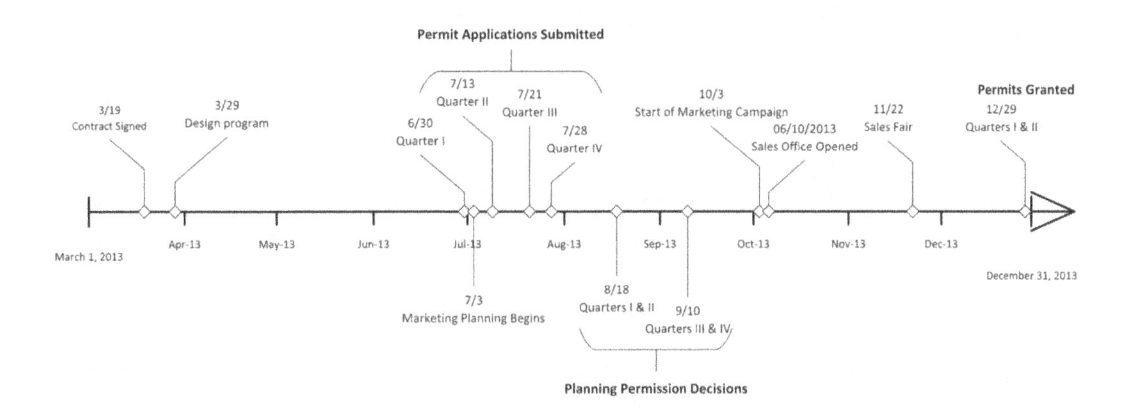

Figure 14.8 Timeline for the process to secure construction permits and begin marketing for the Rosh Ha'ayin project

Reflection and conclusions

In addition to the significantly shortened duration of the design as a whole, the design team members reported achieving record levels of productivity. Some highlighted the fact that this was a unique opportunity for professional growth and learning, by virtue of the close teamwork with other disciplines and exposure to the contractor's construction knowledge.

A major change felt by all the designers was that they were required to model the building to a more detailed level than they would have done either in 2D or using BIM in a mode where each profession worked in isolation. This was a natural outcome of Tidhar's specific requirement that the model prepared for the permit application should be carried over with no rework into the detailed design and construction detailing stages. The design fees across the board were increased to compensate designers for the added effort, yet some designers initially found it difficult to adjust to this way of working. It certainly paid off, however, because construction level models were developed into production drawings very soon after the permits were issued, allowing work on site to begin within one month.

The role of the construction engineering "consultant". Tidhar's construction expert assigned to the project, Mark Starobinsky, provided rich information about construction details, such as the thickness of thermal insulation, sizes of drainage pipes, standard steel formwork dimensions, methods for applying stone cladding to façades and preferred rebar configurations. He prepared a template for an "ideal" construction drawing for each trade, showing the designers how they could best communicate the design to the builders on site. His responsibility was defined as making sure that during construction, drawings would be perfect – no coordination errors, no dimensional errors, excellent readability, etc.

However, he did not use the model directly. This was his first experience in this role and his first encounter with BIM. Although he brought essential knowledge to the team, the traditional way of working to which he was accustomed – reviewing printed drawings, marking them up with red ink and waiting for the designers to fix any errors – meant that the cycle between design, review and correction was slow and needed rework. It also meant that his input was not updated in the model and was inaccessible to other designers and to the design manager.

The lesson learned is that the construction consultant must be well trained and feel comfortable using the model directly. The minimum requirement is to be able to navigate the model, run clash checks and comment on the designs using BIM viewer utility software. The person assigned to this role must have both construction expertise and basic BIM skills. For most construction companies this will mean providing careful, appropriate BIM training for seasoned construction engineers before assigning them to this role within a team using BIM.

The authority of the owners' representatives in the big room must be defined carefully and thoroughly. The big room requires leadership and management, and these are most effectively provided by the owners' representatives. In the Rosh Ha'ayin project, Tidhar was both owner and builder. Maia and Mark were the two key owner's representatives in the Rosh Ha'ayin big room, with the roles of design manager and construction engineering consultant. Both provided input for the designers, and at times this caused confusion. In some instances, Mark's limited access to the model and the resulting disconnect in communicating the rationale behind his advice led to conflict where Maia's and Mark's instructions to the designers were contradictory.

The roles of the owner's representatives and their authority regarding design decisions must be clearly defined and communicated to all the design team. The latter is of course true for any building design team, but the fast pace of work in a big room makes the issue that much more critical. The design manager is responsible for global optimization and must be aware of all the constraints, particularly those imposed by the apartment sales contracts and the building regulations. The construction engineer represents a narrower interest of technical optimization for constructability, schedule, budget and safety. Where these come into conflict, good personal relations and a well-considered process protocol are essential.

BIM in the big room can lead to blurring of authority and responsibility for design. Junior architects and engineers who work in a big room situation find that they are subject to two superiors: their direct team leader in the home office of their design firm, and the project design manager in the big room office. This creates a situation of dual loyalties that can be confusing at times from a professional point of view. The project design manager may be more influential than the firm's team leader if only because the former is always present whereas the latter is mostly absent. Maia Davidson, Tidhar's design manager, learned this in practice when architects in the big room would consult with her on design details, requesting her decision both as the owner's representative and as a more senior architect. She perceived that some of these design decisions should in fact be made by the architectural firm's principal designer because they had implications for the façades or other effects that she may not have been aware of. As time passed, she found it necessary to prepare a set of written guidelines to ensure that the correct people were consulted on each design issue.

Why do we need the design firms? Why not do it all in-house? The design work done in the big room is more technical than creative because the building's overall design concepts are generally set for a project of this type well before the big room is established. The designers allocated to the big room team are typically more junior and are tasked with design detailing work. As work in the big room becomes efficient and effective, some in the owner's organization are tempted to propose that the owner employ the team of designers in-house and work without the influences and constraints of the design firms from which the designers are drawn.

Direct employment of designers might seem particularly attractive where the owner has a work stream that is sufficient to justify employment of a multidisciplinary detailed design team over time, moving from project to project. Tidhar considered this and decided firmly against it. One of the key reasons for rejection was that Tidhar could not foster the same degree of professional development that results from the continuous learning that is only available to a junior designer within a design firm. This has parallels in Toyota's product development practice; while engineers are assigned responsibility for particular products, and may even spend time in the big room, they are never detached from their professional function, which remains their "organizational home" throughout.

Project size matters for use of BIM in the big room. Maintaining the value of the big room requires continuous presence of at least one designer from each main design discipline. Where any designer is absent the benefit of short cycle times for collaborative decision-making on cross-disciplinary design issues is no longer available and the value for those that remain in the big room is reduced. Yet design firms are highly reluctant to send their designers to work in the big room instead of in their home office if their workload does not justify full-time presence. For this reason, only projects that can sustain full occupation for at least one designer from each discipline over an extended period of collaborative work are big enough to make the big room work effectively. Ironically, the more productive designers are as a result of using BIM, the less time they need to spend continuously on any given project.

BIM is particularly effective for projects where there are multiple small-scale differences in local design details across large numbers of spaces that are essentially the same at lower degrees of resolution. There were only eight apartment types (A to H), and only four basic building types (a 9-storey type, a 21-storey type and mirror images of those two) in the Rosh Ha'ayin project. However, the three parking basement floors in each building had quite different configurations due to the different terrain in each quarter, and many apartments and some of the buildings had local adaptations to suit the different sunlight directions (the buildings are arranged in radial fashion around a central park). BIM was very effective in this context. In early phases, the small set of apartment models was reused many times. As design progressed and the level of detail deepened, local changes were made, resulting in far more unique apartment designs than there were originally. The benefit of BIM here is that all quantity take-offs, schedules and drawings could still be produced rapidly and accurately from the model, no matter how differentiated the component parts became.

Lean and BIM synergy

Co-location of designers in a big room with full exploitation of BIM tools is an excellent example of the synergistic application of Lean Construction and BIM. When considering design specifically, we can identify a number of Lean practices at work in a BIM-centric big room in light of Womack and Jones' five-step framework for Lean implementation:[6]

1 Identify value for clients
   - Emphasize collaborative preparation and ongoing development of the brief and programme with the client
   - Set up a concentrated multidisciplinary design team
2 Identify the value stream
   - Identify and remove unnecessary design activities
   - Map the design process and simplify iteration loops
3 Make products/information flow
   - Reduce design batch sizes
   - Concurrent design
   - Transparency, including sharing of incomplete design information
   - Co-locate the multidisciplinary design team whenever possible
   - Change contract incentives to encourage flow and collaboration
4 Create pull from the next upstream activity
   - Pull information flows and decisions from the design team on an as-needed basis
   - Use LPS, particularly reverse phase scheduling and short-term work planning
5 Continuously improve
   - Regularly review work practices within the collaborative design team

Tidhar's implementation of BIM in the big room applied most of the steps listed, but by no means all. There is always room for improvement, in terms of both Lean design practices and BIM technology exploitation, and many improvements have been made as the practice of big room design collaboration has been applied since this first experience.

The BIM in the big room implementation can also be seen in light of the framework definition of the interactions between Lean principles and BIM functionality described in Chapter 1, "Introduction – Growing the Margin". The matrix in Figure 14.9 shows only those interactions from the original matrix (Figure 1.1) that are relevant for Lean design in the specific context of the big room. The principles along the top row are a subset of the Lean principles from the original matrix, and the BIM functionalities are those relevant in the design phase. The cells with a tick mark are the points of positive interaction between Lean design and BIM functionality. For example, by evaluating a building design using BIM – using renderings to visualize the building (functionality 1 in the matrix) or engineering simulations to evaluate performance or behaviour (functionality 2) – one can go a long way to achieve the desired quality right from the earliest phases of design.

The shaded cells with tick marks represent the interactions that are strongly enhanced, and in some cases only achieved, when BIM is applied in the big room. Multiuser editing and viewing of the models (functionalities 9 and 10) are naturally strongly enhanced by co-located working. People can consult with their colleagues ad hoc, using the model as a boundary object for the conversation, and reach design solutions that fulfil the requirements of multiple design disciplines quickly because the communication is synchronous, rather than asynchronous. The technological and administrative overheads of BIM collaboration are greatly reduced.

Another cluster of Lean–BIM synergies that are strengthened by co-location is that of design evaluation (functionalities 1, 2, 3 and 5) and achieving value for clients in terms of conformance to requirements, verification and validation (principles R, T and U). Because design is performed by multiple disciplines essentially in parallel (principle O), many of the problems of negative iteration in design can be avoided. This also means that cycle times and inventories of design information (principles C and D) are significantly reduced, greatly improving flow. Taken together, it is these synergies that allow co-located teams using BIM to reduce overall design duration, consume fewer resources and improve design quality.

In the medium to long term, the big room helps cultivate an extended network of partners (principle X). The social cohesion that develops around commitment to project-specific team goals leads the individuals and their companies to prefer to work together on future projects.

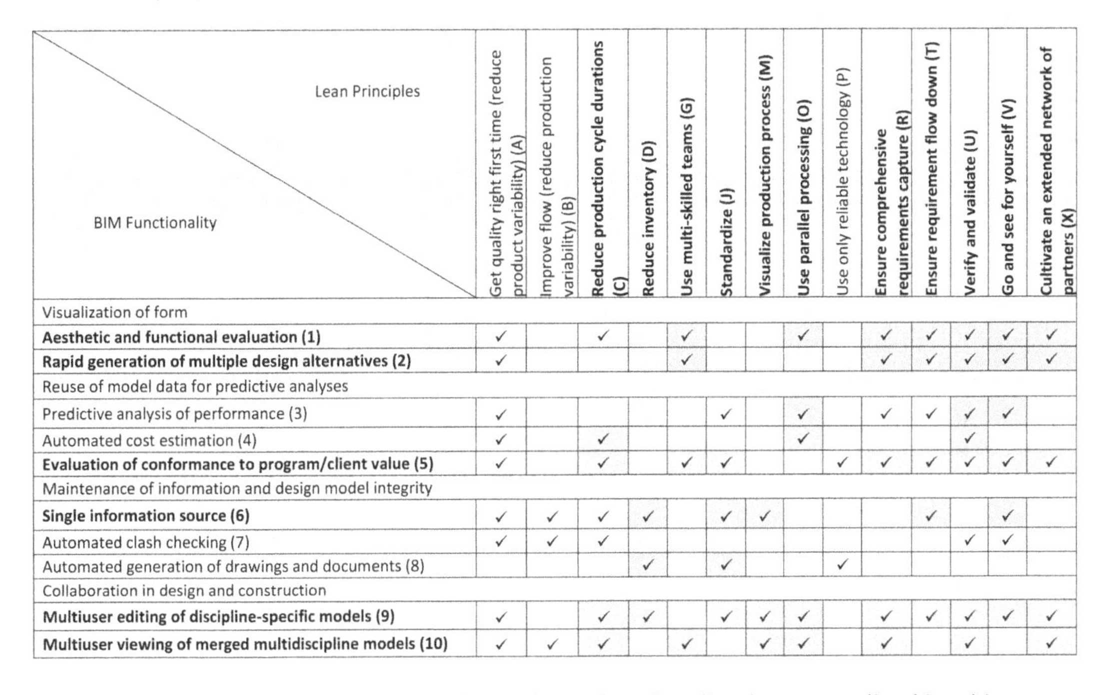

Lean Principles / BIM Functionality	Get quality right first time (reduce product variability) (A)	Improve flow (reduce production variability) (B)	**Reduce production cycle durations (C)**	**Reduce inventory (D)**	**Use multi-skilled teams (G)**	**Standardize (J)**	**Visualize production process (M)**	**Use parallel processing (O)**	Use only reliable technology (P)	**Ensure comprehensive requirements capture (R)**	**Ensure requirement flow down (T)**	**Verify and validate (U)**	**Go and see for yourself (V)**	**Cultivate an extended network of partners (X)**
Visualization of form														
Aesthetic and functional evaluation (1)	✓		✓		✓			✓		✓	✓	✓	✓	✓
Rapid generation of multiple design alternatives (2)	✓				✓					✓	✓	✓	✓	✓
Reuse of model data for predictive analyses														
Predictive analysis of performance (3)	✓					✓		✓		✓	✓	✓	✓	
Automated cost estimation (4)	✓		✓					✓				✓		
Evaluation of conformance to program/client value (5)	✓		✓		✓	✓			✓	✓	✓	✓	✓	✓
Maintenance of information and design model integrity														
Single information source (6)	✓	✓	✓	✓		✓	✓				✓		✓	
Automated clash checking (7)	✓	✓	✓									✓	✓	
Automated generation of drawings and documents (8)				✓		✓			✓					
Collaboration in design and construction														
Multiuser editing of discipline-specific models (9)	✓		✓	✓		✓	✓	✓		✓	✓	✓	✓	✓
Multiuser viewing of merged multidiscipline models (10)	✓	✓	✓		✓		✓	✓		✓		✓		✓

Figure 14.9 Interactions of Lean principles and BIM functionality that are realized in a big room

Lessons learned

Since their first experience in the Rosh Ha'ayin big room, Tidhar has used the approach in other projects and learned a great deal along the way. The most recent project to use an intensive big room for design was the new Microsoft office building, near Tel Aviv. The big room design spaces are no longer housed in temporary site trailers – they occupy dedicated spaces in an office building near Tidhar's head office.

Rosh Ha'ayin was the last of Tidhar's projects in which the consulting engineers worked in 2D. Since then, all the consultants perform their work in BIM. This has removed a great deal of the work of the Tidhar modeller in terms of solving coordination issues, but instead has shifted the focus to model management and design coordination. Over time, and as people have moved from one project to the next with Tidhar, people have learned to work within a team, instead of in their own silos. The consulting engineers' staff, for example, have learned to take direct responsibility for clashes, working directly with one another to avoid the need for extensive and formal clash-checking procedures. This is a shift in the direction that Lease Crutcher Lewis has taken so successfully, as discussed in Chapter 12.

One of the key components that makes this possible is that Tidhar's design managers have formalized the order of design and responsibility among the designers. Interestingly, they now need to invest effort to align the BIM standards that each of the architects and engineers brings from their home offices to the big room. Eli Renzer comments on his experience in the Microsoft big room: "Not only have they learned to model, they have learned to model in their own specific way. Getting them to adapt to the project's BIM Execution Plan is not as simple as it was when they were just starting out".

Some of the key lessons learned include:

- BIM in the big room drastically improves the project design process. It shortens design cycle times, and overall design duration. It removes much of the wastes of waiting and rework, and it improves the quality of design with all that implies for the construction itself.
- By participating in big rooms and working in BIM, designers learn to collaborate with one another in ways that are quite different to traditional practice. Many designers reported difficulty in returning to work in their home offices after their experience, indicating a clear preference for the collaborative and congenial working environment of the project-focused big rooms.
- Social relationships and group cohesion need to be fostered, but the effort more than pays for itself through people accepting joint responsibilities. Teamwork and collective responsibility are reciprocal, each strengthening the other.
- Use of local servers reduces the communication friction of model transfers. Where the BIM software and the exchanges are set-up properly, the technology becomes transparent. Cloud servers and other technologies may fulfil some of these requirements adequately in the future, but not yet.
- Building design and production system design are closely intertwined, and the role of the construction contractor in the design big room is essential. The value of the builder's input in the traditional sense of product design is clear (input regarding the specifications of various construction materials and methods), but the value to the process design is less obvious. The anecdote concerning pipe routing around the tower crane mast, illustrated in Chapter 20, "Pulling it all together", illustrates the need for construction process input during design development very well.
- The big room should ideally be exactly that – one big room. Even simple partitions that create separate cubicles in a shared office space reduce the effect. Rosh Ha'ayin

had three mobile office units, creating three separate working spaces (architects, contractor and consulting engineers) linked by a passage. The Microsoft big room has two big elliptical spaces, with a broad open area linking them, an improvement over Rosh Ha'ayin but still not perfect.

- Big rooms work best where the scale of the project allows consulting engineers to justify the permanent presence of their people in the big room. But even where the balance of workload means that there is not enough work to justify full-time presence for one or more trades, careful scheduling of their time in the shared space leads to excellent results, as was the case in Rosh Ha'ayin. Chapter 19, "The Fira way", introduces the idea of a "virtual" big room and discusses the pros and cons of this approach.

Notes

1 For more information about the role of the *Obeya* in the Toyota product development system, see Morgan and Liker (2006).
2 See Eastman et al. (2011).
3 For more details of this project, see Do et al. (2015).
4 Among them are the BIM Guide documents of the New York District US Army Corps of Engineers, the Singapore Building Construction Authority, Australia's NATSPEC Standard, the Los Angeles Community College District, the University of Southern California, Indiana University and the State of Ohio General Services Division. A detailed review can be found in Sacks, et al. (2015).
5 Earlier generation of information of this kind is exactly the phenomenon described in the MacLeamy curve, which predicts that BIM processes move the curve representing the rate of information generation to the left (i.e. earlier) along a project's timeline. See MacLeamy (2004).
6 The five steps are: specify end customer value; identify the steps along the value stream; improve flow by reducing waste; introduce pull and strive for perfection. For more details see Womack and Jones (2003).

15 What flows in construction?

In construction, we can see materials flowing into a construction site, we can see trade crews and equipment moving through the site, and we are aware of the flow of information. But the spaces we build, the real products that are produced as they flow through the construction process, don't move at all, making the most important flow the hardest one to see.

Why do we need to define a construction equivalent of a manufactured product? If we wish to apply the logic of product flow to construction, including Little's Law and all of its consequent operational logic, we need to define the product that flows. With a sharp definition, it becomes possible to identify the wastes in construction much more easily, because we can measure quantities of work in progress and cycle times, and we can directly observe bottlenecks and their effect on throughput. In this chapter, the first section provides a background to the principles of product flow in manufacturing. The second section answers the question "What flows in construction?" and defines a *constructed space* as the construction equivalent of a manufactured product.

Production flow

The idea of products flowing through a production system is closely associated with Lean thinking. When we think of a production line, the image of car bodies flowing through an assembly plant from workstation to workstation on a moving conveyor comes easily to mind. Even without a moving conveyor, the notion of products flowing along a production line from one work station to another is easily imagined and understood.

When a well-designed production system is functioning efficiently, the rate of flow is constant and production rates are stable and more or less equal at all the workstations. When it is not functioning very well, such as when the rate of production at different workstations varies, there is a tendency for unfinished products to accumulate in front of slower machines, and machines that process parts more rapidly tend to be idle when the upstream flow is inadequate to keep them fully supplied. In his book *The Goal*, Eli Goldratt described vividly how the slowest processing machine in a production line acts as a bottleneck.[1] The "theory of constraints"[2] explains how the bottleneck is in fact a regulator for the production rate of the whole production system – the overall throughput rate can never be greater than the bottleneck rate.

In many systems, the production rates of machines vary, sometimes predictably (when different product types require different processing times) and sometimes unpredictably (when a machine breaks down, for example, or when raw materials are delivered late). These variations mean that the bottleneck may move – what was the slowest machine in yesterday's production may not be the slowest today. When variation in production rates is common, managers tend to compensate for the variation by allowing buffers of extra unfinished products to accumulate behind machines with less predictable production rates, so that downstream machines can continue working even if an upstream machine slows or breaks down. Unfinished products in a line are called "work in progress inventory", or simply "WIP". Where many workstations are unpredictable, buffers of WIP accumulate all along a production line and this WIP – piles of product waiting for processing – can be seen and identified easily. Large amounts of WIP in a factory are generally a sign that a production system has workstations with varying and often unpredictable production rates. The causes may be inherent in the work method or the equipment, or the variations may be caused by the flows of needed materials, information or labour.

The tendency for managers to set policies or select work methods that prioritize local efficiency or productivity measures is another potential source for large amounts of WIP. For example, some machines are designed to process large batches of products in a single operation, or in a continuous run. A large washing machine would be inefficient to run through a full wash cycle unless it were fully loaded with clothes, and so it makes sense to wait until the basket is full before running a wash – but that also means that, on average, clothes spend more time in the dirty washing basket than they would if a smaller washing machine was used instead. The basket of dirty clothes is WIP; WIP is increased where batch sizes are large.

Another common cause for large WIP is the "push" production control policy that is common in many production systems. Push is the traditional approach to managing production when overall production cycle times are long and managers plan and control production according to targets based on estimates of future demand. Under push control, workers are encouraged to produce things as quickly and as soon as possible, under the assumption that if all the parts are working at maximum capacity, the system must also be at its most efficient. In this paradigm, every machine and every worker is set to work as fast as possible. This means that all workstations (except for one – the bottleneck) produce at a rate faster than the bottleneck, which results in accumulation of WIP, sometimes well beyond what is actually needed or will be consumed. Unfortunately, demanding that each worker and each machine work at their fastest possible rate is a very intuitive thing to do, though the result in general is the waste of overproduction. Some of the interim products are never finished and some of the finished products are never consumed.

All of these ideas and concepts – bottlenecks, buffers, WIP, batches, push control – have meaning because we think of products flowing through a production system in much the way we think of water flowing through a system of canals, or of traffic flowing through a road network. A good flow of water is one in which the friction of the canal walls is small, there are no blockages and there is little or no turbulence. Good traffic flow is flow in which all the vehicles travel at the same constant speed, with no traffic jams or traffic lights to impede flow.

But construction is not like production in a factory. In a construction site, the flows are different, though some conceptual parallels can be drawn.

Little's Law

Little's Law is a clear and easily applicable mathematical model for flow in production.[3] It states that throughput is equal to the quantity of inventory in a production system divided by the cycle time needed to produce a single product.

$$TH = \frac{WIP}{CT}$$

TH is the throughput of the system as a whole and is measured as the number of units of a product produced per unit of time (e.g. cars/hour). WIP is the quantity of incomplete products in the production system at any given point in time. CT is the cycle time, which is the time required to produce a single product, measured from the moment the first production activity begins on that product until the moment the finished product leaves the system.

Little's Law explains several features of production systems. If we group products in batches as they flow through a production line, then the time required to produce each individual product grows longer as the size of the batch gets bigger. This occurs because each product in the batch must wait for all of the others to be produced in any given step before it can move on to the next step. Similarly, the smaller the batch size, the shorter the cycle time for the same throughput, because a smaller batch size means a smaller amount of work in progress. This leads to the idea that a batch size of one, i.e. flow of individual products, results in the minimum cycle time.

Box 15.1 Batching production of precast concrete façade panels

By way of example, consider the case of producing a set of architectural precast concrete wall panels for a building's façades. Façades are commonly divided into panels that match the floor-to-floor height and vary within each floor level to accommodate the different windows and other features. Thus, most of the panels that compose a particular floor level are different to one another, but the same pattern is repeated on each floor, so that there are as many instances of each type of panel as there are floors in the building. A 10-storey building with 12 different panels at each floor level will have 120 panels, with ten panels of each of the 12 types.

Batching is often used when a particular step in the process has a long set-up time. To minimize the overall cost of set-ups, it seems to make sense to produce as many pieces of a given type as possible after each set-up. In our example, if it takes a relatively long time to set-up a mould to produce a particular precast concrete wall panel, then it apparently makes sense to pour all of the panels of that particular type in a batch, rather than reconfiguring the mould for each and every panel. With this approach, the plant will produce ten panels of the first type, then ten of the second type, and so on.

Now consider the cycle time needed for enclosing each floor of the building so that interior finishing work can begin. The first floor cannot be fully enclosed until the first panel of the twelfth type is cast. This means that the first floor can be enclosed after producing 111 panels (10 each of the first 11 types plus the first panel of the twelfth type). The construction time of the building as a whole is much longer than would have been the case if the plant had produced one panel of each of the 12 types, thus producing the set needed to enclose the first floor after just 12 pours, rather than 111. Batching of the panels by type lengthens the cycle times for the sets of panels needed for the floors. From the point of view of the building, this is poor flow, resulting in long cycle times. It leads to large quantities of work in progress that must be stored somewhere (all 111 panels must be stored before erection can begin, and many floors of the building remain vacant for extended times because interior work cannot progress in floors exposed to weather). It also results in poor cash flow, as the money needed to produce all 111 panels must be financed before value begins to be created for the client (in the form of enclosed building floors); similarly, the cost of building the structure must be financed for a longer period than could otherwise be the case.

In this example, there is a trade-off between the local efficiency of the precast plant and the global efficiency of the building project as a whole. A local optimization – batching production of panels by type – leads to an inefficient overall construction process. If panels were produced one by one, the construction process could be most efficient.

It may appear that there is some fixed optimum trade-off point between precast panel batch size and floor enclosure cycle time. However, this makes the assumption that the long, expensive mould set-up is fixed, so that one can only improve productivity by making the batch size as large as possible. This is the thinking that Toyota challenged when they realized that they could do the opposite – they could reduce batch sizes and cycle times by actively reducing the set-up times. The story of Toyota's goal of less than 10 minutes per changeover of dies in their stamping presses ("single minute exchange of die" or SMED[4]) has become well known as an example of just this thinking.

In exactly the same way, a precast concrete company that devised a way to reduce the set-up time needed for each mould could reduce the batch sizes for production and thus provide more valuable service to the construction projects it supplied. This was one of the strategies adopted by Malling Precast Products Ltd, a subsidiary of the Laing O'Rourke Group, located in Grays, Essex, UK, when they needed to increase production to supply precast pieces for parking garage structures for the British Airports Authority.[5] They changed their process to allow them to respond to calls from the site for specific pieces within short time-frames by preparing kits of the contents for various precast concrete pieces in advance, then waiting for the customer to call before pouring the pieces. In this way, they reduced the set-up time needed when pieces were ordered, so that they could cast concrete to fabricate pieces "on-demand" and deliver just-in-time, instead of fabricating and storing large buffers of inventory in advance. They also reorganized the plant into production cells, each of which could prepare specific pieces from beginning to end, instead of the previous functional organization in which the products had to be routed through different departments (mould set-up, reinforcing, casting, stripping). This strategy enabled them to reduce batch sizes and cycle times. Among the results reported were:[6]

- *Lead times were reduced for structural precast elements to 1 week, corresponding to a reduction in manufacturing cycle time to 1–1/3 days.*
- *The shear wall production cell had previously averaged 3.2 walls per day, with 12 workers. After application of Lean principles to restructure work flow, 12 workers produced 9 walls per day, an increase in productivity of 181 per cent.*
- *The tees production cell was restructured in a very similar way, resulting in an improvement from a baseline of 9 tees per day to 18 tees per day, an increase in productivity of 100 per cent.*

Another feature of production systems shown by Little's Law is the idea that the throughput of a production system is limited to the maximum throughput of the operation with the lowest production rate in the system. This is the same idea as the bottleneck defined in Goldratt's theory of constraints. Thus, the cycle time and the WIP in a system are governed by the production rate of the bottleneck operation, r_0. It is easy to identify the bottleneck operation at any given point in time in a factory production line – the bottleneck operation is the one in front of which the biggest quantity of *WIP* accumulates. Similarly, there is a minimum cycle time for any given product, CT_{min}. The minimum cycle time is the sum of the net processing times for that product at each machine or workstation. For a product to actually be produced in the minimum possible time, the waiting times for that product between each station must be zero.

A third feature that Little's Law explains is that any production line has a certain "saturation level" of WIP, W_0, which is the amount of WIP needed for the bottleneck machine to reach its maximum production rate (r_0). If the rate at which raw materials or input products are fed into system is less than r_0, then the production rate of the

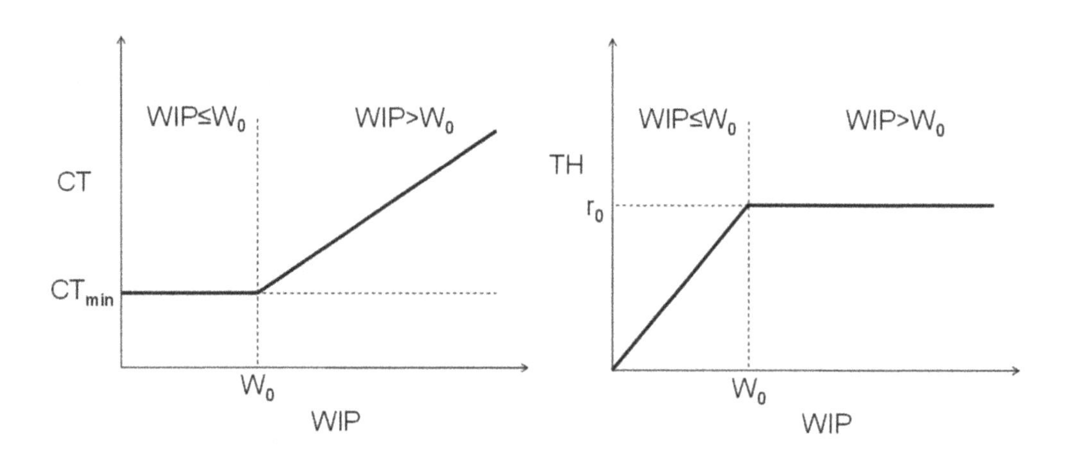

Figure 15.1 Little's Law relationships between throughput (TH), work in progress (WIP) and cycle time (CT)

bottleneck machine and the overall system throughput (*TH*) will be equal to the feeding rate; but when the feeding rate is increased to equal r_0, the overall throughput hits an upper bound and cannot increase beyond r_0. The amount of WIP in the system at this point is W_0, and the relationship $r_0 = W_0 / CT_{min}$ holds when $WIP = W_0$, just as Little's Law requires.

Variability

Variability – whether in product design or type, levels of demand, workstation set-up times and production rates, peoples' levels of skill and effort, or in change orders – invariably reduces the throughput of a production line. In *Factory Physics*, Hopp and Spearman's definitive book on production systems,[7] the authors state this as a set of "variability laws". The first law states:

> **Law (variability):** *Increasing variability always degrades the performance of a production system.*

The second law describes how this happens:

> **Law (variability buffering):** *Variability in a production system will be buffered by some combination of (1) inventory; (2) capacity; and (3) time.*

The effect of buffering for variability with inventory, capacity and time is clear from Little's Law. When variation in operation times leads to longer cycle times it also leads to increased *WIP*, for the same overall throughput. If the time needed to perform a production operation (a particular step) varies significantly from product to product, then some products will be forced to wait between operations (when a following operation takes longer to complete its current product) thus generating an inventory buffer and a time buffer. Also, some operations will be idle from time to time (when the product they are expecting takes longer to be completed by the preceding operation), which represents underutilization, or a capacity buffer. This leads to a build-up of delayed products between operations.

One can also find trade-offs between inventory, capacity and time buffers. Idle machines are perceived to be wasteful, and traditional production managers generally cope with this problem by allowing a small buffer of interim products to accumulate between each operation, so that when a product is delayed in a particular operation, the succeeding operation will have a buffer of products to work on in the meantime. In this way, it is not "starved" of work. This reduces the underutilization, i.e. reduces the capacity buffer, by introducing an inventory buffer.

The size of the inventory buffer or the time buffer will in general be proportional to the degree of variation of the preceding operation, but when capacity utilization is very high, systems become much more sensitive to variation. The reason for this is that when a workstation is always busy, then the higher the variability in either arrival time or processing time, the more likely it becomes that some products must wait to be processed. The phenomenon is illustrated by Kingman's formula,[8] which approximates the expected waiting time in a queue at a single server with generally varying arrival and processing times (commonly called a G/G/1 queue). Hopp and Spearman (1996) adapted

Kingman's formula to reflect the relationship between the cycle time CT_q (equivalent to the waiting time in the queue in Klingman's formula), the degree of capacity utilization, the degree of variability and the mean processing time in a production system, as follows:

$$CT_q = \left(\frac{c_a^2 + c_e^2}{2}\right)\left(\frac{u}{1-u}\right) t_e$$

In this formula, the first bracketed term represents the variability of the arrival rate (c_a is the coefficient of variation for the arrival times of the feeding flow, c_e is the coefficient of variation of the processing time); u is the capacity utilization factor, ranging from 0 to 1, and t_e is the average processing time. There are two important consequences to this, both of which are clearly apparent in Figure 15.2: (1) when u tends toward 100 per cent, the cycle time grows exponentially, and (2) the more variable the production systems, the more the cycle time grows as a function of the capacity utilization factor. Only a perfectly stable system, one with zero variability, can approach full capacity utilization without fear of lead times spiraling out of control. The pertinent conclusions are that production systems running at close to their maximum capacity are fragile, and the more variability they have, the more fragile they are. From an operational point of view, this means that the best way to improve capacity utilization while still maintaining a reasonable lead time is to reduce variability.

A final observation is that when there is variation, its effects are felt not only at the workstation whose production rate varies, but at the downstream workstations too, even if they themselves are perfectly stable. This is because the output of an unstable

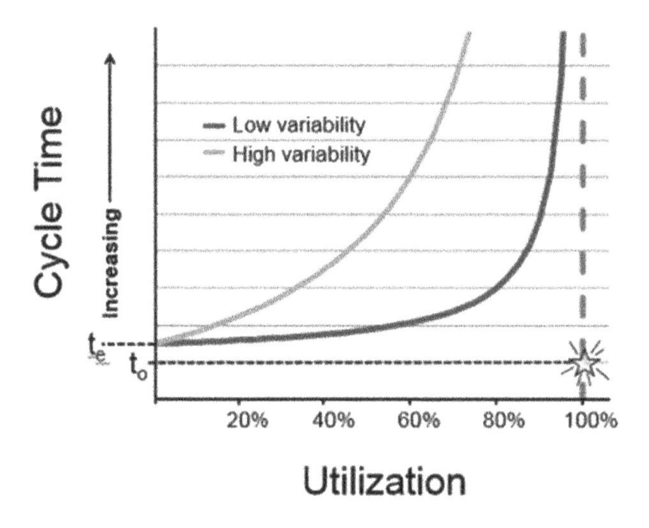

Figure 15.2 The relationship between cycle time, utilization and the degree of variability

Source: Pound et al. (2014, p. 75), reprinted with permission of Factory Physics Inc., Edward Pound.

workstation also exhibits variation and the output of one workstation is the input of the next. The term c_a in the formula above represents the variability in the arrival time of products at a workstation. Even if the processing time variability, represented by the term c_e, was equal to zero, the workstation would still experience longer cycle times than if its feeding rate did not vary. To put this in practical terms: if some products don't arrive when the machine is ready for them, the machine will wait and processing will start only when they do arrive, with the result that when subsequent products arrive the machine will not be ready for them, and so average cycle times will be longer than the average processing time.

Box 15.2 The "parade of trades"

One can see the downstream effect of variation clearly in the "parade of trades" simulation, a computerized discrete event simulation that illustrates the effect of varying production capacity at any of five workstations in a production line. The simulation, prepared at UC Berkeley by Professor Iris Tommelein and Hyun Jeong Choo, is worth trying. It can be downloaded from www.ce.berkeley.edu/~tommelein/parade.htm.

Try setting maximum variability of production rate (9/1) for the last trade while keeping the other four stable (5/5); then, set the first trade to be unstable (9/1) while resetting the last one to be stable (5/5). The difference in overall productivity is striking, but in particular, the potentially damaging effect of instability of the first trade on the subsequent trades is vividly apparent. It is no surprise that downstream trades in construction projects protect themselves from the instability of the upstream trades by allowing buffers of time and inventory to accumulate between them.

Lean thinking about flow

As we have seen, the Lean concept of flow is well defined. It is understood as a path through which a product progresses as it is processed from raw material to finished product (taking flow as a noun) or as the physical movement of the product along the path (as a verb). The flow path is called the "value stream" and the actions performed along it can be classified as value-adding or waste, as discussed in Chapter 8. In Lean terms, the ideal flow state is one in which the product is constantly having value added while it moves through the system, with no waste of any kind.

A Lean system strives to operate at or very near to the boundary where $WIP = W_0$. There is a constant effort to reduce the quantity of materials and unfinished product in the system, and the wastes associated with them, to a minimum. This leads naturally to a concomitant effort to reduce batch sizes with the ideal condition of a batch size of one, a condition called "single-piece flow".

Looking at a production system in this way also helps to explain why a Lean production system works hard to achieve stable and predictable production rates. When production rates for any given workstation are unstable, or when production does not proceed according to plan, downstream workstations must either starve, reducing their local productivity, or allow buffers of WIP to accumulate in order to protect their productivity levels. Buffers of WIP are inherently wasteful, requiring work to maintain them, count them and protect them, place to store them and money to finance them, and as Little's Law predicts, they lengthen production cycle times for every product. This is the reason why a slower but more stable and reliable machine is often preferable to a faster but unreliable machine.

A corollary rule of thumb is that where the production rate of one slow machine is not sufficient, two slow machines working in parallel with small batch sizes are better than one fast machine with a big batch size. Not only is the average batch size smaller, the reliability of two machines in parallel is greater than that of two in series, or one in series, all other things being equal. If the smaller machines have shorter set-up times than the faster, greater capacity machine, then the system will be even better.

Process flow and operations flow

In manufacturing, there are two axes of flow: *process flow*, which represents the progress of a product along a production line, and *operations flow*, which represents the individual actions performed on the product at any given workstation. This distinction was crystallised by Shigeo Shingo, a Japanese production management consultant considered one of the world's leading experts on manufacturing systems, who contributed to the development and documentation of the Toyota production system. Shingo and Dillon[9] defined these two flows as follows:

> When we look at process, we see a flow of material in time and space; its transformation from raw material to semi-processed component to finished product. When we look at operations, on the other hand, we see the work performed to accomplish this transformation — the interaction and flow of equipment and operators in time and space. Process analysis examines the flow of material or product; operation analysis examines the work performed on products by worker and machine.

This is a fundamental distinction and enables us to focus on improvement of the flow of product (process flow), on the one hand, and on improvement of operations, on the other. Process is improved by removing non-value-adding steps such as moving, waiting and inspection, and by minimizing set-up times and rework. Operations are improved by balancing the work of operators, improvements to methods and tools, etc. "To make fundamental improvement in the production process, we must distinguish product flow (process) from work flow (operation) and analyse them separately"; viewing a production process as a single line that includes the operations leads to "the mistaken assumption that improving individual operations will improve the overall efficiency of the process flow of which they are a part".[9]

Figure 15.3 shows these two flows and how they intersect with one another. The process flows shown here for windows and doors produced in a factory are the flow of unfinished products from operation to operation as the raw materials are converted into finished products. The operations flows shown are the cutting and joining operations, which are each performed on multiple types of product. The interdependence between the flows is apparent. For example, considerations of local productivity or efficiency at the cutting operation could encourage a department manager to increase

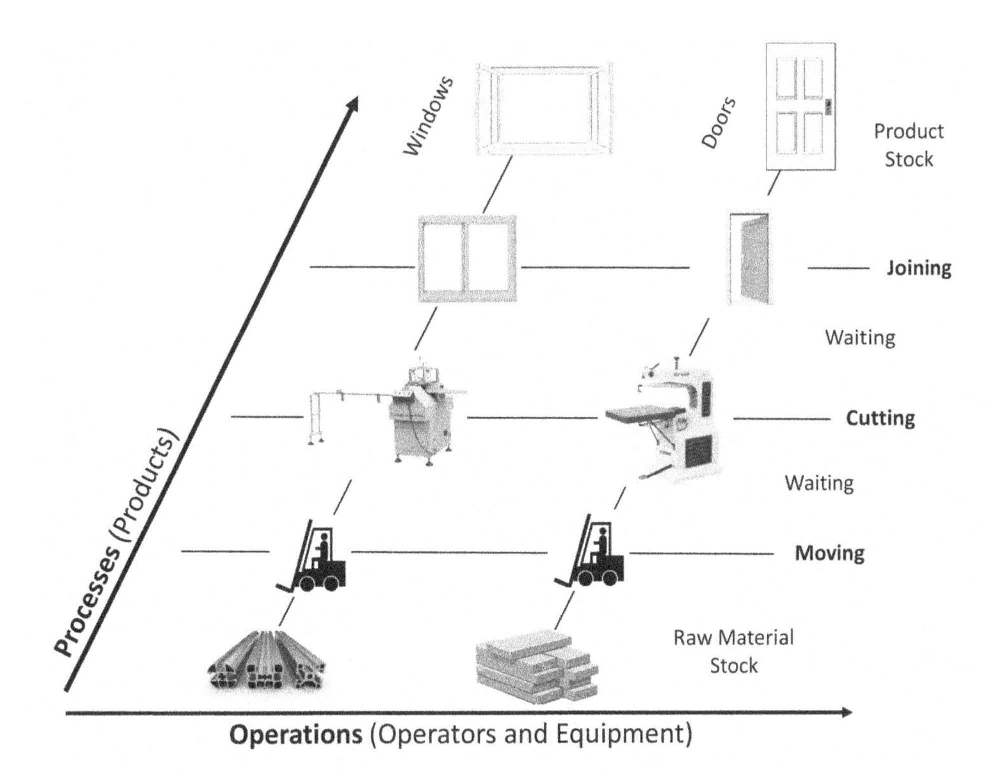

Figure 15.3 Operations and processes in manufacturing, based on "the structure of production" by Shingo and Dillon (1989)

Source: Reproduced from Sacks (2016).

the batch sizes of parts, including batching together of parts from more than one product. Clearly, this would lead to many partial door and window assemblies, and other parts having to wait for the batch process to complete, thus lengthening the cycle times for all products.

Re-entrant flow

The last feature of flow in production that we need by way of introduction, before considering what flows in construction, is the phenomenon of *re-entrant flow*. Re-entrant flow occurs when a product is required to return to a workstation in which it has already been processed earlier in its manufacturing process. Some products – like wafers for semiconductor chips, for example, have many layers, and so they return to the same machines numerous times during their manufacture, each time for a successive layer of deposition of material, etching, lithography, etc. Quality defects are another cause of re-entrant flow, where defects identified downstream in the production process require the defective part to be returned to upstream machines for correction.

If products pass through the bottleneck operation more than once, production control is made that much more challenging because policies that prioritize the use of the bottleneck are needed. Should products closer to the start or the end of their production runs be given priority? LIFO (last-in first-out) or FIFO (first-in first-out) and variations on these are common policies. Planning and control is even more complex when there is a mix of product designs, with each requiring different numbers of returns. A wide variety of approaches to decision-making in such situations have been devised over the years, and many have been applied in factories. Heuristic dispatching rules are easier to understand and to apply than more rigorous computational methods, and they are easier to tweak in response to a system's observed performance.[10]

So, what flows in construction?

The idea of "flow" also lies at the heart of Lean Construction. Along with transformation and value, it is one of the three key ways of thinking about production (Koskela, 2000). However, while it is easy to imagine the flow of manufactured products through an assembly line or the flow of vehicles on a highway, it is more difficult to imagine the same metaphor in construction. The reason is that we cannot see the products moving, which makes it harder to see the batch sizes, to measure the cycle times and to identify the bottleneck operation.

When considering flow in construction, the first things that spring to mind are the flows of raw materials, workers, equipment and information. The first three literally move around from place to place, and most of us are familiar with the idea of information flowing. These flows are captured in Tidhar's "4 × 4" poster shown in Figure 15.4, which shows four flows and four conditions that must be checked before starting work (the conditions are: (1) completion of the previous task, (2) a clean work area clear of obstructions, (3) safe working conditions and (4) external conditions such as the weather, appropriate permits, etc.).

These four flows are part of the *operations* flow axis in Shingo's process/operations model. The trade crews and their specialized equipment are equivalent to the workstations – machines and their operators – in a factory. The general site equipment, such as cranes, work elevators, pumps, etc. are equivalent to the fixed facilities in a factory. Raw material flow is similar in both situations: sites and factories. Information, including both *product* information (which defines what work should be done, i.e. what the product should look like) and *process* information (when the work should be done on a particular product), flows in much the same way in both situations too, as design drawings or model views (product) and work directives (process).

Shingo's model, however, has a second axis of flow – the *process* flow axis, which in a factory refers to the physical product that is being produced. The construction equivalent of a manufactured product is a distinct space in a building. Defining this equivalence allows us to apply the same thinking about flow, including Little's Law and the ideas of a production bottleneck, WIP and cycle time, to construction products as we apply to manufacturing products. The full definition is:

- The construction equivalent of a manufactured product is a distinct *constructed space*.
- A *constructed space* is clearly recognizable to any customer or occupant as a space that has specific utility and value, such as an office, bedroom, hallway, bathroom, classroom, attic, foyer, etc. Vertical zones, like façades or shafts, can also be considered *constructed spaces*.

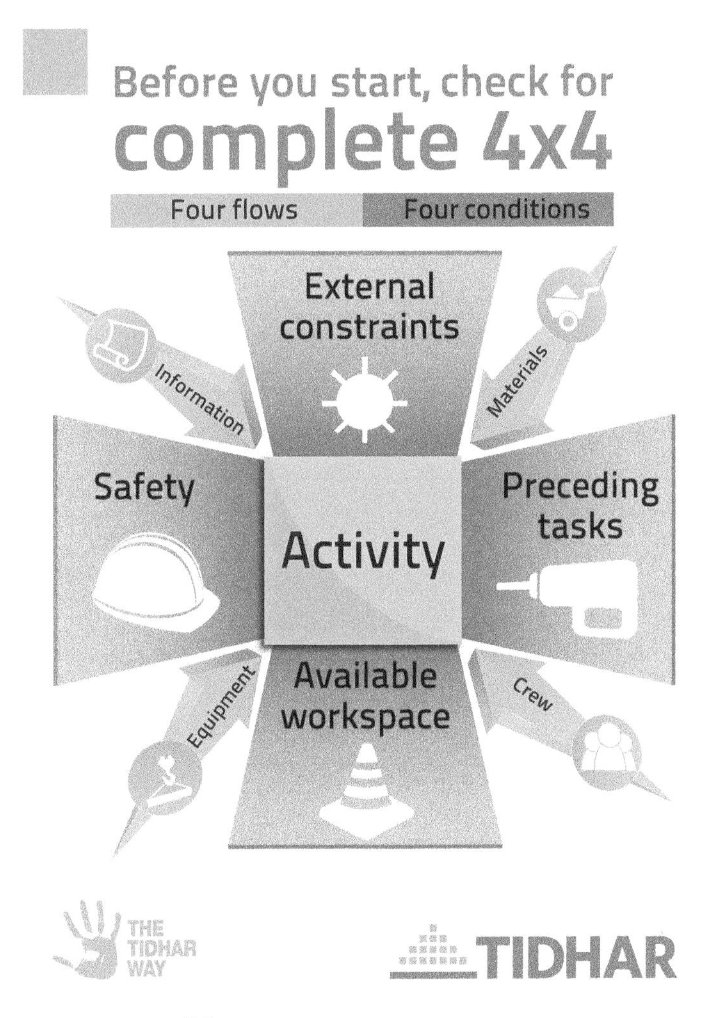

Figure 15.4 Tidhar 4 × 4 preconditions poster

- *Constructed spaces* are parts of larger assemblies: rooms are parts of apartments, which with lobbies and other spaces are parts of floors, which in turn are parts of buildings. We can think of hierarchies of aggregation of constructed spaces.
- *Constructed spaces* are themselves assemblies of components. The parts of rooms are walls and ceilings, doors and windows.
- The *constructed spaces* are the object of planning, so that the level of detail or resolution of *constructed spaces* and their boundaries are defined according to the level of resolution of the analysis applied and according to the construction method at each stage of construction. For a master plan, buildings within a project or floors within a building may suffice; for a weekly work plan, apartments or rooms may be appropriate. Similarly, a breakdown into horizontal structural slab spaces and vertical structural wall and column spaces will be appropriate for structural works, whereas a breakdown into rooms will be appropriate for interior finishing works.

- The common property of all of the *constructed spaces* in a set is that they all undergo a series of operations, performed by a sequence of trade crews, through which they move in time, if not in space. This flow of *constructed spaces* is the construction equivalent of Shingo's flow of products along the *process* flow axis.

Figure 15.5 A 12-storey building under construction at the Tidhar on the Park project in Yavne

Note: Although the structure has reached the eighth floor, no interior finishing works have begun. The WIP of finished floors is seven and the average cycle time for each apartment is growing as the WIP accumulates.

Consider the photograph in Figure 15.5, which shows an apartment building under construction. Although the structure has reached the eighth floor, no interior finishing works have begun. Constructed spaces are added to WIP as soon as their structural work is begun, and they remain part of the WIP until they are handed over to their clients. The cycle time of each constructed space includes the period during which they are part of the WIP. In the building in the photo, the floors are part of WIP, but they are waiting for the next operation and the waiting time is part of their growing cycle time.

One could argue that a particular level in the construction product hierarchy should be the primary product considered in production system design and in construction planning and control, and the right level may depend on the resolution of planning and control and on the phase of construction. Yet whatever level we choose to adopt for any given situation, the statement that "*constructed spaces are the primary flow in construction*, equivalent to the flow of products in manufacturing" opens up a way of thinking about production in construction that is central to Lean Construction.

Peculiarities of production flow in construction

Apart from the fact that constructed spaces have fixed locations (i.e., unless something goes very wrong, they don't move), the other significant difference between constructed spaces and manufactured products is that more than one trade crew can work in a constructed space at the same time. This phenomenon has been called "stacking of trades"[11] and "workstation congestion",[12] and it generally reduces productivity.

In the manufacturing metaphor, this would be the equivalent of a part being worked on by more than one machine at a time. Production flow models assume that this does not occur. The net processing time for a product, T_0, is defined as the sum of the processing times at each workstation in a line, and the minimum WIP is defined as $WIP_{min} = r_b T_0$, i.e. the bottleneck rate multiplied by the net processing time. The principle behind the idea that parts cannot be processed by more than one machine at a time is a consequence of the way production lines are set-up for production flow, and is not necessarily determined by technological dependence between processing steps.

In construction, the processing sequence of constructed spaces is determined by technological dependence relationships for many processing steps, but not for all. Structural works commonly dictate that a lower space must be built before the space above it can be built. However, this is not necessarily the case for many interior finishing and building system trades. Like manufactured products, this means that some steps can be interchanged, but unlike manufactured products, it also means that some steps can be performed simultaneously. One consequence is that T_0 can be less than the sum of the individual processing times, which means that if some of the trade crews can work in parallel in the spaces then WIP_{min} can also be smaller than would otherwise be the case.

Common wisdom in construction planning tends to ignore this reality. In practice, planning for simultaneous work by different trade crews in the same spaces is rare. This stems partly from the perception that productivity is reduced when work areas are crowded, and partly from a tendency on the part of planners to apply "finish–start" constraints between all activities as the default.

Many planners, used to the critical path method way of thinking, also tend to apply finish–start constraints to implement resource constraints (to prevent allocation of more resources than those available). This is a mistake because it unreasonably constrains the critical path method (CPM) solver to schedule task A before task B, ignoring the possibility of performing task B before task A. But more insidiously, this way of thinking prevents planners from considering opportunities where tasks can in fact be overlapped or performed simultaneously, whether by integrating trade crews or simply allowing them to work in parallel in the same spaces.

Box 15.3 Construction completed: a house in 3 hours, 26 minutes and 34 seconds

The world record for shortest duration for building a house is broken every few years, but it was held for some time by a Habitat for Humanity effort that built a single-family house in 3 hours, 26 minutes and 34 seconds in Shelby County, Alabama, in 2002.[13] In 1982, a Californian residential real estate company, Woodhaven Developers, completed two houses, each of 180 m^2, in Sunnymead, California,[14] in 4 hours, 18 minutes and 4 hours, 28 minutes respectively. Each house was built by some 350 to 400 workers, employed by 20 to 30 subcontractors. These projects, although unusual, underline the fact that very short cycle times, with single-piece flow and a WIP of one are possible in housing construction. Trade crews' tasks can be overlapped if the planning is sufficiently accurate and all resources are available with 100 per cent reliability.

In Lean single-piece flow is considered to be ideal. The single piece is a practical lower limit on the batch size. In construction, because the space used for planning is in and of itself an assembly of smaller spaces, one can plan for flow at a level of resolution smaller than the constructed space. More than one trade crew can work in a single apartment (in different rooms or on different systems), and more than one crew can work in a continuous space, such as a façade. The practical lower limit for flow batch size in construction is apparently only restricted by the physical space required for a single worker and/or piece of equipment, and it can be a fraction of the constructed space that is used as the unit of planning.

Batching of constructed spaces

Trade crews always prefer to have as much space available for their exclusive use as possible, because, in general, their perception is that this allows them to maximize their productivity. Large exclusive spaces allow them to:

- work without interference or obstruction from other crews – the constraint of available space to work in is reliably removed;
- maintain a large buffer of ready work that is independent of the variability of the work of the trade crew that precedes them, allowing them to smooth variations in the work content to maintain high productivity of each of their workers – the constraint of complete preceding tasks is removed;
- store their materials close to the work space and for extended times, and in big batches, removing the constraints of material supply;

- maximize the learning curve benefit within their own work;
- offset the set-up times for any equipment or procedures that require long set-ups;
- modulate their resource assignments independently of the other crews.

However, despite the localized benefits for the subcontractor, this practice works against the interests of the project as a whole. Assigning large batches of constructed spaces to any trade crew lengthens the cycle times of the spaces and therefore the duration of the project as a whole. It also increases WIP and management effort.

In some situations, large batches of constructed spaces can also degrade the productivity of the trade crew. When spaces are too large, the distances over which materials need to be moved become long, so that workers spend more time on non-value-adding activities. One can consider a whole project as a unit of production in what we call *portfolio flow*. Trade crews are not limited to a single construction project nor to a particular general contractor. Therefore, there is a loop – project flow impacts trade crew flow.

Re-entrant flow in constructed spaces

In construction re-entrant flow is the rule rather than the exception. When a trade is required to leave and later return to the same constructed space for different process steps, it is literally "re-entering" the same constructed space for a second (or third or fourth…) time. In residential and commercial construction, for example, the return of drywall, plumbing, electrical and other trades multiple times to the same location is an inherent feature of the construction method.[15,16]

A value-stream map of Tidhar's process for interior fit-out of a typical apartment, before Lean processes were introduced, revealed that there were 18 trades and that they handed the apartment over from one to another no less than 44 times. This means that, on average, each trade had to work in each apartment two and half times. Rework to correct defects, to revise work performed prematurely due to "push" control, or as a result of late design change, meant that in reality there were even more returns. Reducing the number of times a trade crew has to return requires changes to construction methods or reassignment of different tasks among the crews.

In a recent project, Tidhar has managed to reduce this to 12 trades and 28 handovers. This was achieved by:

- routing the electrical and plumbing systems through the false ceiling, thus obviating the need for a raised flooring system throughout the apartments, which meant that the system trades could pass through the apartment just once;
- building the interior partition walls using drywall, thus eliminating the need for block masonry work and associated trades such as rendering/plastering;
- self-performing waterproofing, light plastering, window sills and other miscellaneous work with a multi-skilled crew.

The problem with re-entrant flow is that it can easily create a bottleneck that can take construction managers by surprise. Consider the case of construction of drywall partitions or ceilings. This typically requires the following steps:

- construction of the frame by a drywall crew;
- installation of electrical, plumbing and other conduits by their respective trade crews
- closure of the partition by the drywall crew;
- installation of finished end units (sockets, faucets, sanitary ware, etc.) by the system trades.

Now imagine a drywall crew begins work in a residential building, working from the ground floor up. At some point, the first constructed space will be ready for the third step, closing up the drywalls. From the point of view of the project, the best thing to do is to keep the process flowing, with short cycle times for apartments and small buffers, keeping WIP down; this would lead to the shortest possible duration for the project as a whole. So, the superintendent would encourage the drywall crew to go back from whatever floor they had reached, to begin closing up the partitions' second sides. If the drywall trade crew follows this imperative, then they must either draw workers out of the first task, or add additional workers to the crew. If they draw workers down, as is very often the case in practice, then the first task becomes a bottleneck because its rate slows. This has an impact on all the system trades, causing them to wait and reduce their productivity, and also extending the cycle times for the constructed spaces where work is slowed or interrupted.

So, to prevent "starvation" of the subsequent trades, the crews of a trade with re-entrant workflows should be effectively shared between operations that "open up" new constructed spaces for work and operations that "close out" other constructed spaces. Construction managers should be on the look-out for this phenomenon, and be ready to increase trade crew sizes when the first re-entrant steps are reached.

Variety and variation in constructed spaces

Another important observation is that a construction project is not homogeneous in terms of its constructed spaces. Different types of constructed spaces require different approaches to production planning and control.

First, as distinct from manufacturing, the work in a construction project includes establishing the production facility (i.e. "building the factory") on site, where no production facility existed before, and progressively dismantling the facility as the project winds down. This typically includes set-up of cranes, concrete formwork, scaffolding, stores, fencing, offices, etc.

Second, the various parts of a building are different in nature and so are the processes through which they are built. Structural work has a fixed sequence of locations that is dictated by technological constraints because earlier parts support later parts. Building systems (mechanical, electrical, plumbing) do not have fixed technological dependence between their locations, nor do interior finishing works. Exterior envelope works may be part of the structure or they may be independent of it in terms of construction sequence. Constructed spaces belonging to each of these (structure, systems, finishes and exterior) may therefore be classified as different types of constructed space. The implication is that their production flows are also different, and therefore the interfaces between them often exhibit significant buffers of time and/or space.[17]

Finally, there is often variation in the work content of constructed spaces that appear to be similar or even identical.

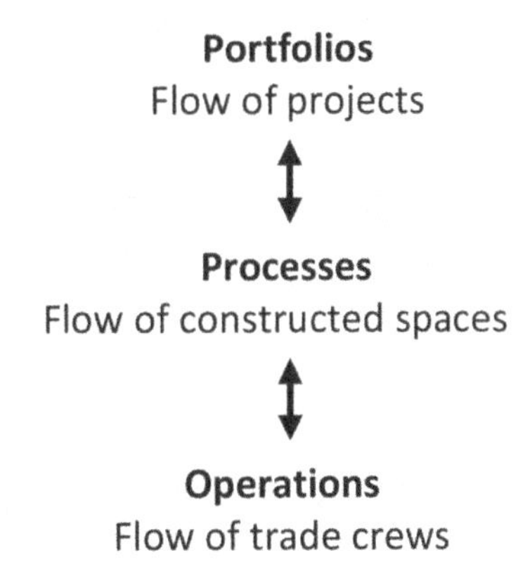

Figure 15.6 Portfolios, processes and operations as a hierarchy

The portfolio, process and operations (PPO) model of production in construction

Up to this point, we have considered two axes of flow – **process flow**, the flow of products in manufacturing and of constructed spaces in construction; and **operations flow**, the flow of workstations in manufacturing and the flow of trade crews in construction. Yet given the explicit "project" nature of construction, we can consider a third flow, the **portfolio flow** – the flow of projects in a company's current portfolio of construction projects.

Adding the portfolio flow view has the benefit of focusing attention on the fact that a company or an owner can consider the flow of projects in a portfolio in much the same way as one considers the flow of products in a production line or the flow of constructed spaces in an individual building project. In this view, the projects are independent products, they can be counted as WIP and they each have a cycle time. The throughput for a company is the number of projects completed in any given unit of time. There are clear and apparent economic benefits to keeping the cycle times of projects as short as possible; applying Little's Law, this means that the number of projects operated in parallel should also be controlled. Where a company plans a certain throughput level, new projects should only be started when the capacity of its own resources and that of its subcontractors allow.

Portfolio flow can also be considered from a second perspective: the portfolio of projects that a trade crew has, in which the projects are the set of operations that the crew performs over time. From time to time subcontractors plan and allocate their resources to the various projects in their portfolio, and their considerations in doing so are quite different to the considerations of the project managers at each project. The flow of trade crews between projects is distinct from the flow of trade crews within a construction site. For the subcontractors, the major sign of waste in the portfolio flow is an idle crew.[18]

Work flow in construction can therefore be understood as functioning on three interrelated axes: portfolio, process and operation. In this model, trade crews are considered to flow not only from constructed space to constructed space within a project, but also from space to space across projects. Subcontractors perform work for multiple general contractors across an economic region, encompassing the portfolios of multiple companies. Opportunistic behaviour on the part of subcontractors, as they shift resources from project to project, introduces instability that restricts project managers' abilities to plan ahead. This reflects an interdependence between operations and projects, so that the ends of the linear hierarchy must be joined in a cyclical dependency relationship. In Figure 15.7 the full circle emphasizes the interrelationship between the portfolios and the operations flows.

In standard line-of-balance charts, locations (constructed spaces, in our terms) are plotted as horizontal strips and the progress of trade crews through the locations is shown using inclined lines that represent the crews' flow through the locations as time advances from left to right. If the vertical width of the strips is proportional to the work content in each location, then the slopes of the trade crew flow lines represent their production rates. Figure 15.8 shows a line-of-balance chart for a single project, with the flows of information, materials and crews into the project.

In Figure 15.9 a project portfolio axis is added, and multiple projects are shown. In this case, the competition for design information, materials and other resources between projects is apparent.

The primary management functions in the list below reflect the three axes, and details aspects of the three axes that are commonly observed in construction projects:

- The **project manager** is concerned with delivery of the project as a whole, uses critical path planning to set milestones and operates through contracts with subcontractors and suppliers.

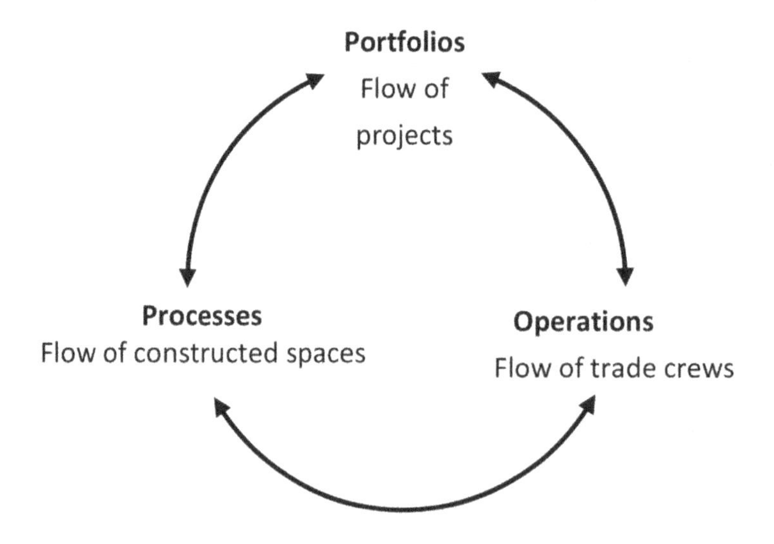

Figure 15.7 Portfolios, processes and operations form a closed loop

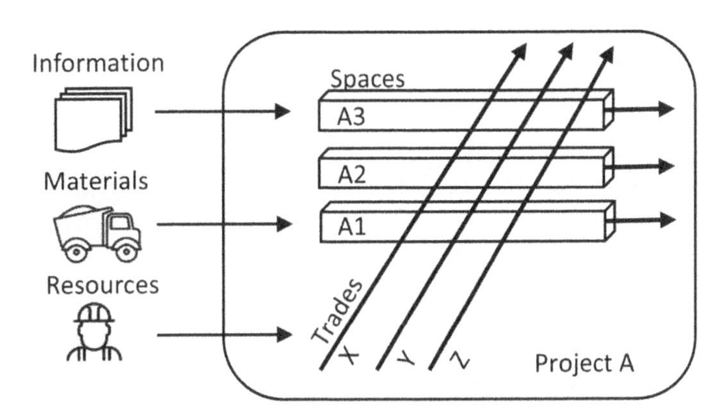

Figure 15.8 Single project line-of-balance chart showing flows of trade crews through locations

- The **works manager** (site superintendent) manages the process, focusing on advancing work to complete the spaces within the project. The works manager uses production control methods such as LPS and tends to build buffers of capacity and materials to ensure continuous work in the locations.
- The **subcontractor trade crew leaders**[19] (operations managers) try to ensure high productivity of their crews through continuous employment, often by evaluating the scope of work likely to be made ready across multiple projects and by allowing buffers of locations ready for their trade to accumulate.

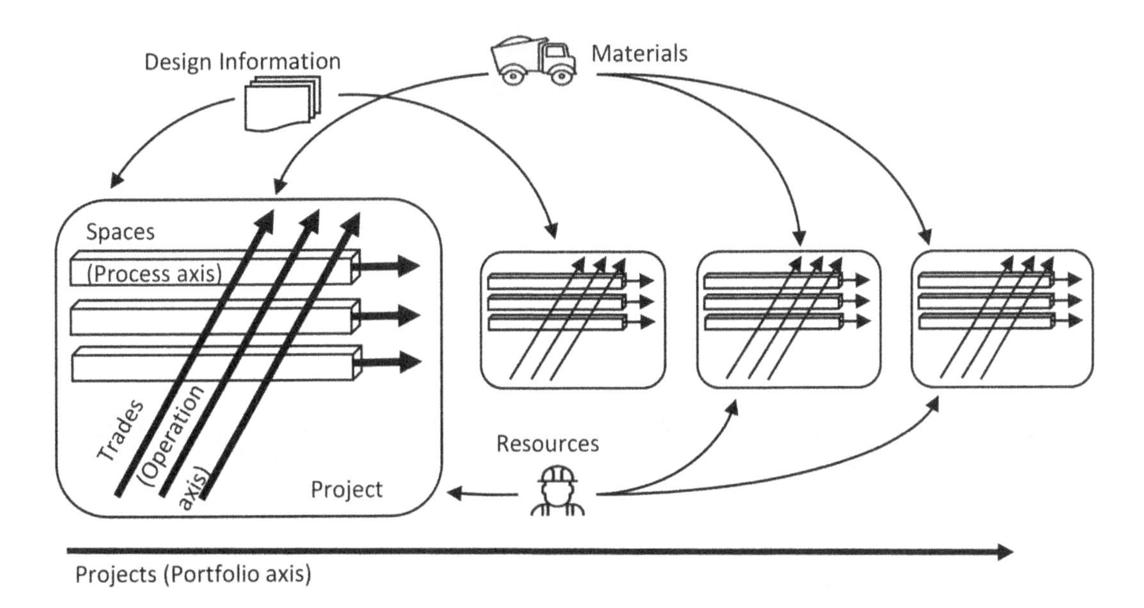

Figure 15.9 Multiple projects compete for resources

Table 15.1 Aspects of the portfolio, process and operations axes of the PPO model

Aspect	*Portfolio axis*	*Process axis*	*Operations axis*
Flow object	Project/building	Constructed space	Trade crew
Cycle time for flow of a single object	Full project duration	From start of work to delivery to client	From first to last day of a crew's work on a project
Optimization targets	Project duration and cost	Flow of locations, reduction of WIP, minimum cycle times, quality	Flow of trade crews in and between locations, continuous work, productivity, safety
Management function	Project manager	Works manager or superintendent	Subcontractor, trade crew leader
Planning and control tools	CPM; contracts	Location-based planning; LPS	Operator balance charts; standardized operations; LPS
Symptom/signs of waste	Budget overrun and/or schedule overrun identified using "earned value" measures; defects; idle crews off site	Unoccupied spaces (spaces with no work in progress); crews absent from site; delayed materials; delayed design information; rework	Idle crews on site; crews waiting for work; small work completion packages; rework
Tactical approach	Contract negotiations; bonuses and fines	Build excess capacity; coordinate across trades	Allow buffers of locations to accumulate before assigning resources; understaffing
Scope of planning and control	Single project	Work or product type (e.g. structure, building systems, interior finishes)	Operations (specialized trade work)

Source: Sacks (2016).[20]

Addition of the project portfolio axis reflects the fact that design and construction occur simultaneously across many projects in any given regional economy. Unlike neighbouring factories, each of which have their own and essentially independent production resources, construction projects in any given economic region are co-dependent on the same subcontractors and their labour.[21] Subcontractors balance their workload across projects, creating a flow of labour between the operations of different projects. Trade subcontractors attempt to achieve unbroken utilization of their crews, even if this requires shifting between projects from week to week or from day to day, resulting in discontinuous flows of space from the project perspective. Designers do the same thing, with the result that the flow of product information (drawings or models) is commonly discontinuous and unstable from the project point of view.[22] In this view, continuity of trade crew flow can be said to be achieved at the expense of the flow of constructed spaces.

Constructed spaces and location-based management

The notion of a space or a location as a unit of planning in construction projects is not new. In the records of the construction of the Empire State Building, for example, a two-storey "tier" was defined for planning, and the progress of the different trades through the building was scheduled to achieve near uniform production rates for each trade in each tier. In 1974, Pe'er formalized the idea of a balanced production line, with trade crews working with equal durations in each of a series of locations, as the "line-of-balance" method; he expressed the view that construction planning must begin by determining a critical operation, and then aligning all other operations with that one, leading to a "line-of-balance" schedule in which all operations become critical, or near critical, by design.[23]

In their book, *Location-Based Management for Construction*, Kenley and Seppänen[24] suggested that the construction metaphor for product flow is the flow of locations, which although they do not move, are analogous to the products moving down a production line, onto which incoming parts and materials are assembled in the operations performed by trade crews. In the same way that a stationary observer can see the flow of physical products pass through a machine in a manufacturing plant, so might an observer moving with a construction trade crew observe the relative flow of locations flow through that trade. Locations in this sense are the same as constructed spaces.

Location-based planning and control calls for construction planners to define tasks as the execution of a work method by a crew in a location. The task dependencies are defined by defining technological constraints between work methods within spaces (like the need to build a partition wall before it can be painted) and between dependent locations (like the need to complete the first-storey structure before proceeding to the second storey), but not by defining them between tasks. In planning and control software based on the location-based approach,[25] the system derives the tasks and the task-specific dependencies automatically, which helps avoid the mistakes made by construction planners who detail tasks directly using CPM. It also makes the location explicit as a resource which should not be over-allocated, avoiding assignment of more than one trade to a location at a time. Perhaps the major benefit of the location-based planning and control approach is that it elevates the concept of the location to the same level as the work methods, the resources and the tasks in the minds of construction planners.

What constitutes good flow in construction?

With a clear idea of what flows in construction, we can ask two practical questions: "*What is good flow in construction?*" and "*How can production planning and control lead to good flow?*"

Using the PPO model as a starting point, it is apparent that achieving good overall construction flow implies simultaneously achieving good project flow, location flow and trade flow. Given the distribution of control over these flows across owners, general contractors and subcontracting companies, collaboration appears essential. Considering the cyclical PPO model, not only should GC project managers take a direct interest in achieving continuous flow for their trade crews, so should project portfolio managers consider the spread of subcontractors across their projects (and across the regional industry as a whole). This approach stands in direct contrast to the neglect of production control in traditional construction practice.

Ideal conditions for good flow in construction

Box 15.4 sets out ideal conditions for good flow in construction for each of the three axes of the PPO model. Just as SMED was a challenge that inspired Toyota's industrial engineers to reduce the time for changing dies from one day to about ten minutes, this set of conditions should be seen as a target to work towards, rather than as a set of realistic goals. They identify the right direction to work towards, not the destination.

Box 15.4 Ideal conditions representing good flow in construction according to the PPO model

Project portfolio conditions (project flow)

- The cycle time for all projects is minimized.
- The work-in-progress inventory in a company's portfolio is kept to a minimum.
- The batch size, measured as the number of distinct projects managed by the same management team, is one.
- Projects move from development to construction at the last responsible moment, in response to pull from customers.

Process conditions (location flow)

- Balanced work: the variation of Takt time across locations for all trades, measured as the standard deviation of the average number of locations completed per unit of time for each trade, is zero.
- The batch size, measured as the number of locations occupied by a crew, is one.
- The sum of the time buffers between trade operations is zero for all locations.
- The number of operations has been reduced to a minimum.
- There is no re-entrant flow.
- There is no rework.
- The work flow is reliable: only work packages with mature constraints are released to operations. This also ensures that "making-do" is prevented (see Chapter 17, "The Last Planner® System", for more detail on making do).
- The number of locations with work in progress is equal to the number of trade crews (i.e. all locations are actively being worked on) at all times.

Operations conditions (trade flow)

- Stable production rates: the variation within each trade's Takt time (multiple of production rate and work quantity per location), measured as the standard deviation of the number of locations completed per unit of time, is zero.
- The operation time for each trade is reduced as far as possible (zero set-up and inspection times as well as minimal non-value-adding time).

Source: Sacks (2016).[20]

For real projects, evaluation of flow must be relative, not absolute, because no actual project or portfolio of projects can fulfil all of these conditions. Flow can be considered to have been improved when the project, process, and operation have moved closer to the ideals listed in Box 15.4. From an implementation standpoint, this requires the ability to measure the quality of flow in construction.

Unfortunately, the PPC measure that is used in LPS is the only measure in common use in construction projects where Lean Construction is practised, and it is an indirect measure of flow at best. A measure of construction flow quality, called the Construction Flow Index (CFI) has been proposed but it has yet to be applied broadly in practice.[26]

Achieving good flow – some advice and words of caution

There are various tools available for improving location flow, some of which naturally deal with aspects of project flow and of trade flow. Lean Construction tools, such as VSM, LPS[27] and Andon boards are among recent innovations in construction. Tools based on BIM also contribute to improved flow: such as model checking applications, hardware and software that provide access to product information using BIM on site, and process visualization and management applications that integrate BIM with project status data.[28] Yet despite the plethora of tools, one should not lose sight of the principles. As Henry Ford commented that the "conveyors are only one of many means to an end", so should Lean and BIM tools be seen as some of many means to achieving good flow.

Achieving good location flow and trade flow simultaneously is difficult, due to variability of the work content and the instability of the supply chains as well as conflicting interests of the participants. Although productivity is a critical factor in income generation, the key objective for subcontractors is the maximization of income per unit time and not the maximization of productivity.[29] This often leads to work out of sequence on site, which then requires making-do or rework, accumulation of WIP, etc. In an example observed on one of Tidhar's commercial fit-out projects, a supervisor decided to allow an otherwise idle flooring crew to begin laying tiles in an area in which overhead gypsum ceilings were not yet installed. When the ceiling crew arrived, the flooring work was interrupted. Grouting of the joints was not yet finished, which meant that the floor had to be protected with boards and the flooring crew would have to come back later (re-entrant flow). In this example, the flooring crew had marginally better productivity and greater income than would have been achieved had they had to wait. However, they came at the cost of waste and reduced productivity for the project. Another example is in the size of the crew that a subcontractor brings to site. Laufer and Shohet found that it often has to do more with the number of seats in the van driving to the job site than with the optimal flow of the locations.[30] Decisions that may appear irrational are often the result of the narrow focus of local optimization and their results can directly conflict with process flow optimization.

Notes

1 Goldratt and Cox (2014).
2 Goldratt (1997).
3 Little and Graves (2008).
4 Dillon and Shingo (1985).
5 Ballard et al. (2003).
6 Ballard, et al. (2002).
7 Hopp and Spearman (1996).
8 Kingman and Atiyah (1961).
9 Shingo and Dillon (1989).
10 El-Khouly et al. (2011).
11 McDonald and Zack (2004).
12 Koskela (2000).
13 See www.prnewswire.com/news-releases/the-house-that-love-built-really-fast----and-just-in-time-for-christmas-kicker--habitat-for-humanity-breaks-world-record-set-by-new-zealand-77236282.html.
14 See www.awci.org/cd/pdfs/8203_f.pdf.
15 Brodetskaia et al. (2013).
16 Construction of drywall partitions or ceilings typically requires the following steps: construction of the frame by a drywall crew, installation of electrical, plumbing and other conduits by their respective trade crews, closure of the partition by the drywall crew, and finally installation of finished end units (sockets, faucets, sanitary ware, etc.) by the system trades.
17 For example, some interior finishes may require that a building be enclosed and protected from weather. If exterior cladding, such as curtain walls, proceeds by discretized elements of the façade, only achieving enclosure once the last unit is installed, the start of interior finishes may be significantly later than would be the case if an alternative exterior cladding system (installed floor by floor) were used.
18 The tactical approaches to avoid this waste include "overbooking" (i.e. commitment to allocation of resources to multiple projects beyond their ability to supply simultaneously), which is common among trade subcontractors and design firms. For details, see Sacks and Harel (2006).
19 A gender-neutral alternative term for "foreman".
20 The table and some of the figures in this subsection are drawn from Sacks (2016).
21 Bertelsen and Sacks (2007).
22 Tribelsky and Sacks (2011).
23 Pe'er (1974).
24 Kenley and Seppänen (2010).
25 Such as VICO software's planning and control module – see www.vicosoftware.com.
26 For more information on CFI, see Sacks et al. (2017).
27 Ballard (2000b).
28 Sacks et al. (2010).
29 Saari (2011).
30 For details of this finding, which highlights the pervasive influence of local optimization, see Laufer and Shohet (1991).

16 Production planning and control in construction

> Production strategy is a critical component of world-class companies, and a powerful source of competitiveness. The competitive advantage comes from how the production system is designed and how the production resources are deployed.
>
> Source: Adapted from dos Santos (1999).[1]

Tidhar began their Lean journey from an apparently simple question: how could they improve their bottom line? Like almost everyone in construction, they were working hard but still not seeing better results, so the typical advice of "just trying harder" was insufficient. They began to understand that they needed to fundamentally change their processes if they wanted to change their outcomes.

Since its early years, Tidhar emphasized process control, publicizing it as part of their value proposition to customers. Arye Bachar, one of the company's founders, had seen construction control as a differentiator between Tidhar and the competition, and the experience gained in the Aviv Gardens project in the 1990s (see Chapter 1, "Introduction – Growing the Margin") had embedded the idea deeply into the psyche of the company's senior engineers. The Control Department was one of the premier parts of the company, employing some of the most talented engineers. For Tidhar, "control" meant making sure the projects proceeded according to schedule and within budget. By taking the time to check the actual progress of the projects against what had been planned (in terms of both time and money), they could identify deviations early. The idea was to avoid "surprises" at the end of the project, for both the company and its customers. In theory, when delays or overruns were identified, corrective action could be taken to get things back on track. In essence, the two tasks of the "control" function paralleled the second half of the PDCA cycle discussed in Chapter 6, checking to see if what had been done matched the plan, and then taking action to try to correct any deviations.

But what Tidhar came to realize was that the way they were doing "check" and "act" was itself deficient. Most projects were checked once a month, and the measures focused entirely on outcome metrics that failed to shed any light on the root causes of the problems. This lead to "action" that was ineffective in creating long-term improvement, since the root causes were often not fully understood and thus not addressed. In this chapter, we discuss the traditional methods of planning and control at Tidhar, which were typical of best construction industry practice throughout the world. We highlight concepts and approaches where Lean Construction differs from the usual assumptions and methods of planning and control, and we describe the steps Tidhar took to improve its planning and control processes.

The work backlog – Tidhar's traditional approach to planning and control

Like most companies, Tidhar would try to plan out the timelines of its projects. The main tool in use was MS Project, a software based on CPM.[2] Tidhar self-performed the structural work in its projects, with most of the work done by a single trade crew. This phase lends itself to CPM analysis because the logical sequential dependence relationships between tasks, such as "finish–start", are crisp and clear. As a result, the level of planning detail was relatively high. Each task – assembly of formwork, reinforcing, casting of slabs and walls/columns and stripping the attendant formwork – was included in the plan, at a level of detail of days.

However, scheduling the finishing works presented a more complex challenge. MS Project was not applicable in the same way as for the structural works, and the plan did not reflect reality. There were too many fine-grained activities, too many trades, too many crews and too many technical and location-based constraints to create an effective and workable plan. Very often, the production rates and the overall durations of subcontracted work packages were unknown. In some cases, like plumbing and electrical work, the subcontractor was also in charge of supplying the materials, which further distanced Tidhar's managers from the in-depth understanding needed to plan effectively; all they knew was how much they were paying for the finished product of the subcontractor's labour but not how long it might take. So rather than trying to capture all the nuances (some of which they did not know) in the plan, a high-level approach was adopted, with each trade appearing as a summary task spread out over the duration of the project. To fill in more detail (both for planning and control), they adopted a "*work backlog*" approach.

The basic idea of the work backlog approach was to set a target quantity of work for each trade subcontractor for every planning period, so that the trade's progress could be monitored and controlled. The target quantity was set as follows: (1) identify and count all of the work spaces (floors or apartments, whatever was the logical choice for that trade) for each trade; and (2) divide the number of workspaces by the overall duration of the trade taken from the summary task on the master schedule.

This gives the number of workspaces the trade must complete in each time unit of the plan. If a plumbing crew had work in 24 apartments and 6 floor lobbies, and it was assigned a 3-month period, then the crew must finish 10 spaces per month. The rate was assumed to be linear, disregarding learning curves, holidays in each given month, the weather in different seasons, or any of the other complex factors that affect work durations. Figure 16.1 shows an example of a work backlog table for a 21-storey residential building from the Rosh Ha'ayin project (see Chapter 13, "Tidhar on the Park, Rosh Ha'ayin").

Once a month, each trade would report how many units they had finished, relative to the amount they were supposed to have finished according to the backlog table. Any leftovers that had not been completed were usually pushed to the end of the project.

Task	Units	Month																
		05-15	06-15	07-15	08-15	09-15	10-15	11-15	12-15	01-16	02-16	03-16	04-16	05-16	06-16	07-16	08-16	09-16
Partitions	92		15	10	10	10	10	10	10	10	7							
Electrical conduits infrastructure	92		15	10	10	10	10	10	10	10	7							
Plaster - walls and ceilings	92		15	10	10	10	10	10	10	10	7							
Plumbing	92		15	10	10	10	10	10	10	10	7							
Kitchen - cabinet order	92		15	10	10	10	10	10	10	10	7							
HVAC - Piping	92		10	10	10	10	10	10	10	10	8	4						
Sealing	92		5	10	10	10	10	10	10	10	10	7						
Windows sills	92			10	10	10	10	10	10	10	10	8	4					
Flooring substrate	92			5	10	10	10	10	10	10	10	10	7					
Flooring	92			5	10	10	10	10	10	10	10	10	7					
Bathtub installation	92			5	10	10	10	10	10	10	10	10	7					
Bathroom wall tiles	92			5	10	10	10	10	10	10	10	10	7					
Gypsum board insulation	92			5	10	10	10	10	10	10	10	10	7					
Sprinklers	92			5	10	10	10	10	10	10	10	10	7					
HVAC - internal unit	92			5	10	10	10	10	10	10	10	10	7					
Sprinklers drops	92				5	10	10	10	10	10	10	10	10	7				
HVAC - ducts	92				5	10	10	10	10	10	10	10	10	7				
False ceilings	92									20	20	20	16	16				
Painting - first layer	92									20	20	20	16	16				
Aluminium installation	92									20	20	20	16	16				
Entrance door installation	92										15	20	20	18	19			
Sanitary fixtures	92										15	20	20	18	19			
Kitchen - cabinet installation	92										15	20	20	18	19			
Kitchen - marble countertops	92										15	20	20	18	19			
Kitchen - wall tiles	92										5	20	20	20	15	12		
Electrical fixtures	92										5	20	20	20	15	12		
Painting - second layer	92										5	20	20	20	15	12		
Internal wood doors	92													30	30	32		
Bathroom - cabinets and faucets	92													25	25	25	17	
Cleaning and delivery	92																45	47

Figure 16.1 A "work backlog" spreadsheet plan for a 21-storey building with 92 apartments, showing monthly production quotas for each activity and trade crew

The work backlog was used in practice to manage projects, but it is essentially a contract management tool and not a production management tool, because it encapsulates the tasks as "black boxes", hiding the details of how production should, could or did in fact work. A good production management tool is one that gives useful information on a frequent basis, but with this method, there was no opportunity to take action other than at the end of the month.

With such a long PDCA cycle time, it was almost always too late to make any significant corrections. Problems remained hidden well after they flared up and began impacting subsequent tasks. This is part of the reason that most of the shortfall was turned into extensions of the target end date; it was too late to do much of anything else. The information, be it about a budget overrun or timeline delay, was like ready-mixed concrete that had been left standing too long before pouring: it was no longer workable or able to provide value. Furthermore, the nature of the information was geared to measure outcomes, not process, which further reduced its usefulness. For example, if the month end came around and it was discovered that the quantity of concrete consumed exceeded the planned quantity, it was impossible to determine, with any certainty, what had caused the problem. Was the amount of waste higher than estimated due to a construction method problem? Inexperienced workers? Was an extra half-centimetre being poured on each slab (which when multiplied over the entire floor area of the building could add up to a large quantity of wasted material)? Was the supplier not filling the mixers, or perhaps some of the concrete being stolen on the way to the site? In this situation, where the cause is not clear, it is very difficult to take effective corrective action. The project manager would be given an order to "fix the problem", but who could be certain what the problem was, when only the results of what was done versus what had been planned were being examined, and that only once a month? No one ever checked how the actual processes that lead to those results were occurring, so it essentially remained an issue of awareness that there was a problem, a line item in the budget overrun, without any effective tools to make course corrections.

Not only was this like driving a car by looking through the rear-view mirror, it was also like trying to correct the direction and speed by working the turn signals. If a trade was running behind and not meeting its work backlog targets, the gut reaction was to add resources. For example, if three plaster workers were not hitting their monthly goal, the project manager would instruct the superintendent to have the subcontractor add another worker. In many situations, this was not only easier than trying to find the root cause by looking into the details of the operation, it was also expedient because it carried no direct cost – the subcontractors were paid per unit of work completed, regardless of the production rate. Coercing a subcontractor to add workers had no directly measurable cost.[3]

The work backlog approach became deeply ingrained in company practice, and through all of Tidhar's Lean and BIM process change, at the time of writing it is still the basis for corporate management's monthly review of all construction projects. While it may be reasonable to take a "bird's-eye" view of project progress at this level of project oversight, at the site level, the work backlog method has major flaws when seen through a Lean Construction lens. First, it creates *push* which tends to lead to accumulation of WIP. When crews are faced with barriers to progress in any given space, they are incentivized to start work in other locations to improve their chances of meeting their monthly targets (targets which only specify how many units are to be completed, not which specific ones). Problems remain unsolved and accumulate, production rates fluctuate and work is performed out of sequence. Large WIP carries the attendant ills of rework, wasted overheads, congestion and inferior flow. Second, where corrective action is reactionary and delayed, short-term solutions are preferred. Corrections do not improve the underlying processes, since they address the symptoms but not the root causes. Continuous improvement is stifled.

Traditional versus Lean concepts in planning and control

Critical path method versus location-based management

CPM models activities as "black boxes" that encapsulate all of the production details and reveal only the total duration. It models the relationships between activities as precedence conditions, hiding all the non-value-adding activities that accompany any core activity from view. CPM does not consider fluctuations in production rates, and it cannot deal with indeterminate or fuzzy dependencies. As a result, it is far from ideal for the typically "messy" production that is common in most construction projects following the structural work.

The typical output of CPM planning is a Gantt chart, as shown in Figure 16.2. One of the problems with Gantt charts is that they often obscure more than they reveal. Each activity is represented as a bar on the chart, quite literally resembling a metaphoric "black box" – disconnected from locations and with no clear information about which trades are responsible for the activity. Grasping information about relative rates of work or work continuity is difficult and far from intuitive.

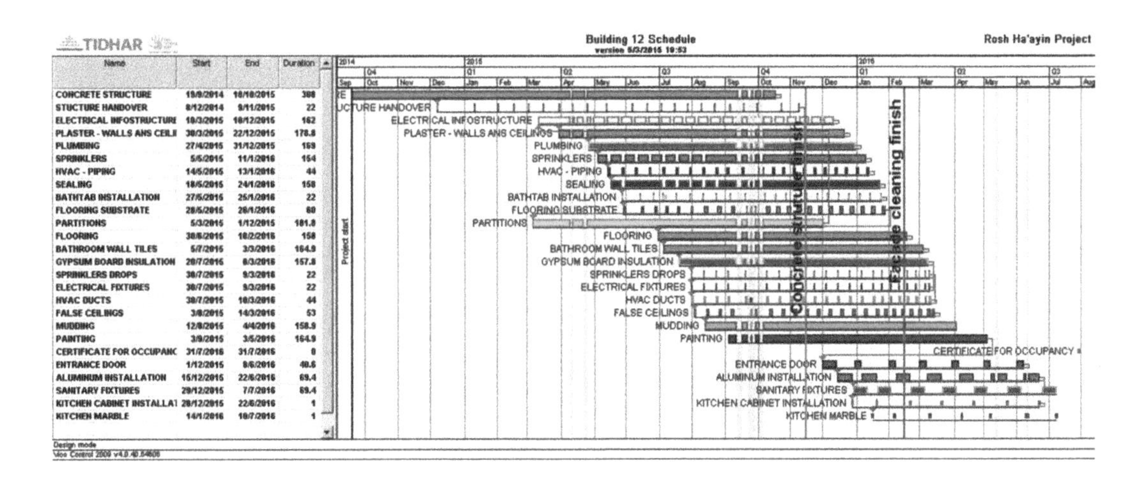

Figure 16.2 A Gantt chart showing part of the activities for a 21-storey residential building

Location-based planning,[4] in contrast, models organizational dependencies between tasks, such as the need for work continuity. In location-based planning, the project is broken down into physical subcomponents (called locations), in which all activities take place. It uses a "location breakdown structure" to model the project as a hierarchy of locations, and associates the activities defined in the "work breakdown structure" with locations. It models the gaps, or buffers, between trades, and it enables the use of highly effective line-of-balance charts, such as shown in Figure 16.3.

Line-of-balance (LOB) charts[5] are far easier to read than Gantt charts. The locations are displayed on the Y axis and the project timeline appears on the X axis. The trades are shown as sloped lines, crossing each location from the start date to the end date in that location. In this way, one can quickly identify waiting times (empty areas on the LOB chart), mismatched rates of work (lines of different slopes) and lack of work continuity (non-continuous lines for a given trade from location to location).

A major benefit of location-based planning and control is that the locations can be considered to be products of the production system. This idea is explained in detail in the previous chapter. Modelling the product is the key conceptual advantage of location-based planning over CPM.

The thorough visual model of the project's locations and activities is in line with the Lean principle of visual management, which seeks to convey information in a visually explicit way so that stakeholders (workers, managers, etc.) can rapidly grasp the status of the production system at any given time. For example, trade crews can understand how their work is interconnected with the work of the other trade crews and with the progress of the project in general.

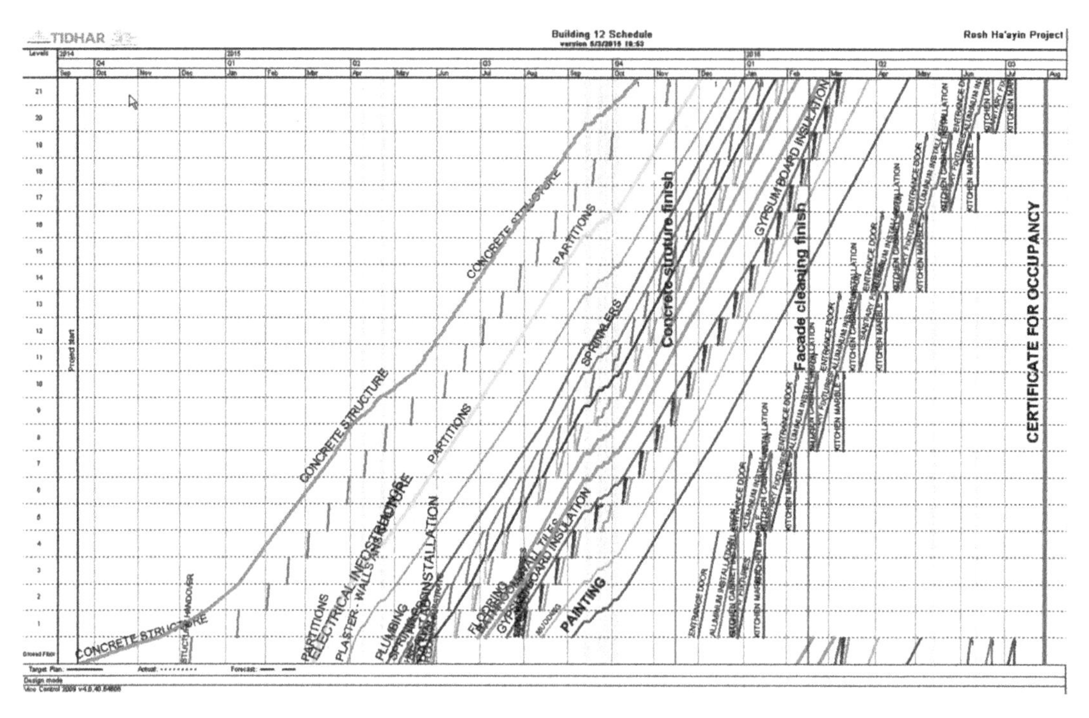

Figure 16.3 A line-of-balance chart for the same 21-storey residential building shown in Figure 16.2

Note: This chart shows all the activities, identifies their trade crew and reveals their production rates, and their locations can be easily identified.

As a planning tool, different scenarios can be displayed and weighed one against the other. There is a greater density of information in an LOB chart than in a Gantt chart, which makes it a more useful tool for improving or optimizing the plan. By specifying and displaying the locations, the project participants are guided to not only maintain a consistent pace of work but also to do so in the correct locations in the project. As a control tool, an LOB chart reveals potential future problems. For example, two converging lines might reveal a conflict where two crews might both try to work in the same location at some future date, unless their current production rates and/or plans are not adjusted. Similarly, situations in which a crew might be forced to make an unplanned break in its work flow can be predicted.

Balance of power between the general contractor and subcontractors

The power balance between the GC and the subcontractors, and the way it plays out in terms of the fine-grained work planning, is another area where traditional and Lean Construction planning and control methods diverge. In traditional methods, the GC's superintendents dictate to the subcontractors what work should be done, paying little or no attention to the constraints placed on the subcontractors. The subcontractors do have the ability to "push back" if they are incapable of meeting a particular requirement, but in general, it is a "top-down" power dynamic. This stands in contrast to the approach adopted in LPS (see Chapter 17 for more details), where each subcontractor is a full partner in production planning, bringing their constraints to the table and committing only to the work that they can do reliably.

Batch size

In the manufacturing world, batch size refers to the number of products that travel together between workstations. For example, moving products in lots of 50 or 100, or however many fit on a forklift pallet. Lean thinkers are always on the lookout for ways to shrink batch sizes, to attain the Lean ideal of "one-piece flow" (make one, move one). Reducing batch sizes reduces waiting, lowers the amount of WIP inventory, reveals quality problems more quickly and in general helps to reduce the amount of waste in the production system.

In construction, the trade crews are the workstations and spaces in which they work to build the building are the products. The batch size is thus the total number of locations that are exclusively assigned to a trade crew to work in, and on, at any given time. Thus "shrinking the batch size" in the construction context means reducing the number of locations assigned as a batch. In the worst-case scenario, a trade crew may be allowed to spread its work throughout an entire project. Instead, restricting a crew to the smallest number of spaces possible is preferable.

The flip side of shrinking batch size is that it creates a more tightly coupled system, which means that if any of the links encounters a problem, the whole chain is affected. Batch size needs to be reduced gradually and in a measured fashion, and is dependent on working to improve the reliability of all parts of the production system (materials, subcontractors, information, etc.). Doing this will shrink the lead-time of the project, since the locations will have less waiting time between trade crews.

At Tidhar, the traditional typical batch size for the structural work and the first wave of interior finishing works (the bulk works – plastering, flooring, partition, etc.) was one storey. The batch sizes for the second wave (installation of components fabricated off site) and the third wave ranged from a storey to an apartment as the spaces became more clearly defined. However, where production rates are high – such as installation of interior doors, gas supply lines, etc. – superintendents tended to allow work to accumulate on numerous storeys before calling in the trade.

At the Yavne project, the batch size for some building systems and interior finishing trades was reduced to half a storey (two apartments), and this significantly reduced the project timeline. Some of the trades objected, explaining that they needed to spread out over an entire storey with their materials, juggling workers around to try to create their own local optimizations.

This issue is still one of contention within the company, but the vision remains clear: the batch size should be reduced to a single product, which in the case of Tidhar's customers, is a single apartment. As we will see in Chapter 19, "The Fira way", single-piece flow can have far-reaching benefits, particularly when one considers flow of products beyond the boundaries of single projects. In Tidhar's case, achieving single apartment flow appears to require significant reconfiguration of the production system, in particular of the alignment of trade crews into multi-skilled teams. This the subject of ongoing research.[6]

Standardized work

As discussed in Chapter 6, "Continuous improvement and respect for people", one of the pillars of Lean is continuous improvement. Continuous improvement means every day, striving to do things just a little bit better than the way they were done the day before. For continuous improvement to be effective, however, the organization must simultaneously work to create standardization in its processes. This is because without a standard way of carrying out tasks, there is no stable base from which to measure the impact of the improvement. Where there is no standard, when one person improves a work method, others continue as before, and the full value of the improvement remains unfulfilled.

In construction, where a GC may self-perform little to none of the value-adding work in the project, there are obvious barriers to creating standardized work and making sure that the subcontractors follow it. Even if the GC establishes a standard, the subcontractor may bristle at the suggestion that the GC thinks they know how to do the work better than they do, particularly if the subcontractor perceives the standard to be a less efficient way of working (which may indeed be the case, since the goal of the GC should be global optimization, while the subcontractor in general favours local optimization).

And yet, if standardization can be implemented (and achieved!), the benefits to the quality of the work process and finished product can be great. A standard represents the best way that the company currently has for carrying out a particular process, the one that yields the best results with the least amount of effort and the one that generates the best overall production flow. If and when a better way of working is found, then the standard should be updated accordingly, so that the better way can become the new standard.

Tidhar set-up a process quality department for many of these exact reasons. The department creates work instructions and procedures that try to capture the years of accumulated "know-how" about the best way to work and the best way to fabricate buildings and apartments. The standards often take the form of checklists or control reports that serve to make sure all the important points have been addressed at each stage of construction. Tidhar hires engineering students to carry out the quality control and tasks them with going through the building under construction and checking each bullet point (these students are part of a special programme at Tidhar where they are hired after graduation; after spending time working in the quality control department, the quality of their work as engineers is much higher). Each apartment and trade gets a quality grade as the result of these inspections, which can then be displayed visually for the entire project. If one of them "fails", they need to demolish and correct the errors (as opposed to smaller errors which do not require demolition).

From project management to production system management

One of the big conceptual leaps at Tidhar has been a change in thinking from "we build construction projects" to "we deliver 2,000 apartments to customers every year". This is a dual conceptual change, because it shifts focus from the transformation view to the flow and the value views. The tools and approaches of project management are quite different from those of production system management. The former emphasize management by contract, while the latter emphasize production flow. The focus on the product (and on the customer) emphasizes value.

To make this leap, they had to address questions such as: what is the product that we build? What is the nature of the production system we want to design and apply? Instead of thinking of each project as a one-off event (which naturally leads to reinventing the wheel each time and much "fire-fighting" and improvization along the way to meet schedule and budget targets), the thinking must evolve into seeing how they, as a company, produce a fairly standard product with lots of repeatability in the system. This leads to seeing how to make system improvements that can pay dividends from project to project. In many ways, this is moving from a "craft" production approach to more of a "mass" production approach (which has the added benefit of being an environment where the tools of Lean can flourish).

Box 16.1 Craft construction, industrialized construction, mass construction, Lean Construction

In technological societies, buildings and other facilities are sophisticated and complex, with a rich variety of architectural forms and technical systems that fulfil the building's multiple functions. They require many different and highly specialized trades for their realization, using sophisticated materials and assemblies, and they are dependent on voluminous detailed information. Construction therefore requires organization, planning and coordination of production. These functions are at the heart of production system design in the context of construction management.

Construction transcends individual projects, and patterns of production system behaviour emerge for different project types. These patterns are typically characterized by the size and scope of commercial companies, preferences regarding specialization or generalization (flexibility) of labour and equipment, location of production (on site or off site) and degree of dependence on information. Each such pattern, or "*construction production system*", is the evolutionary result of the managers of the individual enterprises in the industry learning how to maximize their own utility within the context and constraints of regulation and the economic environment. Three types of construction production systems are commonly discussed: craft construction, industrialized construction and Lean Construction. A fourth system type, *mass construction*, has also been identified.

Recognizing these basic production systems and their interactions is a first step toward devising and applying effective methods to manage the complexities that arise in construction projects. In practice, most modern construction projects are procured in distinct parts whose supply chains are varied, belonging in and of themselves to one of these types. Most project production systems are heterogeneous; as whole systems, they cannot be classified as conforming to any one typical production system. As a result, construction project management can be especially difficult if the ways in which the systems interact are imperfectly understood.

Craft construction

Construction of a unique private home is a good example of craft construction. Craft construction requires highly skilled labour, capable of performing work on site with high degrees of specialization. Communication is usually simple and management provides relatively unsophisticated and short-term directives. Quality is dependent on the craftsman's attitude, skill and experience. Design often evolves with construction as the customer is commonly directly involved and flexible. Logistics are straightforward, mostly involving simple raw materials.

Industrialized construction

Construction of a precast reinforced concrete parking garage is a good example of industrialized construction. With the advent of industrialization, construction could benefit from the ability to mass produce components off site. Some components could be conveniently manufactured off site and installed later as complete units, such as structural steel frame components, doors and windows, precast concrete pieces, etc. Industrialization demands semi-skilled labour, performed off site in factory conditions. Low levels of specialization are required, and the off-site management can provide short-term directives that are based on sophisticated and detailed mid- and long-term production planning. Quality assurance can be applied methodically and systematically. Both manufacturing technology and information technology are central to industrialized construction. Design must be completed in advance and is not flexible to change.

Mass construction

A detailed study of the production system used in the Empire State Building led to the definition of the term "mass construction" to describe the production system that was employed.[7] Like craft construction, mass construction is performed on site, but it has some unique features:

1 multiple uniform and repeated spaces or modules;
2 work flow planned to Takt time;
3 industrial supply chain management;
4 short-term, close monitoring and control of production rates;
5 carefully designed logistic systems to deliver materials;
6 standardized work;
7 minimal variety of parts;
8 careful control of tolerances between parts.

The Levitt towns built in the US soon after the Second World War are another good example of mass construction. These towns consisted of vast tracts of identical single-storey homes, built by a single developer in a single project, with a production system that featured highly specialized crews that marched from house to house according to a strictly controlled Takt time. This was possible because demand far outstripped supply, allowing Levitt to provide no product variety. All of the houses were the same, just as all of Henry Ford's Model T cars were all black.[8]

Lean Construction

The introduction of Lean to the field of construction grew due a realization that there is significant waste in the way construction is traditionally managed. Lean Construction calls on managers to understand the dynamics of the craft and industrialized construction systems, to recognize the waste that results from local and/or opportunistic optimization, and to find ways to collaborate to achieve global optimization and smooth flow, increasing overall utility for the long-term mutual good of all the commercial partners.

Although modern tall office buildings have multiple repeated spaces, uniformity cannot be taken for granted across all floors, and thus standardized work and minimal variety of parts are not necessarily common features. Variety and complexity in the product are one of the reasons for the development of Lean Construction, which aims to provide client-determined value while still maintaining the benefits of flow. Lean Construction can be seen as an improvement not simply on craft and industrialized construction, but also on mass construction. Figure 16.4 shows Lean Construction as derived from the three other types of construction production systems (craft, industrialized and mass construction). It inherits concepts from all three, as well as from Lean production.

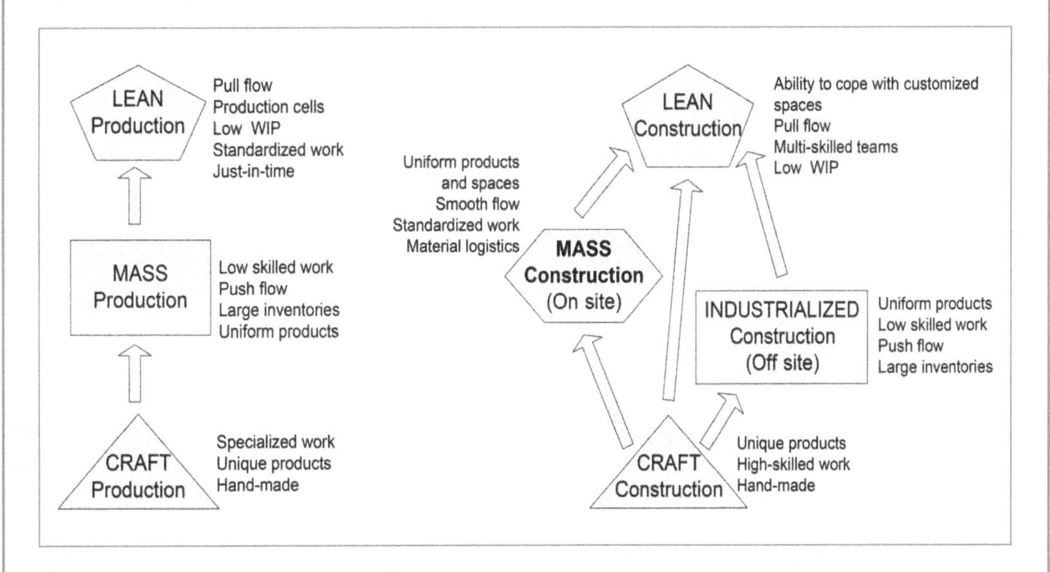

Figure 16.4 Evolution of factory production systems (left) and of construction production systems (right)

Table 16.1 Key characteristics of four different construction systems

System characteristics	*Craft construction*	*Industrialized construction*	*Mass construction*	*Lean construction*
Nature of the building products	Unique components and unique spaces	Uniform components	Uniform components and uniform repeated spaces	Customized spaces
Labour	High-skilled trades	Low skilled but highly specialized	Specialized trades with narrow focus	Multi-skilled teams
Flow system	No consideration of flow	Push – discontinuous flow	Continuous flow (stable due to uniformity of products)	Pull – continuous flow
Batch sizes	Small batches	Large batches	Large batches	Small batches
Inventories	Small inventories	Large inventories of components	Large inventories of components off site, large inventories of spaces	Small inventories
Logistics	No logistics system	Off-site materials logistics system	On-site materials logistics system	On-site and off site logistics system for all resources

The new approach at Tidhar

With sharper focus on flow and value, with better perception of waste in production, and with growing competence in applying BIM and Lean Construction tools, Tidhar's leaders at all levels began to take steps to improve their planning and control processes.

Greater detail in planning and control

As a first step to improving planning reliability they first set out to get a better handle on work durations. This started by simply asking trade crews and superintendents what rates of production were possible in specific projects for each trade. For example, in one project, they asked how many square metres of interior partition walls a mason could build on a good day. The answer was about 17 m^2. A quick check of the BIM model yielded a quantity of about 400 m^2 of interior partition walls on each floor of the building, which meant that each floor required about 23 person-days of labour. Given a desired Takt time of five days for each floor, the crew should have at least five masons.

This is a simple calculation, but until then the construction engineer had refrained from making it, preferring to leave production calculations to the subcontractor. But becoming directly involved in production planning means that one can now consider how production rates fluctuate within the five days, what constraints or preconditions affect productivity, how crews affect one another, what the best batch size is across a range of trades or operations, and what steps can be taken to improve flow from a global optimization standpoint. Opening up the black box of CPM activities is a big step forward for a general contractor that has hitherto assigned the risks of production and productivity to the subcontractors as a matter of policy, if not doctrine.

An added benefit of detailed planning is detailed control, in the sense of being able to check actual versus planned and take corrective action where needed daily, rather than waiting until the end of the month to get bad news. In the example above, checking consisted of making sure that five masons were on site each day and checking not just that 85 m^2 of interior partition walls had been built by the end of the day, but that the correct 85 m^2 were built. Controlling the sequence of work, i.e. which specific partitions were completed, is key to ensuring that the right work can be handed over to the next trade (ensuring flow), and to taking steps further down the line to reduce batch sizes. Any problems become apparent much earlier, so that solutions to systemic problems can be applied (such as adding missing information to the drawings) and gaps can be closed before they spiral out of control.

In essence, Tidhar began investing time and energy in building a production system, rather than just managing contracts: measuring production rates, monitoring planning effectiveness, identifying the constraints that apply to each resource, satisfying the preconditions for working right the first time and generally improving flow for all parties. Chapter 18, "Raising planning resolution – The four-day floors", describes in detail how this principle was applied to reinforced concrete structural work. The next section deals with its application to the building systems and interior finishing works.

Production waves

When the Lean team began compiling value-stream maps of the processes during the Yavne project (see Chapter 7), they realized that the interior finishing works could be divided into three distinct stages. They called the three stages *"production waves"*, adopting a metaphor that saw them as waves that wash through the building, one after another. Each wave represented continuous work, with crews moving from apartment to apartment and one trade following the other within the wave, but with buffers – or troughs – of inactivity between the waves.

Table 16.2 lists the three waves and highlights the differences among them from a production planning and control perspective. In Tidhar's value-stream maps, the first wave had 20 tasks, all done by subcontracted trade crews with three inspection points and a fairly rigid technological sequence; the second wave had another 20 tasks, although many of them were independent and could be done in any sequence; and the third wave had six tasks that were usually self-performed by Tidhar's own multi-skilled teams.

The tasks within each stage were similar to one another but quite different to those of the other stages, and the differences were sufficiently deep that it made no sense to bundle the waves together, under the control of a single supervisor, as they had always

Table 16.2 The three "waves" of interior finishing works

Interior finishing wave	*Typical work content*	*Production location*	*Production type*	*Typical task duration*
Wave 1 **Bulk work**	Partitions (masonry or drywall), plumbing and electrical systems, waterproofing and insulation, flooring and painting	Built on site	Craft work, "wet and dirty", trade-specific	Days
Wave 2 **Installation**	Windows, doors, kitchen cabinets and counter tops, air-conditioning systems and sanitary hardware	Fabricated off site, installed on site	"Made-to-order" products	Hours
Wave 3 **Delivery**	Prepare individual apartments for handover to customers – installing faucets, applying trim around door frames, plastic plates on electrical outlets, touch-up painting, cleaning and polishing, etc.	Performed on site	Delicate, clean work, suitable for multi-skilled teams	Minutes

done. While the bulk work of the first wave is rough by nature, the later production steps refine the product. What might be considered a gross defect in the third wave might be "good enough" for the roughing in during the first wave, just as one might shape a piece of rough wood with a chisel at first and then progress through sandpapers of increasingly fine levels of grit.

The superintendents for each wave need to pay attention to different levels of detail. However, by nature, different people tend to focus on different things. It appeared that requiring superintendents to continuously zoom in and out between the rough and fine works led to less than stellar results as some defects were overlooked. Furthermore, whereas the weekly work planning meetings sufficed for the first wave, the second wave needed a formal make-ready process in which someone would monitor the status of fabrication work performed off site.

Continuous experimentation with a new production system design

To address the problems of bundling all three waves together, as had been the practice up to that time, Tidhar decided to assign the work of each wave to a different superintendent, with formalized customer–supplier relations between them. A detailed standard process map was prepared, defining the tasks and responsibilities for each wave and the customer–supplier inspection and handover points between them. The approach was applied for the first time in the Dan Towers, a project with two 29-storey residential towers with 109 apartments in each. The tasks were set-up for Takt time planning, at a rate of 12 apartments per month, with a production cycle time of two months per wave per apartment.

In theory, this was a good plan, but in its first implementation significant problems arose. Although work progressed more quickly and smoothly, the number of hours worked by Tidhar's own general labourers on rework during the second wave and to prepare apartments for delivery in the third wave rose dramatically. This rework was pure waste, and instead of declining, as had been hoped, it had risen overall, peaking at 300 hours per apartment in the worst cases.

Fortunately, instead of abandoning what had been learned about production system design, they set about identifying the root causes of the overruns. Previously, a superintendent and an assistant had managed all of the interior finishing work for a building of this size. Now, the same two people had separate responsibilities, for the first and second waves respectively. But this had two problems: first, the superintendent was now overworked during the first wave and could not control the quality of work as well as before, and second, in the customer–supplier handover between the first and second waves, the senior superintendent was handing apartments over to a junior assistant who apparently could not demand a complete product from his superior. The first wave work was done, but not "done-done",[9] so that the second wave superintendent had to complete the leftover tails of the tasks with his own staff. These were usually "small" items, like painting a small area of the apartment or finishing a few last flooring tiles, but taken together they were adding up to a big headache.

To address this issue, Tidhar performed another experiment, on two buildings in the sixth and final stage of the Yavne project. By this time, the wave approach was standard practice at Tidhar. In the experiment, the second wave superintendent was not assigned any general workers. Rather than repeatedly fixing problems after each installation subcontractor during the second wave, all of the rework was left to the third wave. The assumption was that some of the items that were being fixed in the intermediate steps were damaged by, or required more work as a result of subsequent tasks. The results were encouraging; rework was reduced from the Dan Tower peak of 300 hours of general labour per apartment to 80 hours.

Waste was reduced but not entirely eliminated, of course. Yet this was a sufficient improvement to continue implementing the wave solution. The next experiment was performed in the Rosh Ha'ayin project. As described in detail in Chapter 20, "Pulling it all together", at Rosh Ha'ayin the responsibilities for each of the three waves were clearly assigned to two different managers, one for the first and third waves and the other for the second wave, with no overlap. The installation manager, as the superintendent of the second wave was now called, was not assigned any general labourers, thus avoiding any possibility of performing rework. Clear handover requirements were defined for the first wave, to try to reduce the quantity of rework that would remain. A checklist was compiled to make sure each apartment was up to standard before being transferred from one wave to the next.

As in the parable of lowering the water in the Lean "production stream" where reducing WIP reveals the rocks (production system problems that were always beneath the surface) on which boats can founder, these steps improved matters but also revealed new problems. For example, some of the first wave superintendents prioritized starting work in new apartments at the required rate over completing and handing over apartments to the second wave; they pushed rather than allowing pull from the second wave. This was an error in the design of the control system, where the wrong parameter was being measured and applied for control. And so, experimentation and improvements continue...

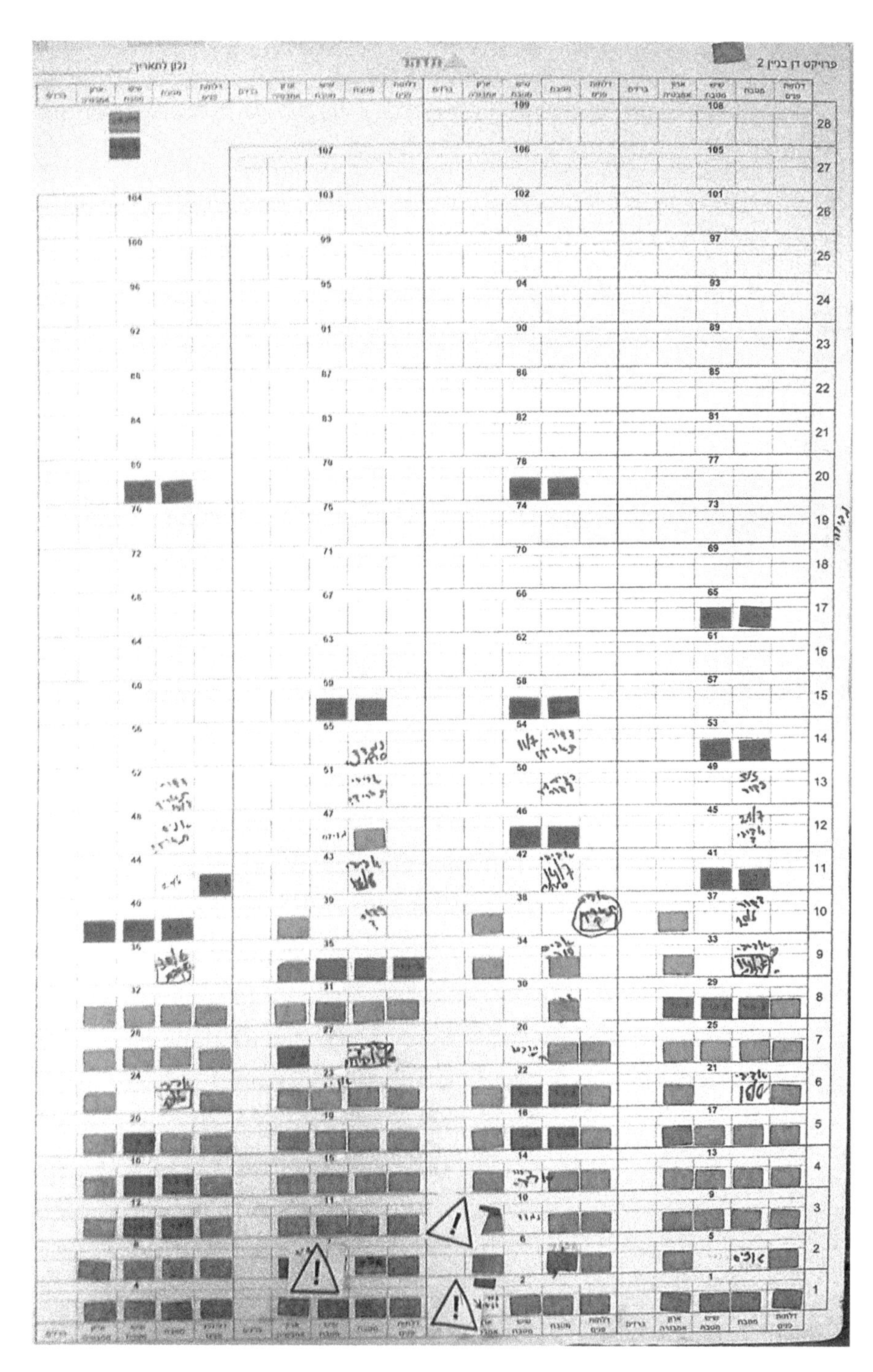

Figure 16.5 Visual display board for the second wave, Dan Towers project

Note: The chart shows the storeys on the vertical axis, and four apartments each with five installation tasks (doors, kitchen cabinets, kitchen marble counter top, bathroom cupboards and sanitary fittings) on the horizontal axis. Annotations include green, red or alert stickers, and text.

Working with the supply chain to improve the second wave

The second wave consists primarily of installation on site of building components fabricated off site. Kitchen cabinets exemplify the issues that arise in this process. They are long lead-time items because each individual kitchen must be configured to meet customers' preferences (layout, location and quantity of plumbing and electrical feeds, type of fixtures and counter top, etc.) and because measurements of the as-built kitchen dimensions must be taken on site before fabrication can begin in the factory.

Installation of the prefabricated cabinets on site is quick, but suppliers generally demanded some three months of lead time between order and installation. The minimum time from measurement of the kitchen until delivery was 45 days for cabinets with no customization and 60 days for customized kitchens. Tidhar had to plan the sequence of work carefully to enable suppliers to measure as early as possible and to fill the intervening downtime with other tasks. In practice, promised delivery dates were more often missed than met, and site supervisors dealt with suppliers through a mix of coercion and supplication. They had never attempted to delve in and try to understand the suppliers' problems – with most of the work performed off site and distanced by a supply contract, opening up the "black boxes" that second wave suppliers represented was not considered.

The genesis of a shift in thinking began following implementation of LPS in buildings 10 and 11 at the Yavne project (Chapter 17 describes the implementation in depth). Conventional wisdom held that the "wet and dirty" works of the first wave (plaster, masonry, flooring, etc.) were responsible for most of the delays since this wave comprised the bulk of the work volume. Application of LPS reduced the duration of the first wave trades from six to three months through a combination of production planning and detailed phase scheduling. But despite this success, the project still ran late, as six months more were needed to complete the remaining "dry" works of the second and third waves. It seemed that the first wave could be more easily improved because the workers were on site, in view of the superintendent. In contrast, the off site production was out of sight and out of mind. Apparently, superintendents could only hope that the items would arrive on schedule, with no window into their progress until it was too late. Now, with clear definition of the waves and identification of the barriers to flow in each wave, it became clear that the next step to improving the installations was to follow Toyota's example and invest time and effort to improve processes at their principle supply chain partners.[10]

Tidhar arranged a series of visits to kitchen suppliers, to visit the factories where the cabinets were produced and meet the people who handled the orders. The suppliers were more than happy to have Tidhar visit; rather than feel "invaded", they welcomed the opportunity to explain their constraints and look for ways to improve together. Tidhar's team included Ilan Nachman (vice president for procurement), Vitaliy Priven (process engineer), Ortal Yona (production engineer), Ronen Barak (R&D) and Cheni Kerem (Lean consultant). They began with a straightforward request: "Show us everything that happens to our order from the moment we send it to you, until the kitchens are delivered to our site." Using VSM, they walked through the process from beginning to end.

Almost immediately, they found three issues that delayed delivery:

- The supplier's worker responsible for taking measurements would wait for a number of kitchens to accumulate in a batch ready for measurement. Because the measuring task itself was brief, the long "set-up time" of travel to the site and climbing to the apartment encouraged the worker to wait until there was more than one kitchen ready to be measured before going to the site. Large batches naturally resulted in longer cycle times for each kitchen.

- The sequence in which kitchens were measured on site did not match the sequence in which customer approval was sought (see the next bullet point) and so some kitchens had to wait even if customers had signed off on their selections (the next problem).
- The supplier started production only after a customer visited their showroom and signed off on their customization selections, avoiding rework by ensuring that they were informed of any last-minute changes of fixtures, colours or layout. However, the supplier could not control the customers, and much waiting time was introduced into the process.

Once a customer order was finalized, manufacturing and assembly were completed within days, but Tidhar had been unaware of these causes of delay. The solutions were fairly straightforward:

- For all but the most complicated kitchen designs, the site supervisor would receive instruction about what to measure and would provide the information to the supplier.
- Tidhar facilitated the customer sign-off through their own customer change representatives, who developed a personal relationship with each of the customers.
- Tidhar began monitoring and controlling the process including the status of work at the supplier for each of the sets of kitchen cabinets. The process gates were clearly defined – customer sign-off, measurement, release of work orders to production, production itself, delivery to site and installation – and could therefore be monitored.

Tidhar opened up the black box, shone light inside and got involved in the process. As a result of the corrective actions, 95 per cent of kitchens now arrive on time, as opposed to the less-than-50 per cent on-time arrivals that were typical before the improvement work. The process was made visible, so that any issues that arose could be dealt with in real time, rather than waiting for a missed delivery date. Most importantly, the new process enabled Tidhar to reliably plan the work that followed kitchen installations.

Not surprisingly, the problems by and large were not in the production of the kitchen but in the supporting information processes. As was the case for Toyota and many, many manufacturing companies, it appears that major gains can be achieved in the construction industry too if one looks at the production performed by suppliers off site as an integral part of the overall production system.

Lessons learned and ongoing challenges

The nature of Lean continuous improvement is such that the work is never finished; there is always more to do. Tidhar's engineers have come very far in changing their perceptions of best practice in planning and control, but they would be the first to acknowledge that they could do better. For example, they have not yet addressed the production planning of non-repetitive parts of projects, such as underground parking garages, lobbies or penthouses. Although these locations are not actively managed, they do eventually get built, but the process is far from optimal. They could benefit greatly from detailed planning, as a lot of waste is caused by managing the work in batches that are too large for proper planning and control. Good flow is hard to achieve when trade crews are given large work areas and left to their own devices. This reflects the problem discussed at the beginning of this chapter: when the level of detail is low, the ability to control and correct is also low. It is also the rationale for raising the resolution of planning in structural work, as told in Chapter 18, "Raising planning resolution – The four-day floors".

Careful analysis of the nature of the production operations can reveal inconsistencies in the way in which work is bundled. This was the case for the interior finishing works, and definition of the three "waves of production" that wash over every apartment enabled improvement through better production system design.

Finally, Tidhar's leaders have learned that applying an improvement may not necessarily bring good results on its first implementation, but applying the controlled experimentation approach of continuous improvement can lead to eventual success. With each additional step to learn and understand their processes in more detail, and with active engagement as opposed to passive observation, Tidhar's people improved the quality of their planning and control, and with them the project outcomes.

Notes

1 Dos Santos (1999).

2 CPM was developed in the 1950s, following research for the US Navy. It was subsequently adopted by the construction industry (see Fondahl, 1962) and became the standard method for scheduling construction projects. It lies at the core of all construction scheduling software, including recent tools that implement location-based scheduling, such as VICO Control (VICO, 2016).

3 For a discussion of the interaction between general contractors and trade subcontractors, see Sacks and Harel (2006).

4 The definitive book on the subject of location-based management is Kenley and Seppänen (2006).

5 In some countries line-of-balance charts are sometimes called "flowline" charts.

6 Korb and Sacks (2016).

7 Partouche et al. (2008).

8 "A customer can have a car painted any color he wants as long as it's black" (Ford and Crowther, 2003).

9 Greg Howell, one of the founders of the International Group for Lean Construction (IGLC), was fond of recalling his experience as a construction engineer when a trade crew leader reported to him that a flooring activity was done. When he pointed out that some of the trim panels were missing, the crew leader responded "Oh, I thought you were asking if we were done. I didn't realize you were asking if we were 'done-done'." Some background to coping with this using the "language-action" approach can be found in Macomber et al. (2005).

10 See Chapter 17 of Liker (2004), which discusses Toyota's approach to working with their supplier to help them improve to mutual benefit.

17 The Last Planner® System

The Last Planner® System (LPS) is a production planning and control method in which planners review work packages prior to releasing them to trade crews to ensure that their preconditions are fulfilled. It is a countermeasure to the unstable nature of traditional construction management in which trade crews are expected to meet predetermined schedules without due consideration for the conditions that govern their work. Stability is achieved in part through direct involvement of the trade crews or subcontractors who actually perform the work in setting short-term work plans. LPS consists of master planning, phase planning, look-ahead scheduling, weekly work planning and monitoring of outcomes using a "per cent plan complete" measure.

> Thanks to the Last Planner® Weekly Work Planning meeting, I got 30 per cent of my time back. Rather than running around trying to figure out where we stand on the project and directing people where to go next, I suddenly have time to actually plan the work and invest more time in quality and site organization. Instead of retroactively fighting fires, I have time to proactively control the work.
>
> (Ahmed Hassin, Superintendent, Tidhar)

Closing the gap between plan and reality

In every industry, one can find front-line workers with complaints (often fully justified) about the disconnect between management directives and the reality they confront on a daily basis. In Lean terms, the white-collar managers and engineers are divorced from the Gemba.[1] Construction is no exception in this regard, and construction planning, scheduling and control is a prime example of this detachment.

Preparing a schedule is one of the first tasks in any construction project. The project is decomposed hierarchically using a formal work breakdown structure (WBS) and the relationships between the tasks defined are modelled using CPM. The schedules are computed with CPM software, such as Primavera or MS Project, and the highly detailed and technically very impressive – and colourful – Gantt charts are printed on giant pieces of paper to be hung in the construction trailer. Hours of work by highly trained engineers go into creating the construction timeline, and the result is indeed formidable.

But the sad reality is that very often the ink is barely dry on the schedule chart before reality throws the proverbial wrench into the works, and from that point forward, the plan is not much more than an expression of good intent. Project managers work valiantly to return the project to schedule, but the day-to-day control is improvised. The schedule printouts become little more than trailer decorations, divorced from reality and having no practical use as production management tools.

For anyone experienced in the construction industry, the above paragraphs offer no surprises. The situation described is seen as a "necessary evil" inherent to the industry and so pervasive as to be barely noteworthy. Clients demand schedules and contractors prepare them, but they are not very useful for actually running the job.

From a Lean point of view, the scheduling process as outlined above is simply waste that adds little value to the customer. If the CPM schedule adds no real value to the control of production, then why take the time to create a schedule?

In the early 1980s, Greg Howell was working with time-lapse photography techniques in construction projects. He used the films to identify opportunities to improve labour productivity, but he noticed that there was a low degree of correlation between the scheduled work and the actual work tasks underway. He discussed this observation with Glenn Ballard, Alexander Laufer and others, who had noticed the same phenomenon. Working with Howell and Ballard, Laufer wrote about the uncertainty inherent in construction and the challenges this posed to planning.[2] When they compared notes, the disconnect between planning and doing became clear.

Ballard and Howell's response was to suggest that there was essentially no effective planning and control of production, which resulted in a great deal of waste for construction workers – waiting, rework, overproduction, inventory and the associated negative cash flows – all of which contributed to projects being unpredictable. They realized that the planning, scheduling and control process must be altered to serve the production process directly, ensuring that the right people would have the right tools, materials and information to do the right work in the right place at the right time. This line of thinking led them to develop and test LPS.

Box 17.1 Construction management: traditional versus bureaucratic methods

In 1982, Herbert Applebaum reported on a fundamental disconnect between two groups of people on construction projects: the "modern" managers, equipped with large staffs and computerized tools, and the more traditional contractors who were fully enmeshed with the project, the subcontractors and the work actually being done.

The article starkly shows the incompatibility of traditional CPM planning and scheduling, on the one hand, and day-to-day production control, on the other hand. Applebaum published his observations in an insightful article entitled "Construction Management: Traditional versus Bureaucratic Methods" in the journal *Anthropological Quarterly*.[3] One wonders if the article had appeared in the *Engineering News Record* instead of in the *Anthropological Quarterly*, Lean Construction might have got off to an earlier start.

The problem with traditional methods of scheduling

Before describing the LPS in more detail, let us first dive deeper into the root causes of the problem with traditional scheduling methods. The problem is that overall construction schedules are only useful for monthly management reporting and not for day-to-day management of the work on site. Box 17.2 explores the root cause using a "five whys root cause analysis".[4]

Box 17.2 "Five whys" analysis addressing the problems with traditional schedules

Problem: the pre-prepared schedules are not used by construction professionals for managing production.

1 Why are the schedules not used for day-to-day production management?

Because the construction professionals don't derive any utility from the schedule (the schedule doesn't help them do their job).

2 Why doesn't the schedule offer any utility to the professionals?

Because the schedule does not reflect the day-to-day reality of the construction project.

3 Why does the schedule not reflect the day-to-day reality of the project?

Because the subcontractor's constraints are not included by the planner in the plan.

4 Why aren't the subcontractor's constraints taken into account?

Because when work packages are delegated to subcontractors, the work is considered a black box, with no interest or attention to the subcontractor's constraints. The work described in the schedule (including work sequences, durations, sometimes even the work content itself) does not match neither the actual work that needs to be done nor the work that is actually done on the project, and it is not updated to reflect the delays that have occurred. It does not help us understand when things can be done, and it's not reliable.

5 Why is the work that is delegated to a subcontractor considered a black box?

Because the management method is to focus on results and not on process. Their assumption is that by transferring work to a subcontractor, they can also transfer the risk and all of the responsibility for achieving the results, absolving the GC of the need to take responsibility for work flow, or to pro-actively make the work ready.

The uselessness of the schedule can be traced back to a lack of information about the project during the planning of the schedule and to the lack of reliability of the plan as it is not updated, coupled with a misplaced focus on results at the expense of process. A plan that is too detailed represents only one possible outcome among a combinatorically large number of possible outcomes due to uncertainty at each step. The further into the future it goes, the greater the uncertainty. People work with information which is inherently uncertain, which naturally leads them to make assumptions. In fact, the schedule itself is one big assumption that is the amalgamation of many, many smaller assumptions. Assumptions of how long it will take to paint a wall or lay piping. Assumptions that materials will show up on time and in the quantity required. Assumptions that no complications will be discovered on the site when excavation begins. Assumptions that subcontractor crews will even be available to start work on the date required. Every one of these is uncertain and can have many possible outcomes.

When viewed this way, the schedule looks no more stable than a house of cards. If any one assumption is not borne out, the whole house comes crashing down. With neatly divided work packages with predefined durations, CPM does not respond well in a dynamic environment, which is why the schedule it creates has a near-zero probability of being realized. All of the tasks that are on the "critical path" are scheduled to start as soon as possible. This means that there is no buffer. What is the probability that all of the assumptions about duration will be correct? Essentially zero. Therefore, the probability of the schedule remaining accurate also approaches zero.

In the real world, there will always be something unexpected. Systems are complex and interdependent, making it essentially impossible to predict how any one change will affect everything else, particularly where human behaviour is involved. There are aspects of any project that cannot be known when creating the master schedule. Lack of knowledge and/or experience can further complicate the lack of information in the common situation where the person who creates the schedule is not deeply familiar with the intricacies of the processes they are scheduling. They may set arbitrary time durations and sequences that do not reflect the true nature of the work as it is performed on the site.

An external scheduler doesn't always take into account the interests of the subcontractors who are tasked with actually performing the work. Subcontractors value work continuity and having work ready for them when they arrive on site very highly, since this ensures that the time of their crews can be used productively. If a schedule provides a subcontractor with just three days a week of work content, the subcontractor will tend to delay its arrival on site to allow a buffer of work packages to accumulate. But this delay will inevitably lead to further delays down the line as the effect of the inconsistency ripples through the system.

Box 17.3 Project managers and subcontractors stuck in a stable "lose lose" equilibrium

Sacks and Harel examined the dynamic of a GC interacting with subcontractors using game theory.[5] They modelled the behaviours of each, given the GC's commitment to moving the project ahead and the subcontractors' attempts to maximize profits across all of the projects they are involved in. As may be guessed, this leads to a situation where the GC is incentivized to "over-promise" how much work will be ready and request excess work crews from the sub, while the subcontractor is incentivized to "under-deliver" by committing fewer workers to a given site. They found that by increasing the quality of the information about the status of the project (one of the key goals of LPS), the situation can be improved for both players in the scenario. When the subcontractor is trying to determine how to best allocate their resources, they place a high value on continuity of work. This is why information quality is key; it allows them to properly plan for continuous work.

In Chapter 16, "Production planning and control in construction", we explained how traditional methods of scheduling typically assume that work packages are discrete elements that can be neatly boxed into a Gantt chart (see Figure 16.2 for an example). In reality, the work content is much more fluid and dynamic. For example, if only 70 per cent of the material arrives on site for a particular trade, it may be possible to do 70 per cent of the work. Or it might be possible to do 0 per cent of the work. But in the scenario where some forward progress is possible, then the rectangle on the schedule has been split into two. In the meantime, the subcontractor might route some of the crew to start doing another type of work. So now a future rectangle has been pulled forward and overlaps with the previous one. These problems with the CPM/Gantt approach to scheduling have been addressed by Koskela, who described the problems with the decomposition procedure that is inherent in the "transformation" paradigm.[6] First, there is more than one way to decompose, and decomposed parts can overlap. Second, tasks are not as independent as decomposition assumes. And third, the basic assumptions behind the initial decomposition change over time (work is structured and assigned differently, design details change, etc.), so that static decomposition does not reflect reality. All this leads to the conclusion that the transformation paradigm by itself is not a good basis for production control in construction.

Finally, one of the fundamental problems with traditional scheduling, and one that leads directly to the uselessness of the finished schedule for production control, is that the scheduling process is typically linear: the scheduler creates the schedule and hands it to the construction team to execute. There is little possibility for feedback into planning to adapt the plan as the situation changes – there are sometimes no feedback loops, or the loops are too long to have any effect on the control level. As any designer of complex systems can testify, a system without feedback loops to correct behaviour during operation will fail.

At Tidhar, prior to implementing LPS, the situation was similar to that described above. Highly detailed schedules were prepared (primarily focusing on the structural works, which were self-performed), but they were not used by the superintendents. Project managers would occasionally open the schedules to report status, but that was the extent of their use.

CPM is not completely irrelevant to the construction process, as long as its limitations are understood and no attempt is made to exceed them. It is useful at the very beginning of a project, for preparing a rough outline of the schedule to get an idea of the project duration. The planning resolution here (see Chapter 18, "Raising planning resolution –

The four-day floor", for a longer discussion of this concept) may be at the level of complete months. In this context, CPM is not a planning tool, but rather an estimating tool. It can help set overarching goals, those results needed by the owner, the company or the project itself, but it will not help with the process of actually building the project.

The Last Planner® System

The main idea behind LPS is to stabilize the flow of work, making plans predictable and reliable, thus reducing waste and adding value to the construction process, and ultimately to the customer. It enables a team to proactively reduce uncertainty by (1) reducing the number of assumptions that must be made ahead of time to a minimum through the make-ready process, and (2) increasing the level of detail of planning by involving the "last planners", those actually tasked with performing work. LPS is a registered trademark of the Lean Construction Institute, which maintains *The Last Planner® Production System Workbook,*[7] the canonical guide for implementing LPS. LPS has been adopted in construction projects around the world, with successes reported from many countries.

LPS entails planning at four levels: master planning, phase planning, look-ahead (or make-ready) planning, and the weekly work planning. The first step is "**master planning**" to define the project at a high level, i.e. with low resolution. The master plan is prepared in the early stages of the project, and is intended as a rough framework with the project milestones to determine if the project can be delivered within the constraints of the customer or project. It is typically carried out by the GC in a centralized manner, as a point to begin the conversation of the next step, phase planning.

Each major stage of the master plan is broken out into its component tasks during "**phase planning**" (sometimes called "reverse phase pull planning"). The phase plan sets the standard process for the phase and a base for the detailed schedule. The method implicitly recognizes that "over-planning" ahead of time by going into minute detail before the project is at the stage necessary to plan accurately is a waste. Phase planning is conducted together with the main subcontractors, taking advantage of their experience and input to create a workable plan. The focus is on how the deliverables flow between different subcontractors, to make sure that each trade has what it needs at each step. This focus on handoffs between trades is a key component of this step, which is typically carried out once per project phase. By starting at the final delivery and working backwards to determine which prerequisites have to be completed at each step (reverse phase pull planning), the team makes sure all bases are covered.

Next, "**look-ahead planning**", also referred to as the "make-ready process", takes a more granular approach to the work packages, ensuring that all prerequisites are complete for each work package before it is released to the weekly work planning meeting. The look-ahead plan identifies the preconditions for each task and supports action to make the tasks ready by removing constraints.

Finally, the plan for the coming week is developed in a "**weekly work planning**" (WWP) meeting. In this weekly meeting, the foremen of the crews who will actually perform the work meet to review the current status of the project and develop the plan for the ensuing week. They commit to perform specific tasks in specific locations and at specific times during the week, and in so doing, they become the eponymous "last planners". Only at this meeting is it finally determined who will do exactly what on exactly which day, considering the status of the work and all preconditions, as well as the resources needed. Delaying this commitment to the last possible moment increases its degree of reliability, since the plan is much more likely to be carried out.

Unlike traditional planning, the plan is prepared not by a single individual who may not have full information, but rather with the active participation of all those who hold information about the preconditions for each task – the trade crew leaders who will actually be doing the work as well site engineers, purchasing, etc. The goal is to create the most reliable plan possible by basing commitments on the most reliable information, assigning only the work that "can" be done to the list of work that "will" be done. Detailed planning is delayed until the "last responsible moment" to keep assumptions to a minimum, avoiding as far as possible the snowball effect of plan failure where incomplete tasks from a previous planning cycle make it impossible to fulfil the tasks assigned in the next planning cycle. If done right, it shields the work crews from half-baked tasks that are unworkable.

Figure 17.1 A WWP meeting at Tidhar on the Park, Yavne

Another key component of the WWP meeting is the review of the plan and commitments that were made the week before, to see how many of the tasks were completed as planned. This measurement is called the PPC, and it is tracked over time as a measure of the reliability of the plans. The goal is to improve PPC over time, which tends to happen when it is measured each week for the team to see. It is not intended as a tool to browbeat non-compliance as much as a tool to support improvement by identifying weaknesses. The reasons for particular work packages not being completed are tracked to identify areas for correction. Often, the reason for not completing work as planned is beyond the control of the crew to which it was assigned: materials that didn't arrive, a work space that was not cleaned and readied for them to begin, or changes to the design. These are issues that the GC staff need to be aware of so that they can improve their make-ready performance. The analysis of reasons for not achieving promised work packages is a feedback loop that enables PDCA improvements, hopefully leading to corrective action being taken. The constant flux in the schedule, particularly during the look-ahead/make-ready step, is also a feedback loop: real-time information from the field drives adaptation of the schedule.

While a week is a natural time-step for most common building construction projects, a one-week cycle is not the only possibility. There are types of projects that require either longer or shorter intervals; for example, on critical infrastructure rehabilitation projects, intervals of hours have been used, like the Channel Tunnel rail link.[8]

Finally, we note that while the discussion above focused on the construction phase, LPS is also effective for planning and controlling other phases, such as design.

Tidhar adopts LPS

Tidhar's first practical experience with LPS occurred within the framework of a research collaboration with the Technion – Israel Institute of Technology. In early 2011, before Tidhar began their Lean journey, a three-week experiment was carried out to test the impact of the KanBIM[9] system on the completion of work packages by subcontractors. In order to delineate the work packages that would be completed, the researchers (led by Rafael Sacks and Ronen Barak) conducted two WWP meetings. In the framework of this research, Ronen developed planning boards and sticky notes that were used to graphically portray which work packages each of the subcontractors had committed to for that week. But once the research was completed, no more WWP meetings were held.

The next experience with LPS took place in 2012, for another Technion-driven experiment. By this time Tidhar had formally embarked on its Lean journey, and had decided to collaborate with and fund a Technion Lean Construction research project. Vitaliy Priven, then a PhD student (and subsequently a Tidhar employee), proposed to study the impact of the WWP meetings on the social networks among unrelated construction crews. The research goal was to explore the mechanisms by which LPS worked, trying to establish whether and to what extent strengthened social relationships enhanced coordination among crews, and in turn, whether this contributed to the success of LPS implementations.

The plan called for implementing the WWP for the entire interior systems and finishing stage of construction at a sample of high-rise apartment building projects, while monitoring a second sample set, without intervention, as a control. Ronen Barak, who had recently been hired to manage Tidhar's R&D effort, prepared a presentation outlining the method and its expected benefits that he presented to the Tidhar management. The research was given the green light, with buildings 10 and 11 in the Tidhar on the Park project, in Yavne (see Chapter 7) selected as the pilot location.

Tidhar was well positioned to benefit from implementing LPS, since they suffered from many of the problems described at the beginning of this chapter. High-level schedules were developed by an industrial engineer who was divorced from the work and the workers, with many approximations. The interests of subcontractors were not considered, so that schedules often contained discontinuous work for some of the trades. Discontinuity is anathema to subcontractors, which means the schedule was already setup to fail. The rule of thumb at Tidhar was that the actual duration of an activity would be roughly twice the theoretical net duration, with the result that many buffers were included in the committed schedule. There was no detailed schedule at the level of daily work, and the impact of this lack of production planning and control was reflected in the apparent chaos that was typical of many project schedules.

Using the "work backlog" approach (see Chapter 16, "Planning and control in construction"), superintendents were given a target benchmark number of storeys to complete per month. That was the extent of the planning – results were demanded without any process guidance. No one cared how it was done, as long as it got done. In practice, superintendents acted like "air traffic controllers", trying to coordinate between subcontractors by calling each in turn to coordinate start dates and get progress updates. Different trades had different rates of production, meaning some subcontractors would find themselves underutilized and thus spend time waiting, while other trades failed to meet the work backlog target. Team members didn't really know when each crew was supposed to start, since it was hard to know when predecessor tasks would finish.

Thus, more often than not, projects quickly became acrimonious. Communication was not aided by the proliferation of languages; a given site might have a mix of Hebrew, Arabic, Russian, Chinese, Turkish and others, depending on the subcontractors involved in the project – a veritable tower of Babylon. And yet, this wasn't viewed as a problem that could be solved, but rather simply the way in which construction work was done. Indeed, despite all the problems, Tidhar had built thousands of housing units in this traditional fashion, mostly on time and on budget (in the standards of the market). "Process control", such as it was, entailed counting how many storeys were completed each month and yelling at people who failed to meet their goal. Superintendents would be disciplined by senior managers, and the subcontractors would be disciplined by the superintendents in turn. Gaps would inevitably open up from the planned schedule, and once this happened there was no return to the planned timeline. Ironically, delays could be caused by trades with production rates both faster and slower than the target rate: slower trades failed to meet the (for them) unrealistic deadlines, while faster trades delayed start to allow a buffer of work in progress to accumulate for them, which would delay the start of the next steps after their work.

The Technion team experienced this viscerally when they arrived on site for the first day of experimental set-up. They wanted to measure the PPC, a measure of how well the project team was meeting their planned objectives. This required knowing the plan – but the site superintendent informed them that there was no plan, per se. He held the entire work sequence in his head, and he doled out tasks to trade crews as they came due. When pressed, he described the coming week's plan in approximate terms, but there was clearly no way to measure how well or poorly the plan was being carried out. Consequently, there was no way to address issues at their root causes, since there was no gap that could be measured to identify them. There was the requisite CPM Gantt chart posted on the trailer wall, but it was irrelevant to the project that was actually being worked and built. Thus the "current state" at Tidhar on the eve of their Last Planner® implementation was typical of many companies who are used to planning and working in a traditional fashion.

Works begins in Yavne 10–11

Buildings 10 and 11 in the Yavne project were two of the 12 built in stage 1 of the project. They were chosen for the pilot since the superintendent seemed open to and interested in LPS. The timing was right, since the structure was almost complete and finishing work was just about to begin. The finishing works were of specific interest for the research since the coordination among the many different subcontractors involved in this stage was perceived as a problematic area that could be improved.

The first steps Vitaliy took were a form of phase planning with the subcontractors. "I met with each subcontractor in order to go over the order of the work their trade had to do, including the work rates and interdependencies for each one", he relates. "They were sceptical, and they didn't always have the information at hand or weren't used to thinking of such things in a formal way. They hesitated to 'committing' to particular work rates that they might not be able to achieve consistently." He captured the information in an LOB schedule generated using VICO Control software. Quite apart from the need to compile a useful phase plan, both Ronen and Vitaliy saw this as an opportunity to change the approach to work planning. The next step was to begin the WWP meetings, using the boards and sticky notes developed previously. Ronen ran the WWP meetings at the start.

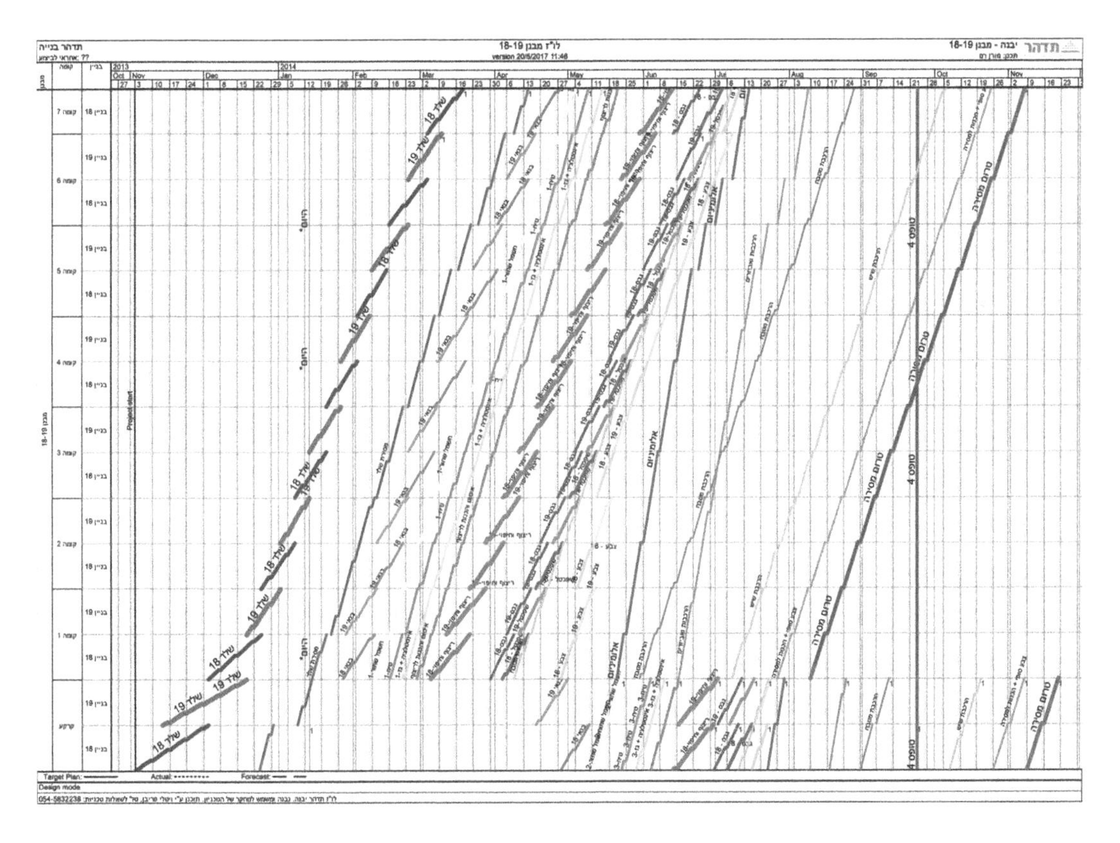

Figure 17.2 Line-of-balance phase schedule for the interior and finishing works of Yavne buildings 10 and 11, prepared using VICO Control software

"Everything clicked in Yavne 10 and 11; it was the right time, with the right people from Tidhar. The project was the right size and scope, with 32 apartments in each building, which kept things relatively small and manageable. You could keep the whole project together in your head", said Ronen. At the start, the participants' reactions were mixed, with some reservations. "We thought it was some sort of child's play, with sticky notes and smiley faces in different colours", said one of the subcontractors. But as they began to see results, the impression became more positive. Ahmed Hassin, the hitherto relatively anonymous site superintendent whose quote opened this chapter, suddenly found himself in the spotlight when LPS gained steam and started showing positive results.

The subcontractors also experienced a dramatic change. Instead of being yelled at for missing artificial deadlines, now they were being asked what they needed and for their ideas about how to improve the schedule and the work for their benefit. Rather than being scapegoats, their insights about how to develop the schedule together with the GC were valued, and their needs to get the work done efficiently were being catered to. Each week at the WWP meeting, they were given a platform to raise problems and issues if conditions were still not ideal. For example, a lack of electricity connection points, or of water supply for mixing materials on each floor, or raw materials that hadn't been delivered to the job site on time: issues that had always bothered the subcontractors and interfered with their work, suddenly came to light.

From the subcontrators' point of view, the GC was finally listening to them and paying attention. By involving them in the planning process and asking each week what work they planned for themselves in the coming week (and just as importantly, asking at the

Figure 17.3 The sticky note box brought to each WWP meeting

Note: Each trade has a distinctive colour for its tasks, and there are notes for problems or delays.

end of the week what tasks they had completed among those they had promised), the level of reliability of the plan soared. Allowing the subcontractors to determine the work they would do led them to take ownership of the schedule and deliver what they promised. And for the first time, subcontractors were talking to each other, rather than each of them interacting only with the GC. According to the electrician:

> In the old way, when I needed to know when masonry work would finish up so I could get in, I would ask the GC. But now, since it wasn't the GC who was promising that the prior work was going to be ready for me, I started talking directly to the mason, who could give me the most up-to-date and accurate information.

There was, admittedly, a strong Hawthorne effect (in which participants under observation in an experiment change their behaviour as a result of exactly that observation, often leading to productivity improvements), and this helped the pilot succeed. Everyone in the company and the subcontractors involved wanted the experiment to succeed. As a new effort that had some early success, it enjoyed heightened attention from management. At least one member of top management was present at most weekly planning meetings. This further increased the sense of importance the participants had for themselves and the success of the project, particularly because their commitments made at each WWP meeting were voiced before the visiting members of the Tidhar management (they didn't want to look bad in the eyes of those managers). Management attention was a supporting success factor in the pilot.

All participants were required to arrive at the meetings on time, in order to respect each other's time. The project manager reviewed the previous week and highlighted key focus areas for the upcoming week. Next, the superintendent would go to the prior week's plan board and check all tasks that were completed as scheduled. Uncompleted tasks were marked with an "X", and the reasons for non-completion were discussed briefly. Next, a special "score board" was filled out, in which each trade and the GC were given a green-yellow-red score (represented by smiley faces in the appropriate colour) for each of five categories: planning, production rate, safety, quality and site cleanliness/organization. The score for planning was in fact a PPC score, calculated as the number of check marks versus Xs. The overall production rate was evaluated by plotting the work completed on the LOB chart of the phase plan. The score for safety was assigned by the project's safety officer, and quality was determined by the superintendent. The score for site cleanliness/organization was given by each subcontractor to the preceding trade in each location: it measured how clean the work area had been left.

According to Adi Brayer, the project manager:

> Even though in the beginning it felt like we were in arts and crafts in kindergarten with the colours and sticky notes, we saw that there was a real transformation that happened in terms of people's behaviour and how much responsibility people were taking for their work, particularly since they saw how other members of the team were counting on them.

The next step in the weekly meeting was to take down the sticky notes from the prior week so that the next week could be planned. In the first few weeks, the subcontractors approached the board one at a time to plan their work by posting their notes. Once

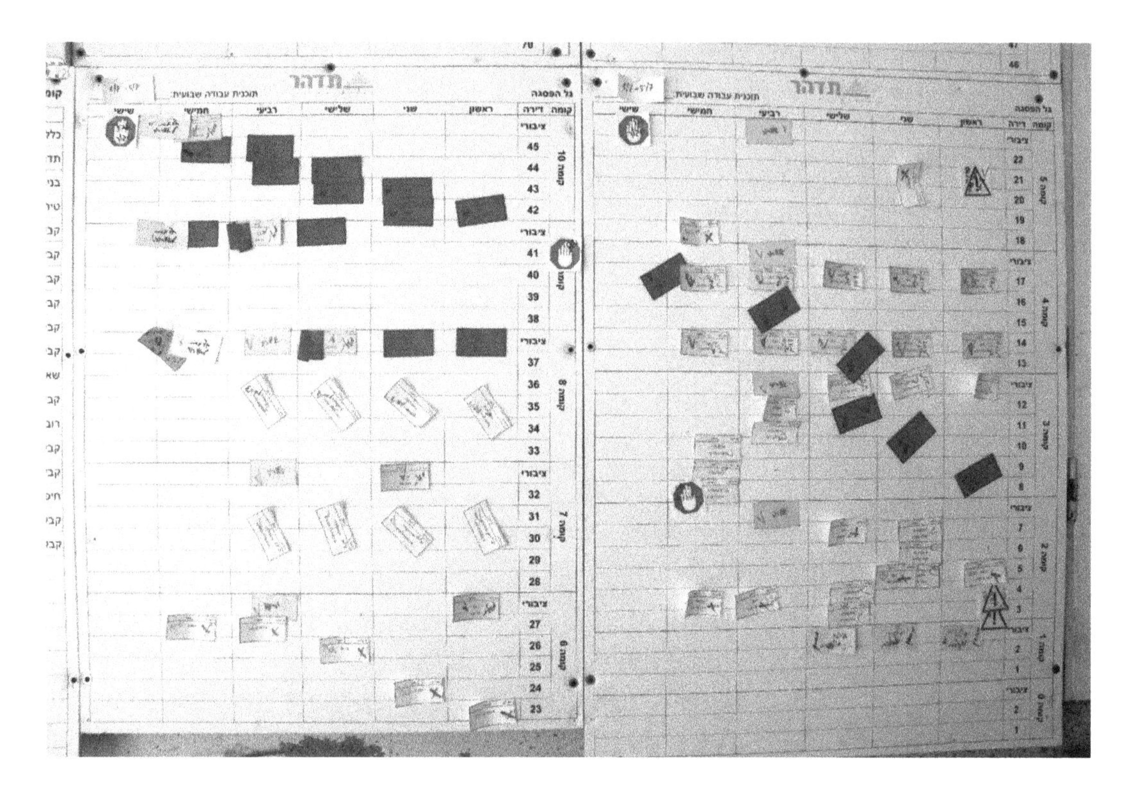

Figure 17.4 The WWP board, including all the works planned from the preceding (concluded) week and the upcoming week

teams gained experience with the process and learned to communicate amongst themselves, everyone went up to the board together to plan, discuss, agree and commit to the next week's plan. In the last part of the meeting, everyone returned to their seats and reviewed the entire plan together. Then the meeting leader went around the circle to ask each person if the plan as portrayed was acceptable to them and if they did indeed commit to it.

The results of this first LPS pilot were clear-cut and very impressive: the first wave of finishing works was completed in only five and a half months instead of the usual ten or eleven. Equally impressively, they finished exactly on schedule. Quality was higher than usual, and the work flow was stable – instead of the usual stress people felt, work was completed in a calm and measured fashion. Everyone at every level knew what was supposed to happen and when; there was a palpable sense that the project was "under control".

The subcontractor trade crew foremen were the most satisfied customers of LPS, since instead of being yelled at, people listened to them (this has remained true for the successful implementations of LPS at Tidhar through to the present). Workers too, not only their crew leaders, experienced improvement in their work climate, since the project was better organized, meaning that the work would be prepared for them before they arrived: materials, work locations, prior team cleared out of the work space.

Table 17.1 *Results from pilot project in Yavne Buildings 10 and 11*

	Experimental group		*Control group*	
	Bldgs 10 & 11	*Bldgs 13 & 14*	*Bldgs 12 & 15*	*Bldgs 16 & 17*
Begin finishing works	20/12/2011	05/07/2011	10/09/2011	15/11/2011
Finish finishing works	28/05/2012	02/07/2012	25/07/2012	10/10/2012
Finishing works duration	5.5 Months	12 Months	10.5 Months	11 Months
Number of man-months of clean-up/correction work	29	58.5	52.5	48

Source: Priven (2016).

Given the results discussed above, company managers came to see the "magic" first-hand. Each visitor from the corporate office who participated in the meetings had his or her eyes opened. The success of the pilot proved to Tidhar that the WWP meeting was the real thing and that it worked, even without the other more intricate layers of LPS.

In May of 2012, only five months after starting the LPS pilot in Yavne 10–11, a meeting was held to review the results of the experiment. Ahmed Hassin, the superintendent, spoke forcefully in support of the method: "I suddenly have more time in my day. I spend less time yelling at people and putting out fires, which leaves me more time to review the plan and step back to see how things are coming along." The subcontractors also reported their satisfaction; their concerns were being taken into account, which allowed them to simply do their work and be more productive. The project manager said he had fewer headaches since everything was going so much more smoothly, and the regional manager also noted that subcontractor reliability had greatly improved. Subcontractors were fulfilling their own commitments to complete work packages. At the same time, the conditions for them to succeed were now more frequently in place. Reliability of the work preconditions is a fundamental aspect of LPS, more so than success obtained through will and commitment.

Ronen Ben-Dor, the regional engineering manager, saw first-hand how the method helped him manage and understand what was going on in real time. He decided that all of the other projects within his purview would implement LPS in order to gain similar benefits. Soon afterwards, Tal Hershkovitz (CEO of Tidhar Construction) issued a directive: all Tidhar projects would now use LPS, both current projects (which would have to learn quickly and start using LPS mid-project) and future (which were expected to be run according to LPS from day one). Around the same time, Ronen Barak was assigned the task of supporting the rollout of LPS to all of the Tidhar projects. He began with the Lagoon and the See-Unik projects, both located in Netanya, and, in parallel, prepared a Hebrew translation of the LPS guidebook.

At the Dan Towers project in Petach Tikvah, Tidhar began extending the use of LPS to the whole construction process, going beyond the first wave finishing works, which had been their focus at the start of LPS implementation,[10] to include the upstream structural work and the downstream waves (installation of components fabricated off site and the final finishing work in preparation for delivery). The "production waves" reconception of the work flow enabled a clearer delineation of work packages and defined handoffs between the trades that were involved with each wave, which reduced the amount of rework and re-entrant flow.

LPS the Tidhar way

With any new method, implementing organizations tend to customize and adjust the method as part of the process of "growing into it" and making it their own. Tidhar was no different in this regard, and their adaptations reflect their particular environment and organizational culture. Tidhar created a PowerPoint "implementation guide" for LPS in Hebrew and used it to train new workers and prepare project teams to work with LPS.

The official LPS, as described by the Lean Construction Institute, includes four levels of planning: master scheduling, phase planning, look-ahead/make-ready planning and weekly work planning. At Tidhar, the focus is primarily on the WWP level. Some phase planning is done: at the beginning of each project phase, a Tidhar employee prepares a draft phase schedule that they present to the subcontractors. The subcontractors give their feedback and ultimately the team commits to the revised plan.

One area where Tidhar has departed significantly from the "pure" LPS method is look-ahead planning. This stage and the make-ready process within it is intended to filter work packages so that only fully ready work packages are presented as options at the WWP meetings. After experimenting with this stage and finding it cumbersome in terms of the resources required, the decision was made to simply rely more heavily on the status knowledge of the WWP participants in avoiding unready tasks. Part of the reason

Figure 17.5 The tradesmen gather at the board to plan the upcoming week at Tidhar on the Park, Yavne

for this, and for its unlikely success, lies in the nature of the work. The majority of the materials in the first wave are bulk materials. Their availability could be checked at the floor or building level, rather than at the individual task level. In addition, the tasks are not dependent on heavy equipment, such that two key preconditions for make-ready are less relevant.

In Lean Construction, the opposite of "make-ready" is called "making-do", which Koskela has identified as a form of waste particular to construction.[11] Making-do happens when work packages are not fully ready – one or more of the prerequisites (materials, work space, tools, workers, information, etc.) is unavailable or incomplete. Starting work in these conditions requires some degree of improvisation. While the ingenuity required to make-do would seem to be a boon for the project, making-do frequently leads to quality problems and usually harms the readiness of subsequent tasks by consuming resources or materials intended for them. Participants in the WWP meetings at Tidhar are aware of this problem, so to compensate for the missing formal make-ready process in most projects, they do it in their heads on the fly during and around the WWP meetings. For important tasks, they do it verbally, or walk through the Gemba to see what the status is. But without a thorough make-ready process, some work packages that are not actually ready slip through to the weekly plan, which means that PPC ratings are degraded.

When projects become too chaotic or risk missing deadlines, Tidhar has shown that its people are capable of buckling down and implementing a thorough make-ready process. This was the case in the Herzliya Marina project. The company was building a Ritz Carlton hotel, but the project encountered difficulties and was at risk of missing the deadline for opening the hotel in time for the winter holiday season. The metaphorical red lights and sirens were going off, so the company took drastic action: a team of Ra'anana head office managers took a break from their day-to-day jobs, set-up shop in the parking garage of the hotel and pitched in to get the project completed.

Ronen Barak was assigned to manage the look-ahead/make-ready process. He used a control database built ad hoc in Excel to monitor status and make sure that only fully vetted and completely ready work packages were released to the construction crews. The effort succeeded and the project was finished on time, to everyone's satisfaction. However, while the Marina story ended well, it is indicative of and indeed reinforced a reactive "fire-fighting" instinct, rather than inculcating a culture of proper forward planning.

Another aspect that Tidhar focused on as they made LPS their own was the dynamics of the WWP meeting. Their standard practice is to hold the meetings within the building under construction (as opposed to an exterior location or a construction trailer), because they prefer to be as close to the Gemba as possible. Tidhar tries to strengthen the sense of teamwork and commitment by having the subcontractors post sticky notes themselves onto the weekly planning boards, which helps to make their commitment more visceral. Participants stand up during this phase and, in typical Israeli fashion, the interactions are lively, but the outcome is shared with any WWP conducted the world over: shared understanding and public commitment to the schedule for the coming week.

Ronen and Vitaliy trained new Last Planner® and WWP participants when "pulled" to do so by individual project managers. They coached and monitored project teams, making sure the LPS processes were run by the book. Other projects implemented LPS without help, and most succeeded. Superintendents who achieved and observed the benefits of LPS by working with it did not go back, but others, who didn't really "get it", remained set in their old ways.

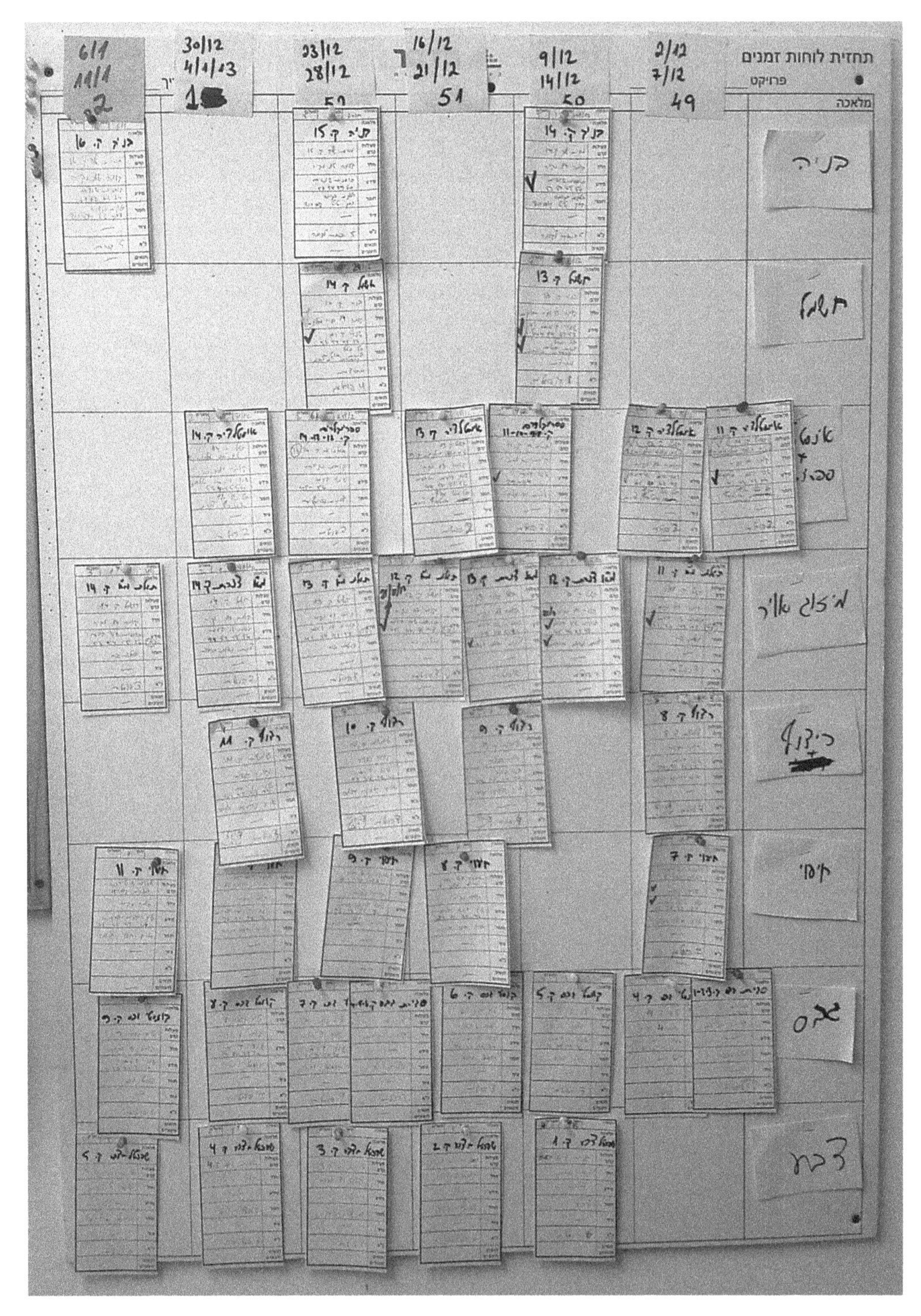

Figure 17.6 The project make-ready board, with each task and its preconditions listed on one sheet, grouped by trade (rows) and date (column), at the Ritz Carlton Hotel project

Box 17.4 The "social subcontract"

Tidhar has shone in its openness to experimentation. As will be told in Chapter 21, "Reflective practice and practical research", the company has collaborated with academia (the Technion) both during their implementation of Lean Construction and Last Planner® and also in pursuing new avenues of research.

In particular, Vitaliy Priven pursued "action research" at Tidhar, looking at how interpersonal connections and social networks impact and are impacted by LPS. He measured the strength of the social networks in various projects and over the course of the implementation of LPS, and developed a tool known as the "social subcontract".[12] This tool was designed to promote trust and mutual responsibility between the GC and subcontractors as well as among the subcontractors themselves. The eventual goal was to improve PPC (and the project performance).

The researchers learned a great deal from the application of the social subcontract tool, and published award-winning articles on the basis of this work. The social subcontract is based on the philosophical concept of a social contract, which governs how different people get along together in a society. The idea is to build connections among the various members of the project to create a shared sense of responsibility for the project, and in particular, to create relationships between

Figure 17.7 Tidhar feedback board from social subcontract meetings

the various subcontractors. The participants create a set of behavioural guidelines that they all commit to, and each week they give feedback about how each participant is delivering on his/her commitments to the project and the other participants. The GC is an equal party to the agreement and the constructive criticism, which means that, when need be, subcontractors are free to tell the GC that the GC is failing to support them and their work – a far cry from the typical power dynamic in a project, but one that is designed to improve project outcomes.

Chapter 21, "Reflective practice and practical research", provides more detail about the research.

LPS at Tidhar today

At the time of writing, six years into the LPS journey, all Tidhar sites have some form of LPS implementation. This aligns with the CEO's edict that all of Tidhar's projects be run according to LPS. Yet the degree of adherence and the results are not uniform across the projects; some projects have tried it half-heartedly, in a "going-through-the-motions" fashion, and given up when they saw insufficient benefit to justify the costs of people's time.

At the start of each new project an initial master plan is prepared in broad strokes. Next comes the phase scheduling step, in which all of the major subcontractors come together with Tidhar to plan the order of operations required to meet the master plan and build the required product. The group decides which trade will do which tasks in which sequence, defining the duration each one requires and delineating the work content for each. Once they do this for one floor, they copy it for the entire building to check if it all works out and meets the required schedule and the needs of each participant. This occasionally leads to adjustments, as in the example of a "quick" task that may jump back and forth between two twin buildings in order to maintain continuity of work for those tradesmen. The project plan is computed and portrayed in an LOB chart using VICO Control software. The plan is the "local" standard and it is expected to be improved upon during the lifetime of the project execution.

Schedules prepared in this way are typically a little longer than the "traditional" method (usually by around 20 per cent), but the probability that the schedule will be met is much higher using this method than the old way. Instead of preparing an overly optimistic plan that is doomed to fail, the phase scheduling method allows all of the parties involved to develop a more realistic and achievable plan. Another big improvement has been in work continuity, which is crucial for most subcontractors. For example, a plan may call for flooring workers to work only half the week, then wait half a week. But no flooring subcontractor is willing to work this way. In practice, such a crew would hold off beginning to work until a sufficient buffer had built up, then come and work on the buffer continuously. When extrapolated out to all of the trades and the entire project, subcontractors opening up buffers of work in this way was a major source of "unplanned" delays (though when explained this way, it seems completely logical). When subcontractors became involved in the planning, they planned for continuity, thus creating a plan that is set-up a priori to be more reliable.

Today there is no process owner for LPS at Tidhar. After the initial successes, there was talk of appointing or hiring an LPS champion, but as the system seemed to be self-propagating, the idea was dropped. In his role as Lean facilitator, Ronen Barak works to support new projects, attending three to four WWP meetings at the beginning of each project to make sure they get started on the right foot, but the LPS implementation is the responsibility of each project.

In those projects where WWP meetings are successfully implemented, the subcontractors are the biggest fans. The quality of implementation can be roughly measured by the punctuality and attendance of the subcontractors; at those projects where LPS is succeeding, the subcontractors arrive before the superintendent. The benefits to them are stability and work flow, and a platform to raise problems that they need Tidhar to fix in order to do their work better. The benefits flow in both directions. Having to report in the presence of their peers on the progress (both successes and failures) of the work packages to which the subcontractors committed has a positive effect on the overall progress and reliability of the project.

Although PPC is measured every week in every project, it is not reported up the chain to Tidhar management. It is used only at the site, relying on the pride of the people involved to motivate them to improve it. One of the main benefits of LPS to the GC (beyond the productivity improvements and reduced lead times) is that it makes the process of managing the construction much easier. It improves the quality of the information about the project status. This was seen already in the original Yavne buildings 10 and 11 pilot, where the superintendent reported that LPS freed up 30 to 35 per cent of his time. This allowed him to spend the recovered time doing things that he previously couldn't, such as proactive quality issues assurance activities.

Challenges

Despite the successes at Tidhar, there are inevitably some areas where they have failed to hit the mark. By reviewing both those areas where not everything has gone right and the difficulties encountered, it is possible to learn from their example.

The decision to skip the make-ready activity, discussed above, is a clear area where Tidhar has missed the mark. Often, when work tasks are not completed on time (which the WWP will discover), a "five whys" root-cause analysis points to a lack of readiness (i.e. a necessary prerequisite wasn't complete before the work began). A make-ready process would address this problem, but project managers claim that they don't have the time or resources to engage in effective make-ready. Yes, they admit, rework also costs money, but that happens after the fact, when the managerial overhead is harder to quantify (and as yet, no one has tried to quantify it). No experiments have been conducted to compare sites that proactively make work packages ready versus sites that do not. Instead, they prefer to pay indirectly for the consequent damage. It is likely, although unproven, that this costs more than the equivalent time that would be required for "making-ready", not to mention the headaches associated with the rework, schedule delays and any quality problems discovered during the warranty period when a customer already occupies an apartment. Ronen explained to a new project team:

> A particular problem that I have encountered is that most project teams have no trouble implementing the Last Planner® System by the book, but some teams just

> don't get the "spirit" of the system. When I run a weekly work planning meeting, I try to flatten the hierarchy. I try to empower everyone to speak their mind. Many superintendents do understand the potential power of the WWP. They use it wisely to build collaboration and understanding of the needs of each party, and the results naturally follow: PPC of 85 to 95 per cent. But other superintendents can't let go of the habit of browbeating their subs, so they just go through the motions of LPS – these projects will more likely have a PPC of 60 per cent. In truth, the "supplier–customer" relationship between the GC and subs is constantly flipping in every project; each has different things it must supply to the other (information, materials, prior work and final work). It's hard to get a handle on this, but you must remember that you also serve the subcontractors! It's not a one-way street where the subs only serve us. We have to remember that we also need to serve the subcontractors!

Not only is there no Tidhar LPS process owner, there is also no governing mechanism to ensure compliance with the CEO's "LPS in every project" dictum. If a particular project manager or supervisor does not apply LPS, there are no repercussions. LPS requires an investment of management time, but not everyone is willing to make the investment. Some simple projects are relatively stable, and for them the investment is perceived to be unjustified. The result is that despite clear benefits to the company, LPS adoption has not been 100 per cent and the lack of management follow-up (in the terminology of PDCA, the "check" step) is likely responsible.

Lessons learned

Lessons learned include:

- The Last Planner® System is a new way of approaching project planning, one designed from a Lean point of view, that attempts to minimize the wastes associated with traditional planning.
- By working through the steps of master plan, phase scheduling, look-ahead planning and weekly work planning, construction projects can greatly improve the adherence of the project to plan and to schedule.
- Part of the power of LPS is its involvement and empowerment of subcontractors, who take an active role in planning in a way that addresses their objectives and constraints, asking in return that they make personal commitments to deliver what they have promised. This increases plan reliability.
- If properly applied, LPS can free up time from project leadership, such as superintendents, who can then use the time to invest in quality and improvement.

Notes

1 "Gemba" is the "shop-floor" where value is added to the product for the customer; see Chapter 9, "Learning to see", for a discussion of the concept.
2 See Laufer (1996).
3 See Applebaum (1982).
4 The "five whys" method was developed at Toyota and is a key Lean tool for solving the problems that are encountered in organizations by digging deeper than the level of symptoms

to determine root causes. If problems are imagined as weeds growing in the organization, the goal is to dislodge not only the above-ground portion of the weed, but also the root, so that the weed won't grow back later. By asking the simple question "why?" at least five times (with the answer of each preceding question becoming the next "why?" to be asked), it is possible to probe deeply into an issue. Very often, the root cause is a fundamental issue with the company, one that may have implications in areas far afield from the symptom that were initially observed.

5 Sacks and Harel (2006).
6 Koskela (2000).
7 Hamzeh et al. (2007).
8 Koerckel and Ballard (2004).
9 KanBIM is a BIM-based workflow information system, that uses on site BIM model to graphically display the status of the work and help make scheduling decisions. See Sacks et al., (2013).
10 The three waves of interior finishing works are detailed in Chapter 16, "Production planning and control in construction", in the section entitled "Production waves".
11 Koskela (2004).
12 The social subcontract (Priven and Sacks, 2016a) is based on Jean-Jacques Rousseau's "social contract" concept (Rousseau, 2010).

18 Raising planning resolution

The four–day floors

Focus on process is more effective than management by results. Planning can be done at different levels of detail. In construction, as the resolution increases, abstract planning gives way to production system design. In production system design, the "black boxes" of abstract planning are opened up and the intricate interactions and movements of workers and equipment must be thoroughly thought through.

Ronen Barak's immediate reaction when asked what started the whole process of changing production rates for the structural work at Tidhar on the Park, Rosh Ha'ayin, an initiative that left its mark on the company, was:

> "The WIP is way too high. Starting all 20 buildings in parallel is crazy!" That was the first thing that came to my mind when I looked at the master schedule for the Rosh Ha'ayin project. It would be almost impossible to achieve standardization or a good learning curve. From a management standpoint it seemed to me that the toll on the company, the project, and the project team would be unbearable, even if we could mobilize all of the resources we would need. We had better come up with something better.

Tidhar standards

Around 80 per cent of Tidhar's projects are typical residential towers from 8 to 30 storeys and four to five apartments on each floor. The size of the floors varies from 600–700 m^2 and the quantity of reinforced concrete needed for each floor is typically 280–320 m^3. Tidhar uses the "Baranovich" construction method to directly attach the exterior stone cladding when casting the external walls. The work cycle for a floor involves nine steps in two stages:

A: Vertical elements (exterior wall, interior walls and columns)

1. In a ground floor "factory" the natural stone cladding is cut, drilled with at least four anchor holes using a special-purpose machine, and set on the exterior steel formwork panels. Window and door stoppers are added, and a steel mesh is tied to the stone cladding.

2 The exterior formwork panels are lifted by the tower crane to their exact positions along the exterior walls on each floor.
3 Reinforcement is tied and the formwork is closed with interior steel panels. At the same time, any interior walls are formed, reinforced and closed.
4 Concrete is poured using the tower crane (for the lower floors, a mobile concrete pump is sometimes used, which reduces the workload on the tower cranes).
5 The formwork is stripped. The exterior panels are returned to the ground, and interior panels are transferred to the next section.

B: Horizontal elements (slabs, stairs)

6 Slab formwork and shoring is prepared.
7 Rebar is tied and any necessary MEP fittings and conduits are installed.
8 Concrete is poured using the tower crane.
9 The formwork is stripped and moved to the next location.

Stage A is repeated across four to six sections of the vertical elements per floor to balance the flow of operations with a concrete pour at the end of most days. Stage B requires less labour and is typically performed in at least two zones so that distinct teams of workers can work on the horizontal and the vertical stages in parallel. Where two buildings are built in parallel, each one serves as a zone. In the case of a single building, the floor must be divided into at least two zones.

Once the structural frame is complete, workers use hanging platforms to clean the exterior stone surface and the reveals between the stone panels. The Baranovich method results in walls with mechanically secured stone cladding without the need for scaffolding. Tidhar designs and fabricates its own steel formwork panels for the walls and the columns at a central workshop. The design of the panels depends on the positions of the joints between concrete pours, which in turn is a function of the number of pour

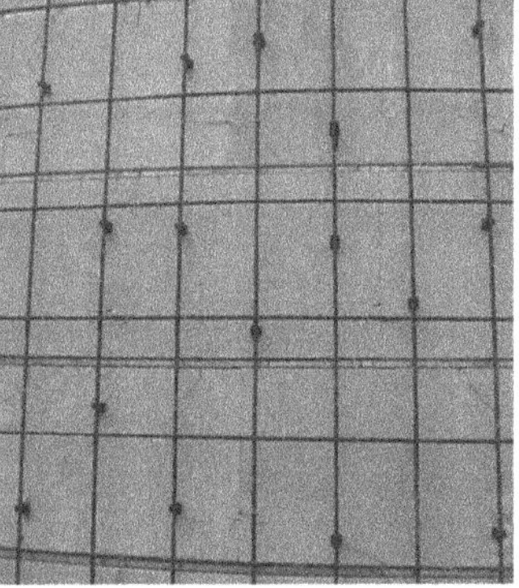

Figure 18.1 Preparing the "Baranovich" system formwork panels at the ground floor "factory": (left) applying stone cladding to a formwork panel; (centre) a window stopper on a wall formwork panel; (right) steel mesh is tied to the back of the stone cladding to ensure adequate fixing of the stone to the concrete wall

cycles that will be used for each floor. Thus the pour cycles must be planned before the panels can be designed and fabricated.

In Tidhar's standard practice, each building has one tower crane to serve both the structural and the finishing works. More than half of the projects have adjacent pairs of buildings that have either identical or mirrored floor plan layouts. If a pair of buildings shares an identical floor plan, one set of formworks is provided to serve both buildings, and each floor is cast in at least two sections, so that the cycle of vertical walls followed by horizontal slabs can be performed alternatively on each building. In the case of mirrored geometry, some of the panels will be shared, but additional panels are provided where the geometry makes it necessary.

The reinforced concrete construction crews typically number from 25 to 32 labourers. One crew builds a pair of buildings in a pair simultaneously. The crew is divided into teams according to professions and building elements: a cladding factory team, a vertical formwork team, a vertical elements reinforcement team, a slab formwork team, a slab reinforcement team and a service team (signalman, cleaning, transportation). To achieve better productivity, the work is synchronized so that when the vertical teams are working on one building, the slab teams work on the other building, obviating the need to split the slab formwork in two on a floor, which is necessary when two buildings are not built in parallel.

Up to 2014, Tidhar was generally considered in the local industry to be among the fastest construction companies with respect to the rate of construction of the reinforced concrete frames of buildings of this kind. When working on two buildings simultaneously, the record rate was four floors per month per crew (the equivalent of two floors per building per month); when working on a stand-alone building, the record stood at three floors per month. Because the perception was that they were far ahead of their competition, Tidhar's people were complacent. They felt no drive to explore ways to improve the processes and, in turn, to improve the company's results. When you're the head of the pack, your processes must be pretty efficient. In the minds of top management, there wasn't much waste.

"The Rosh Ha'ayin Underground"

Tidhar on the Park, Rosh Ha'ayin, had 1,036 residential units spread over 20 buildings: eight 21-storey buildings and twelve 9-storey buildings. The project was arranged in four similar quarters, each of five buildings. The project layout and the make-up of the buildings are shown in Figures 13.1 and 13.2. At a sales event, 800 apartments were sold in one day, also described in Chapter 20, "Pulling it all together", which lead to a very tight schedule for the whole project.

The initial master plan called for two-month intervals between the beginning of work in each quarter (in the sequence 455, 454, 453 and 456). Each quarter would require two months of excavation and earthworks, 14 to 15 months for the structural works, a further ten months to complete the interior works, and two months to obtain certificates of occupancy and hand over the apartments to the customers. The overall schedule – from the start of excavation on quarter 455 to the delivery of the last apartments in quarter 456 – was 40 months. When examining the master plan, it became clear that from month 10 to month 21, the WIP for structural works would be 100 per cent. This would mean four sets of resources working in parallel in a fairly small area. It also meant that although the quarters, and indeed the buildings, were almost identical to one another, since they were working in parallel achieving standardization would be a major challenge, compounded by the fact that each quarter would have a different project management team. The opportunity of building on the lessons of one quarter to improve the others would be severely curtailed.

Cheni Kerem, a Lean consultant working with Ronen to improve production at Tidhar, suggested reducing the WIP by increasing production rates and staggering the structural works in the four quarters:

> What if we could reduce the WIP by allocating two sets of resources per quarter, cutting the cycle time to half the time, and instead of erecting the whole project at the same time, we do the structural work in two quarters in parallel and the other two quarters after the first two are done?

Ronen's first thought was simple: "Impossible". Tidhar's record production rate, standing at four floors per month, was one of the best in the country, so doubling it to eight seemed crazy. However, if it could work, it would definitely improve matters – half as much equipment, half the number of crews, reduced cycle times, earlier start of finishing works in the first quarters and earlier deliveries. Most importantly, there would be a much better opportunity for learning and standardization. Ronen decided to give the idea another chance since in some ways it sounded pretty reasonable, but he also knew that he could not pitch such a far-reaching idea to management until the solution was well planned and demonstrably feasible. He put together a working team to develop the idea further, including himself, Cheni, Vitaliy Priven (at the time, a PhD student pursuing research with Tidhar) and Ortal Yona (a young engineer who was slated to be the production engineer at Rosh Ha'ayin).

The team called themselves "The Rosh Ha'ayin Underground", recognizing their feeling that they had to work under the radar to develop the idea without management interference. At the same time, they realized early in their work that they needed to pursue the Nemawashi[1] approach in order to build consensus around the ideas. The Underground's goal was to compile a comprehensive production plan for the construction of the buildings' structures that would achieve a production rate of eight floors per month. They began by focusing on the pairs of tall buildings in the four quarters, eight in all, each 21 storeys tall. The idea was to achieve a rate of four floors in each building per month, which would mean that they could both be completed within seven months (the ground floor and the roof typically take one month each).

The team realized early on that none of them knew enough about the structural work to start compiling a feasible plan. They therefore approached three leading structural work superintendents to learn the way they think and plan the work cycle in a typical floor. Each of the superintendents was asked to create a production plan for the structure as they normally would plan it. The session with each superintendent took about two hours, and by the end of the process, the team had documented a traditional plan. The plan was portrayed in two parts: a floor plan drawing with each stage highlighted in coloured markers, and a day-by-day schedule in a table, seen in Figures 18.2 and 18.3, respectively. The marked-up drawing shows five sections for the vertical elements, each requiring three days in overlapping steps, as shown in Figure 18.3. The detailed pour plan, an example of which is shown in Table 18.1, also shows the quantities of concrete to be poured in each section. This traditional plan calls for pouring each floor slab as a single pour on the last day of work on the floor, so that the volume of concrete on those days (117.9 m^3) is much more than the daily average. Even without these peaks, the daily volumes for vertical pours fluctuate a great deal, from days with no pour at all, to days with two pours with twice as much (62 m^3) as other days.

The first thing the "underground" team noticed was that the building geometry strongly influenced the way in which the vertical sections were subdivided. There was very little variation among the plans created by each of the three expert structural superintendents. The only differences were in the internal walls and the order of the pours. The schedules were mostly identical as well since the rule of thumb was that each pour section should take roughly two to three days. The team decided not to question these assumptions, agreeing to accept them as the basic principles for the plan they would develop. They wanted to avoid antagonism with a plan that people could not readily understand or agree with. The magnitude of change the team was suggesting was big enough without engendering additional push-back.

Another output from the sessions with the superintendents was data concerning resource allocations and crew size. Here too, the three superintendents had similar opinions regarding the type and operation of the tower crane, selection of the formwork and structure of the crew, with almost the same total number of workers.

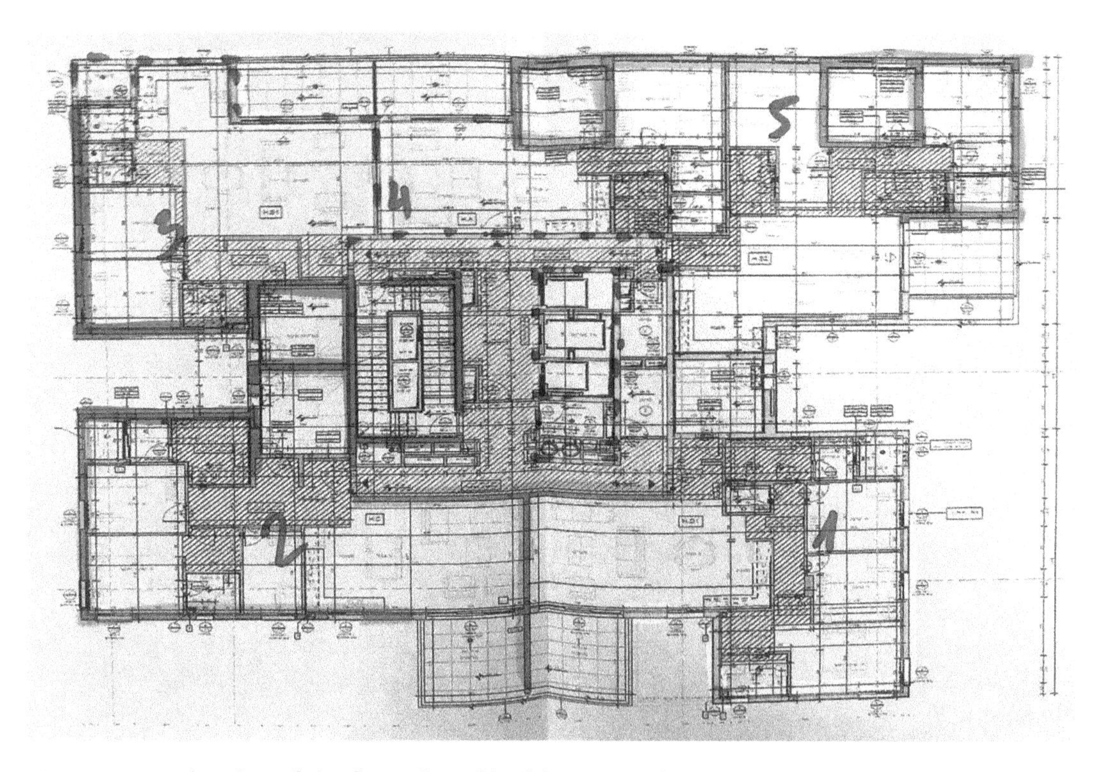

Figure 18.2 A drawing of the floor plan of building 12, Tidhar on the Park, Rosh Ha'ayin

Note: The marked-up drawing shows five sections representing the steps for construction of the vertical elements.

Bldg #	Cast Unit	Concrete Volume (m^3)	Days																							
			1	2	3	4	5	6	7	8	9	10	11	12	13	14	15	16	17	18	19	20	21	22	23	24
11	Cast 1	37																								
	Cast 2	36																								
	Cast 3	19																								
	Cast 4	34																								
	Cast 5	28																								
	Slab	118																								
12	Cast 1	37																								
	Cast 2	36																								
	Cast 3	19																								
	Cast 4	34																								
	Cast 5	28																								
	Slab	118																								

Figure 18.3 Traditional construction plan prepared by an experienced structural works superintendent for two identical neighbouring towers (buildings 11 and 12) in Tidhar on the Park, Rosh Ha'ayin

With much better understanding of the current practice and know-how, the team started working on a new production plan. The goal was to double the production rate from four floors per month to eight using the basic principles of production planning for the structural works and through minor modifications to resource allocations where needed. The quantitative goal was: "One floor every six working days on each of the two buildings". This change in language was the essence of their work. According to Ronen:

> We wanted to change the way people think of a production plan. When the production rate is quoted as X floors per month, the control resolution is too low to achieve improvement. Once the goal is set with a resolution of days, the ability to conceive the plan is greater, and with it, the ability to act. In other words, you don't have to wait a whole month to know whether you have reached your goal. You will know every day if you are on track or not, and if you're not, you will know what needs to be changed and improved.

The team started by building a structural model in Autodesk Revit software. All the concrete elements were modelled. In addition to the structural geometry, all the temporary elements, such as scaffolding and formwork, were added to the model. The idea behind using BIM tools for production planning was to enable analysis of as many plan alternatives as possible with all the precise data needed, including pour volumes, crane lifts, etc. BIM gave them the ability to visualize the plan itself, using 4D simulation. This was valuable for the underground team, but even more so for explaining it and mobilizing support at all levels.

Once the team had all of the information, they started examining solutions to the challenge – how to double the production rate without adding resources. The main resources in use during the structural work were cranes, formwork, construction crews and the site management team. Their first proposed solution was as follows: since the production rate was to be doubled, it meant that half of the project (two quarters) would be erected first and the second half thereafter. This time-shift would allow them to commit all of the project's resources to the first two quarters during the first period, and they could then progress to the third and fourth quarters during the second phase. Figure 18.5 illustrates how this was intended to work. The resources from quarter 455

Table 18.1 Concrete pour quantities derived from the BIM model for five vertical pour sections and the slab section for a typical floor of buildings 11 and 12 at Rosh Ha'ayin

Element	*Cast#*	*Volume [m³]*	*Surface area [m²]*	*Length [m]*	*Work platforms [unit]*	*External formwork*		*Internal formwork*		*Formwork total units*
						Length [m]	*Units*	*Length [m]*	*Units*	
Walls	1	36.8	170	60	5	25	5	94	28	33
Walls	2	35.6	154	62	6	41	10	78	25	35
Walls	3	19.4	98	36	0	0	0	72	19	19
Walls	4	34.0	155	59	5	23	5	90	27	32
Walls	5	27.7	136	58	8	52	14	59	24	38
Walls total		**153.5**	**713**	**275**	**24**	**141**	**34**	**393**	**123**	**157**
Slab	1	56.6	315							
Slab	2	61.3	329							
Slab total		**117.9**	**644**							
Concrete total		271.4								

would move to quarter 453 in the 14th month, and the resources from quarter 454 would move to quarter 456 in the 16th month.

At first, this looked like a straightforward solution. By doubling the production capacity, they could double the production rate. However, on closer examination, it became apparent that it would not work: there wasn't enough room on the site for two cranes for each building, nor could two sets of formwork be used effectively for each quarter. The number of engineers and superintendents needed for this concentration of effort would also be cumbersome. The team decided that each quarter would have only two cranes (one for each building), 1.2 sets of formwork (the buildings were very similar, but not quite identical) and one large construction team.

Vitaliy reflected:

> We had the quantities, we had the resources, and we had a good high-level plan, but it wasn't enough to make the change. We knew we had to deepen our understanding of every task in the process. For example, what is the order in which the external formwork is lifted, or for that matter, taken down? How much crane time is needed for casting or fixing scaffoldings? How is the overall crew divided into teams for each of the tasks? We knew that we would have to identify and remove the wastes, so that we could be more productive with the same resources.

Unfortunately, this kind of knowledge was not immediately available. The company had never addressed production issues to this level of detail before. When asked, project teams gave all sorts of "rule of thumb" answers, expressing the accepted wisdom across the company: one crane can cast no more than 600 m³ of concrete per month; exterior formwork needs to be on the ground in the stone cladding workshop

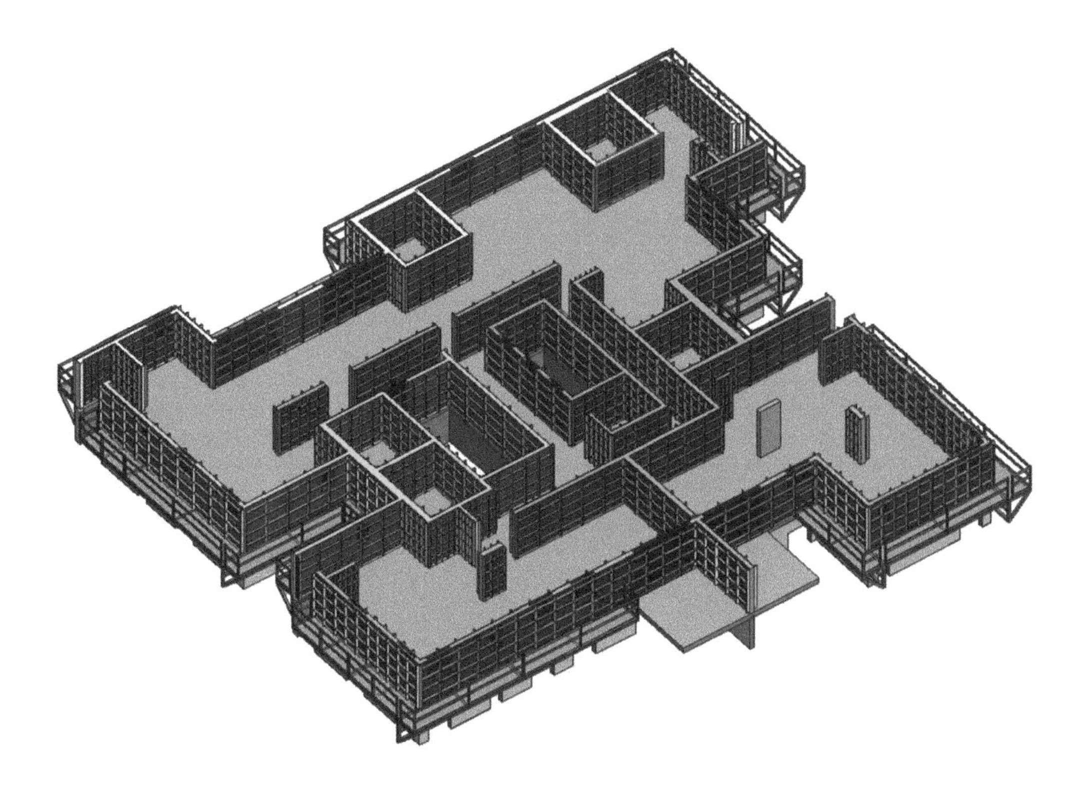

Figure 18.4 BIM model of the reinforced concrete structure of a typical floor of buildings 11 and 12, Tidhar on the Park, Rosh Ha'ayin

Note: The model includes all of the structural elements as well as the formwork and scaffolding.

at least three days at a time for fixing the cladding before it can be hoisted; and it takes between 30 and 60 minutes to unload an 8 m^3 capacity concrete mixer. Not only were these rules of thumb not helpful, they were actually limiting, for two reasons. First, they placed upper limits on people's thinking; if anyone came up with a plan that required or promised better performance than which the common wisdom said was possible, they ran the risk of scorn by their seniors, who were themselves captives of

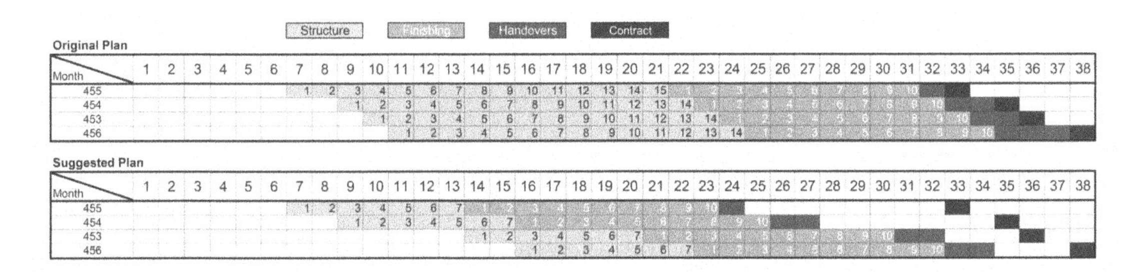

Figure 18.5 The first proposal for reconfiguring the production plan for the four quarters

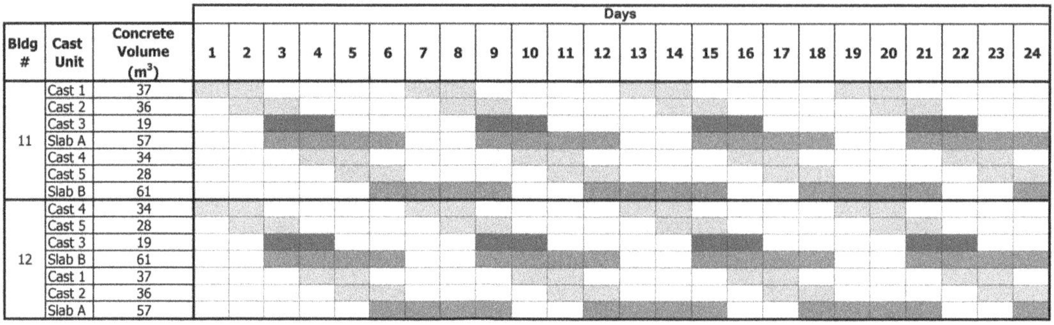

Bldg #	Cast Unit	Concrete Volume (m^3)	Days																							
			1	2	3	4	5	6	7	8	9	10	11	12	13	14	15	16	17	18	19	20	21	22	23	24
11	Cast 1	37																								
	Cast 2	36																								
	Cast 3	19																								
	Slab A	57																								
	Cast 4	34																								
	Cast 5	28																								
	Slab B	61																								
12	Cast 4	34																								
	Cast 5	28																								
	Cast 3	19																								
	Slab B	61																								
	Cast 1	37																								
	Cast 2	36																								
	Slab A	57																								

Figure 18.6 The proposed plan with the erection of two buildings simultaneously

Note: Each building has a dedicated team, one crane and 1.2 sets of formwork. Compare this with the traditional plan in Figure 18.3.

these limits. Second, they were too superficial to provide the detailed information that the team needed in order to properly evaluate the wastes in the process. They represented black box encapsulations of the operations. The Underground team would have to open up the boxes and look inside them to get the information needed to prepare a production plan at the resolution they intended. The information was gathered mainly through site observation. For about one month the team spent much of their time on site timing each of the eight activities on the production line of casting walls and slabs.

One of the activities that was explored was the fixing of the cladding to the exterior formwork panels. The rule of thumb was that it took three days to prepare the forms. But the team's observations revealed otherwise. When working continuously, one worker could prepare one typical formwork panel (5 m long and 3 m high, for a total of 15 m^2) in less than three hours. The average durations measured for each step and the cumulative duration for such a panel are shown in Figure 18.7.

Given that the average number of formwork panels in a vertical element batch was about eight and that the team in the cladding fabrication workshop typically had two to three workers, the cladding team could prepare a whole set in about one to one and a half days. The reason behind the three-day rule of thumb was that the work had always been done in bigger batches. Bigger batches mean more WIP, and more WIP means longer cycle times for the same rate of throughput. To make matters worse – or better,

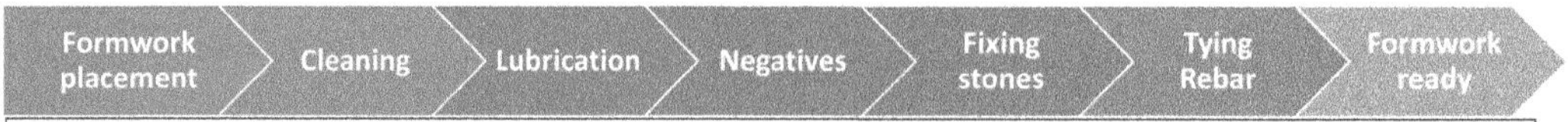

Formwork placement	Cleaning	Lubrication	Negatives	Fixing stones	Tying Rebar	Formwork ready
10	10	10	10	90	30	Task time (min.)
10	20	30	40	130	160	Cumulative time (min.)

Figure 18.7 Measured task durations for preparing formwork panels with stone cladding in the ground floor fabrication workshop

Table 18.2 Work durations in the stone cladding workshop (note that cast unit 3 includes interior walls and has no stone cladding)

Cast unit	*External formwork*	*Total time [min]*	*Total time for one worker [hours]*	*Total time for two workers [hours]*
1	5	800	13	7
2	10	1600	27	13
3	0	0	0	0
4	5	800	13	7
5	14	2240	37	19

from the point of view of the Underground team – larger batches meant more floor area for the workshop, which also meant longer walking distances for workers carrying heavy stone cladding pieces, and thus less efficient work. It appeared that reducing batch size would improve productivity by reducing waste and not only reduce cycle time.

After collecting and analysing all the information, a detailed plan was prepared for each of the vertical pours. The resolution used was now two hours. Each of the seven tasks of a typical cast unit was sketched out, as shown in Figure 18.8. This detailed plan showed the team that the two-day duration for one cast unit was feasible and that the one-day buffer between two identical cast units was sufficient. The high-resolution plan informed the construction team, managers and crews exactly what work needed to be done in each time slot throughout each day. The plan called for completion of two floors every six days, one in each of the two buildings, equivalent to a production rate of one floor every three days. With 24 working days in a month, this meant that the calculated production rate would be eight floors per month.

With their fully detailed plan substantially complete, the Underground team performed a risk analysis. Cheni remembered:

> We knew we had a good plan but it was still far from being proved feasible. We also knew that it would not work and people would not believe in it if we didn't address the major issues of concern carefully: the crane, the formwork, and the construction crew.

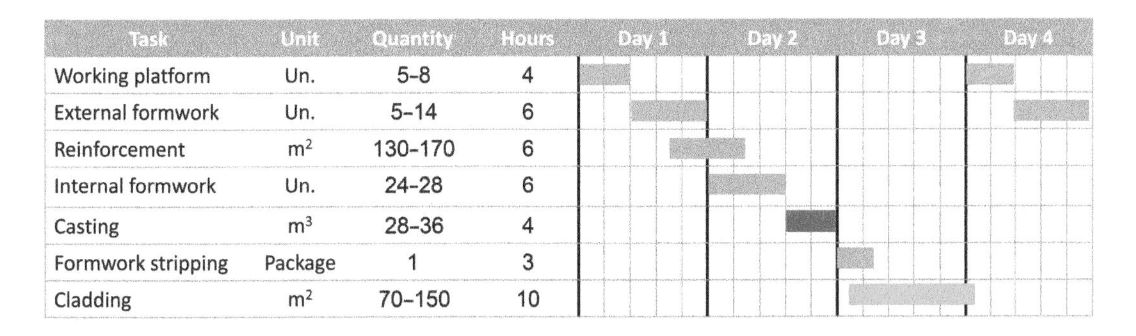

Task	Unit	Quantity	Hours	Day 1	Day 2	Day 3	Day 4
Working platform	Un.	5–8	4				
External formwork	Un.	5–14	6				
Reinforcement	m^2	130–170	6				
Internal formwork	Un.	24–28	6				
Casting	m^3	28–36	4				
Formwork stripping	Package	1	3				
Cladding	m^2	70–150	10				

Figure 18.8 A detailed Gantt chart for the seven operations to be performed on a single vertical pour section

The plan's biggest problem was indeed the resources, especially the crane and the formwork.

The crane utilization analysis showed that for two of the eight days in the six-day cycle time of each floor, each crane would have to operate for 14 hours a day. This is two hours more than the accepted practice of 12-hour working days. Another big issue with the crane was downtime. Looking at previous projects and interviewing superintendents revealed that cranes in the company, whether rented or owned, had an average of one to three days per month of downtime, for scheduled maintenance and for breakdowns. Under the proposed plan, the two buildings were built simultaneously with two cranes (one each), but they would be interdependent because formwork was shared. If one of the cranes were to break down, the specific formwork panels needed for the consecutive section on the other building would remain "trapped" on the first building, disrupting the production cycles of both buildings, even if the tower crane of the second building continued to operate normally. This issue highlighted the impact of the decision to share a set of formwork: sharing a formwork set made the buildings dependent on one another. Even without regard to the tower crane, any damage to or failure in a proactive program of the formwork would bring the production of both buildings to a halt.

To cope with the crane maintenance and breakdown issue, the team suggested adopting a proactive program of "on site maintenance", similar to the Lean-related Total Productive Maintenance approach. Given that the project as a whole had six to eight cranes at all times, it made sense to employ a full-time technician and have at least one set of spare parts available for the full duration of the project. For the formwork, the team suggested expanding the set with several extra formwork panels that could act as spares. The team didn't come up with solutions for all of the specific risks arising from the plan that they identified, but they believed that if they could recruit the extensive experience and strong motivation of Tidhar's people, most of the minor technical challenges that arose could be solved.

The last part of the plan was the economic aspect. Ortal said what was on everyone else's mind: "It would be much easier to sell the plan if we could show that it not only saves time and reduces WIP but also saves money." The team decided they would calculate only the costs directly related to the structural work, which no one could dispute. These costs consist of the management team overhead (project engineers, superintendents, foremen and crane operators) and equipment costs (cranes and formwork). The team assumed that the costs of material would not change. The cost analysis started by calculating the monthly costs for the traditional way of working, to set a benchmark cost. This benchmark could be compared with the suggested plan cost. The analysis showed a direct savings of NIS7.5 million (about US$2 million at the time) which was 7.5 per cent of the total above-ground structure cost. The team knew that the true value of their plan was much higher than that, because of the better quality, lower WIP, better flow and better organizational learning that would result from building the quarters in sequence rather than in parallel, with lower WIP. However, they found it very difficult to quantify the values of these benefits in a way that would persuade people who were still rooted in the "transformation" paradigm and so these added benefits were left out of the economic analysis.

Management commitment (or lack thereof...)

The plan was ready: a detailed schedule, a thorough plan for equipment utilization, full risk and cost analyses. It was almost time to present it to management. The underground team thought it would be a good idea to continue engaging in Nemawashi before presenting the plan to the company's senior management. The team identified the key players: Tidhar's chief engineer, the logistics manager, the intended regional engineer and the CEO of Tidhar Construction. The team met with each of these managers one-on-one to present the plan and get feedback to hone their case. The managers raised many issues and, especially, fundamental concerns about feasibility.

One of the concerns was the interlaced cycle of two buildings. In order to use only one set of formwork, the plan was that the cycle in each of the two buildings would start at the same time but one of the buildings would cycle clockwise while the other would cycle the vertical pour sections counter-clockwise. This sequence was new. It had never been tried before, and its implications were hard to foresee. With the BIM model in hand, Ronen decided to create a 4D simulation using Autodesk Navisworks Timeliner. The team already had a detailed schedule and a detailed model with all the formwork and scaffoldings, which made connecting the model and the schedule straightforward. The impact of such a visual tool was huge. People could see the plan in action (virtually), and theystarted to relate to the fine details of the plan and give excellent feedback. Following the round of individual meetings with all of the managers, the team revised and finalized their plan to accommodate the comments and concerns that were raised, in true

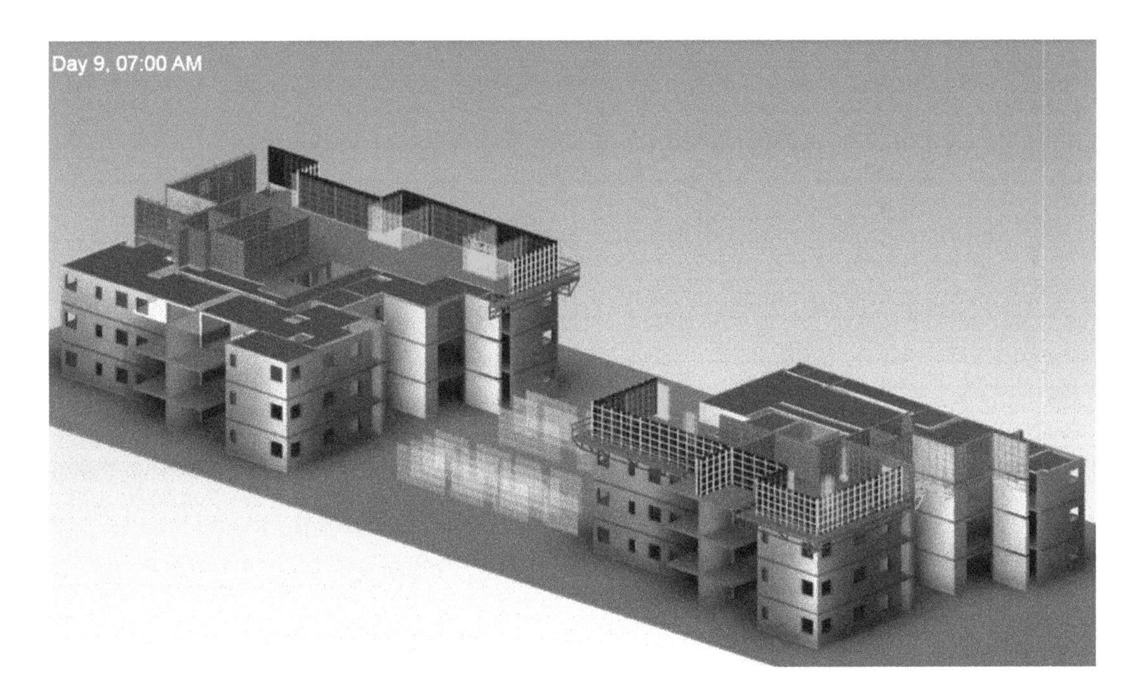

Figure 18.9 A single frame from the Navisworks Timeliner 4D simulation of the Underground team's plan for completing a floor every four days in a pair of buildings at the Rosh Ha'ayin project

Nemawashi fashion. They were now ready for the formal presentation, with the goal of obtaining approval, or at least, for the plan to be considered for further investigation.

All of Tidhar's top management, including Gil Geva, the CEO and Tal Hershkovitz, Tidhar Construction CEO, gathered in the conference room in Tidhar's head office in Ra'anana. The team presented the plan they had developed, using all of the data and the visual tools they had so painstakingly prepared. They brought the structural superintendents with whom they had consulted with them to the meeting. Those present asked questions about the plan, its advantages and its risks. As is common in Tidhar, a passionate, at times heated, discussion developed. On the one side, there was the Underground team with the superintendents; on the other, the company's most senior engineers. Tal explained his position:

> The plan is well formed and the analyses are thorough, but it's too radical. Too ideal. To make it happen, everything will have to work like a Swiss watch! The working culture and the ethics of the crews we work with are far from being sufficiently precise or reliable. Cranes will break down; workers will fail to keep up with the pace over time; and superintendents will not be able to control the production at the fine level of detail that the plan requires. It's just too risky, and to be honest, it's unrealistic in our context. It cannot be done.

Dudi Nasi, one of the structural superintendents, responded:

> The plan is fragile, but it is doable with the right team, the right spirit and the right management commitment. I can make it happen on the ground, but I can't do it without your support. Give us the chance to prove we can do it!

Chaim Kirshon, the chief construction engineer, argued against the plan:

> What if you have problems, and you're delayed by a month or more on the first two quarters? Under your plan, all of the resources will be tied up there so we won't even be able to start the second two quarters. We will never be able to catch up. There is no buffer, so we're exposed to anything and everything that can go wrong. We will end up paying penalties for late delivery – Tidhar never delivers late!

"Yes", the team members replied, "but we will have a savings of NIS7.5 million, some of which can be invested in addressing and reducing these risks." The debate continued in this vein, but the team began to see that they were up against a proverbial brick wall.

The fragility of the plan, due in particular to the interdependence between the pairs of buildings and the quarters, along with the novelty of the approach, meant that management was unwilling to accept the risks they perceived. These issues and attitudes were not new to the Underground team; they had heard them repeated frequently throughout the development process. But, as Dudi had said, the plan depended on management supporting them to solve the day-to-day issues that would arise. In retrospect, perhaps the managers' aversion was grounded in their doubt that the superintendents and the production control team would succeed in guiding (or coercing) the crews to perform the work in the strict and specific sequences that would be needed. Fine-tuned production control was not a part of Tidhar's construction management culture, and it was most certainly not a part of the culture of the construction crews available in the local construction industry.

The meeting was adjourned without committing to the new plan. The team was told to go back to the drawing board. They felt frustration and disappointment: not just because their plan hadn't been approved, but because of the lack of motivation they had seen to make a leap of faith and invest the personal and corporate efforts needed to make the admittedly aggressive plan a reality.

A few days after the presentation, Tal met with Ronen. Tal tried to explain why he felt that the organization and the market were not ready to make the change that Ronen's team had proposed. As he said during the presentation, he believed that the plan was too fragile and that the modes of behaviour that were common in the industry would make it a non-starter. However, Tal did recognize the potential to improve the process:

> The Baranovich method is very efficient as it is, and the way we perform it is good and solid. Yet at the same time, I agree that it can be done better. If your team can come up with a plan – based on the current practice, with the same crew structure, cranes and formwork cycle – that produces 6 floors a month, I will embrace it and do all that I can to make it happen.

The team accepted the challenge wholeheartedly. At first, it looked like it would be easy. They had already figured out how to increase the production rate from four floors a month to eight, so the target of six floors a month should have been a cinch. Looking deeper, however, revealed a different story, because the starting post was different. The plan they had presented, with eight floors per month, called for a separate crew and a separate tower crane for each building in the pair – only the formwork set was to be shared. Current practice for this set-up – with one crew per building – was three floors per month. Since the eight floor per month plan was for a pair of buildings, the target rate was four floors per building per month. The improvement they had planned was from three floors per month to four floors per month, a rate increase of 33 per cent. Yet what Tal had proposed as a compromise target was to achieve six floors per month with one crew for two buildings, instead of the current record rate of four floors per month. The new plan they would devise, to meet Tal's target, needed to deliver a gain of 50 per cent in the production rate for a typical crew!

A production rate of six floors per month meant that the construction crew would need to cast a slab in one of the buildings every four days. Since the cycle time of both buildings is the same, the overall duration of each floor, i.e. walls and slab, was eight days. The team started by dividing the walls into five pour sections, as in the original plan. The plan they came up with is shown in Figure 18.10. Under this plan, each pour section required two days to complete and there was a one-day overlap with the next section, meaning that the total duration of the walls was six days per floor (the first section was completed on the second day, and the following four sections were completed at consecutive one-day intervals, making a total of six days). The slab was to be cast as a single unit with an overall duration of five days.

To ensure an even work flow for the slab crew, the crew would start to raise the slab formwork from the previous floor to the current floor after two wall sections had been completed (i.e. on day four of the cycle, as can be seen in Figure 18.10). This plan met the eight-day cycle time for the whole floor, but it was one day short of the four-day goal

Bldg #	Cast Unit	Concrete Volume (m^3)	Days																							
			1	2	3	4	5	6	7	8	9	10	11	12	13	14	15	16	17	18	19	20	21	22	23	24
11	Cast 1	37																								
	Cast 2	36																								
	Cast 3	19																								
	Cast 4	34																								
	Cast 5	28																								
	Slab	118																								
12	Cast 1	37																								
	Cast 2	36																								
	Cast 3	19																								
	Cast 4	34																								
	Cast 5	28																								
	Slab	118																								

Figure 18.10 The proposed plan to produce six floors a month with a single crew working on two buildings

Note: On days 5–6 and 9–10, highlighted in the timeline, the resources were insufficient because three wall cast units were scheduled to be built simultaneously.

between slabs because the slabs required five days, which meant that only five full floors could be completed in a month.

The only way to reach the goal was to push the beginning of each sequential floor one day back. But this gave rise to a new problem. The wall crew could deal with only two vertical pour sections at a time, closing one section and opening the next one on the same day. The suggested plan showed that on four out of the six days, the wall team would need to work on three pour sections: two on one building and one on the other (days 5 and 6 and days 9 and 10, highlighted in Figure 18.10). The only way to overcome this problem was to increase the size of the wall crew, but they would then experience an unbalanced workload because on the other days there would not be enough work for the larger team.

For Ronen's Underground planning team – who were now working in plain sight, given Tal's direct challenge and promise of support – this wasn't necessarily a bad solution if it meant that they could achieve their overall goal. However, the apparent drop in productivity would be a "hard sell" for the construction crew. In the local industry, the work crews are paid per cubic metre of concrete cast, not per hour. This arrangement generally suits them well, as it results in relatively high monthly earnings. In the case at hand, enlarging the crew but not the total amount of work would ostensibly mean they could earn less money. This is counterintuitive because by following the proposed plan their overall income would increase due to improved total production per unit time (despite the discontinuity in their activity). Nevertheless, Ronen's team feared that instead of waiting when the plan required them to, they would reduce the number of workers in the crew, thus upsetting the flow of the plan. However, neither the crane utilization analysis nor the formwork downtime showed any causes for concern. If the issue of the wall crew size could be somehow overcome, then the plan was feasible.

The production planning team turned their efforts to selling the plan, starting with the superintendent and his team. As expected, their main concern was the two days of underemployment, but they did understand the overall gain of the plan and they were willing to test it. The plan was also presented to Tal, who requested it, and to a small group of managers. This time the "no-longer-underground" team got the approval they needed and wanted, and everyone was set to go.

Creating a visual plan

The key element of the plan was organization. The team understood that the main problem with the current practice was not simply the production rate, but the wastes in the interfaces between the different phases of the process. They recognized three major sources of waste:

- The crew in the cladding fabrication workshop would often prepare the wrong wall formwork panels. Under the existing controls, their priority was to complete as many panels as possible, regardless of the sequence of panels that was needed for the floor production process. This would have to be changed – they would need to prepare the panels in exactly the sequence that was needed.
- There was considerable variation in the time and effort needed for releasing wall formwork after the concrete pours. Forms that had not been greased sufficiently would become stuck and window negatives could be difficult to extract. When this happened, and it often did, the crane would be (quite literally) "tied-up", tethered to and supporting the formwork panels that were being stripped while workers struggled to get them free. The equipment would have to be made more reliable so that the work flow could be more stable.
- The duration of work for the slab team varied because of the large batch size (in each cycle, there were four wall pours but only one slab pour). Monitoring the progress of the slab crew was therefore less well defined, which meant that they were not able to consistently maintain the correct rhythm. When they fell behind, it affected the sequence of the wall crews.

To solve these issues the planning team realized that the plan must be broken down, as before, into smaller-grained work packages. After talking to the construction crew, it became clear that providing an hourly plan, like the one made with Navisworks, would not serve the purpose and might just make things worse by confusing the crew. Instead, they would create a daily plan for the walls crew that detailed five operations-specific work packages: exterior formwork, reinforcement, closing formwork, casting concrete and stripping formwork.

The production planning team struggled to find the right tool to create the necessary plan. They were familiar with a number of different software packages, but they feared that the outputs would not be readily accessible to the construction crews. They also didn't have all the information at hand, and they again met with Dudi the superintendent and his construction crew leader, Zhu Hongxi. "Let's just print out eight copies of the formwork plan on an A0 poster sheet (one for each day), hang them on the walls, and mark each of the work packages for each day in a different colour", Ortal suggested. All team members embraced the idea and they were set to go.

It wasn't an easy task. Dudi and Zhu weren't accustomed to thinking through the process at this level of detail. They were results oriented, rather than process oriented. They felt far more comfortable setting overall production targets, in units of area of formwork or volume of concrete per person per day, than detailing precisely which piece of formwork or which section of wall was to be cast on each day, and in which sequence they were to be cast. Creating a structured daily plan was frustrating. Ortal recalled:

It was like teaching someone to breathe in a different way. Dudi and Zhu were very professional in casting concrete. In the field, they knew exactly what needed to be done and how to do it. They were very good in reacting to the emerging reality they faced in each minute of the day. Although it was all in their heads, they found it very difficult to explain in words or to put down on paper how they went about deciding what should be done on the site, or whom should do it. I tried so many approaches to work with them to compile our plan: going element by element, hour by hour, going forward, going backwards, but nothing gave us the plan we were looking for. Eventually, we decided to go with what they knew best. Results. We asked a simple question: "What do you want to achieve from each of the crews on every day of the casting cycle?" That did the trick and we got what we wanted.

It took the team five two-hour sessions to generate 12 daily target sheets for the two buildings (the buildings were not quite identical).

The paper plans with all the mark-ups had to be computerized. Ronen explained:

We could have done it in Autodesk Revit and Navisworks, but since this was the first time we were doing this, we decided on a quick and simple approach. We used Microsoft PowerPoint, plain and simple. We embedded the formwork plan as a picture and started to draw on top of it. The second reason we did it this way was that we didn't have the reinforcement modelled. We could have created a mock-up element representing the rebar, but again, it was time consuming and we were not sure that it would have value.

The result was a daily production target sheet showing all the tasks for the day. Each operation was represented in a different colour, and to provide context, the previous day's work and the already existing elements were also highlighted, each in their own colour. To complete the picture, Cheni suggested adding a crane utilization table with all the elements, quantities and the planned duration for each lift. Figure 18.11 shows an example of the daily production target sheets that were prepared.

"We had arrived at a really good plan for the wall team", Ortal complained, "but what about the slabs? We still needed to organize a plan for the slabs and create similar daily production target sheets." She was right, of course. The team realized that they had been focusing so much on the vertical elements that they had neglected the slabs entirely. Even in their own master plan, the slab had been encapsulated in a single five-day long "black box" activity. Not only did they not know much about production rates for the different tasks, the actual process sequence itself was unclear. The bigger problem was, they soon learned, that nobody could give them any additional advice or information. The slabs had always been considered as one cast unit, usually between 600 and 700 m^2 with approximately 150 m^3 of concrete. The team did know the task breakdown: set-up formwork, install bottom layer reinforcement, install electrical and plumbing conduits and fixtures, install top layer reinforcement, pour concrete and strip the forms; but they did not know how these tasks could be broken down into daily work packages with distinct locations. There was no apparently obvious location breakdown. It was clear that the batch size wasn't the whole floor, but they couldn't pin down what it was exactly. Unlike the walls, where the work breakdown structure was clear and sharply defined and constrained, once the slab formwork was fixed, the whole

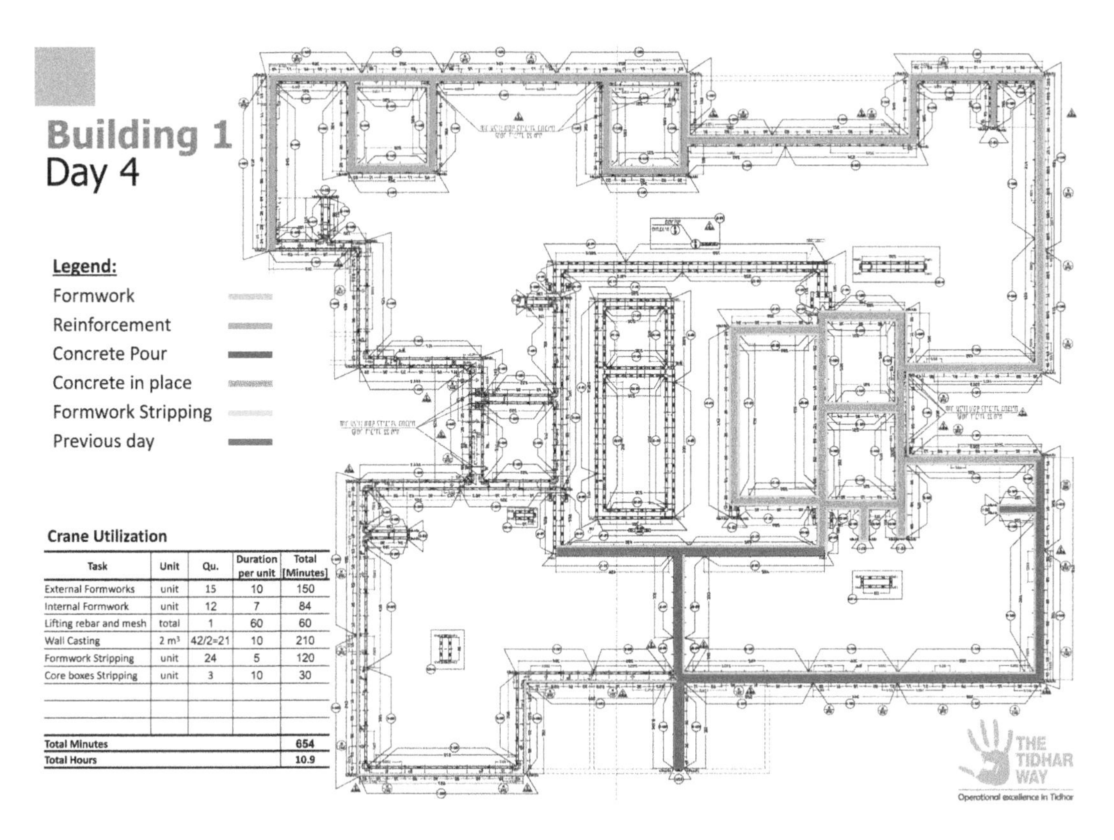

Crane Utilization

Task	Unit	Qu.	Duration per unit	Total [Minutes]
External Formworks	unit	15	10	150
Internal Formwork	unit	12	7	84
Lifting rebar and mesh	total	1	60	60
Wall Casting	2 m³	42/2=21	10	210
Formwork Stripping	unit	24	5	120
Core boxes Stripping	unit	3	10	30
Total Minutes				**654**
Total Hours				**10.9**

Figure 18.11 A typical daily production plan sheet for the construction crew building the vertical elements

slab became available to all the crews. Even after hours of observation, the planning team still found it difficult to establish some sort of coherent process. It seemed that every worker on the slab was doing his business in a different location, going back and forth from location to location. It was like an ants' nest. The planning team became progressively more frustrated as the knowledge they were seeking was nowhere to be found. None of the superintendents they were interviewing knew the answers and even the crew leaders couldn't help. Unfortunately, the slab "black box" could not be cracked open and the team had to move on with the plan they had, with daily target sheets for each day of the vertical elements and just one five-day target for the slab. The only things they knew for certain were that the overall work on the slab should take five days and that the slab formwork should commence as soon as the second wall section on each floor was stripped of formwork.

The planning phase was over. The team had a solid visual plan for a typical floor, and the time had come to put the plan to the test. The structural work for the parking basements and the first two (non-standard) entry floors in both buildings was almost complete. Their production system plan was due to come into effect from the third floor of each building, as these were the first of the 19 standard floors. Now they faced a new problem – the only way to synchronize and start the process correctly was to stop the work entirely on both buildings once the second floors were done, instruct the crews and begin "cleanly". But the prevailing culture of "getting things done", of not stopping the work, was a hurdle they would have to overcome. To make matters worse, the subcontracted crews were paid per amount of work done, not for their time, and so any planned stoppage was greeted with hostility. Vitaliy reflects:

> We almost started a war, it was against every principle the company used to work by: the work never stops. There is no time to waste. Thing can go wrong so we should always start as early as possible. We had to convince everyone from the project manager, through the superintendent and the crew leader why it was in their best interest to start the third floor with a clean slate. It took hours of explaining that we can't start working unless we know for certain the formwork will flow the way we want, how we can't start until we're sure that every crew member will be available to perform the planned tasks, and what the implications of deviating from plan would be on the interfaces between the walls and the slabs.

In order to convince everyone, the planning team calculated the downtime that would be needed to synchronize the two buildings and all the people involved, and found that it would be less than one week. Since the construction team had remnants of unfinished work in the basement (such as pouring stairs and landings), the delayed start would actually have minimal effect. Dudi, the site superintendent, was the first to agree and commit to the new production plan. He was the only one who realized the potential of working according to a detailed plan, one that defined hour by hour what everyone should be doing. Best of all, he was really eager to break the company record of number of floors per month. His personal best was five floors, a badge of pride in a company that fostered competition among its superintendents. The team assured him that the plan, if executed in full, would bring him to a record six floors per month: *one complete floor – walls, columns and slab – every four days*. Dudi gave the green light and 11 January 2015 was set as the starting date.

Putting the plan into action

The eleventh of January arrived and everyone was anxious to start work. Dudi and his crew were on the third floor and starting to lift formwork panels according to the plan. The planning team, Ortal, Cheni, Vitaliy and Ronen were also there to see their plan being realized. It was an exciting day, and spirits were high. After lunch, five hours after the work started, Ronen recalls returning to the floor:

> I remember on my way back to building 11, looking up and seeing a formwork panel lifted, and I thought it was to the wrong place. I wasn't sure. I climbed up to the floor and I opened the daily target plan, which I had in my pocket, to check. The formwork panel they were lifting was part of day three – they were not supposed to lift it that day. I approached Dudi and asked him why the crew lifted that panel. At first, Dudi didn't understand my question. "Why does it matter?" he asked "We need it there anyway." It was a defensive answer. He thought I was criticizing him; he didn't realize I was just trying to understand the problem. Maybe it was the right formwork panel and we got the plan wrong? After explaining myself and insisting that we try to

understand the problem, Dudi told me that he knew it was out of sequence, but that particular panel was stripped from building 12 and delivered to the cladding workshop at ground level earlier than expected. The workers had already finished fixing all the cladding stone on it, and it was ready. Not only that, it was occupying space in the workshop that was needed for other formwork panels. So they hoisted it.

This was a big revelation. Never before had the planning team or the crew paid any attention to the formwork stripping process. The production plan focused on creating value by lifting the right formwork to the right place at the right time, but it ignored the stripping and fabrication operations that were a fundamental part of the flow. The crew had pushed the wrong panel because it was the next one ready, not according to the plan, but still ready. Now, they realized the importance of stripping the formwork panels in the right order. Dudi too, was on board, and devised a practical solution to controlling the flow – he sent his foreman to supervise the stripping process in building 12. Cheni recalled:

We were ecstatic: we could see the process for the first time, we could see the problems as they occurred, and we could find solutions not on a piece of paper but right there on the floor. The potential was huge, and we were only five hours into the process. It was the first reward for our hard effort.

The next few days were no different. The crew tried to work according to the daily plans, and many other problems arose throughout. Both Ortal and Ronen spent hours at the Gemba observing the work, interacting with the crew and monitoring progress:

At first, the workers did not understand our job, our place in the management hierarchy, nor what we were doing. We asked so many questions, and they felt our role was unnecessary interference. Since the crew was made up of Chinese workers, the culture differences and the language barrier played a big role in miscommunication.

It was a long process and Ortal and Ronen had to find creative ways to communicate with the crew and explain the benefits of the production plan. Ortal reflects:

Eventually the crew got used to our presence. We became part of the team and worked closely with them. We encouraged them to develop an understanding of the way they worked, of the flow and of the problems they encountered, and they needed us to help them solve problems and become better and faster. It was only with a lot of respect for the crew's hard work that we could succeed in what we were trying to achieve.

The first cycle took nine days instead of eight, but it was considered a huge success. At the end of the cycle, as a token of appreciation, Dudi and the planning team organized a small party on the working floor with cakes, pastries and cold beers. Dudi reflects:

It was very fulfilling. After six months of planning and exploring different options, we could finally see the first results of the process. We needed to let our partners,

> the workers, know they were part of the success. There was still a lot of work to be done and we needed them on board for the rest of the way.

Dudi made a short speech that was translated from Hebrew into Chinese, explaining what their ongoing achievement meant to him. He in turn was surprised a short time afterwards when he received a call from Gil Geva, Tidhar Group CEO and co-owner. Gil was familiar with the long planning process and the change it was about to bring. He called Dudi to congratulate him and his team for their effort in executing a well-orchestrated plan. Dudi was overwhelmed by the conversation with Gil. It gave him the confidence to continue with the process in the knowledge that he had earned respect and support for his effort from the highest level of the company.

After finishing the second complete cycle, the team realized that the third wall pour section could be eliminated by distributing its elements among the other four pour sections. This was the smallest section, with just 20 m^3 of concrete. The improvement was supposed to cut a full day out of the cycle and help the team achieve the eight-day target cycle time. The reason the team could consider this improvement at all was the plan itself.

Figure 18.12 A well-earned celebration – some of the construction crew on the fourth floor of building 11, Rosh Ha'ayin, marking completion of the first complete cycle of the "four-day floor" production plan

The plan gave the crew a standardized process and helped them understand every step of their work and the effect it had on the whole process. They also knew the production rates of the different tasks with precision, and so they could calculate the effect of expanding the work package with extra formwork and concrete to pour. A new plan was prepared and the crew started to work accordingly. The wall casting cycle with four pour sections in five days was stable and was maintained until the end of the project.

Going to the Gemba

Ortal and Ronen understood the importance of going to Gemba, and they made themselves available to the work crew at all times. They decided to embrace the Deming circle of "plan-do-check-act" and to use it on a daily basis. The "plan" and the "do" resolution was daily and the "check" and "act" were synchronized accordingly. This meant that either Ortal or Ronen checked on the progress of the work every day, compared progress to plan, identified any deviations, analyzed what the problems were and analyzed solutions with the crew. Monitoring every day was essential because it prevented small problems from growing into big ones.

Visual controls

As work progressed, it became clear that processing the formwork panels in the right order was crucial to achieving the correct flow. Each panel flowed through steps of stripping from the concrete, lowering to the workshop, cladding with stone and hoisting to their next pour location, and so the sequence of panels within each of these operations had to be the same as the order for the whole process. Ideally, the batch size would be one, with single-piece flow, rather than a batch of panels being processed without any particular internal order. The production planning team came up with the idea of attaching large signs on top of each of the formwork panels, numbered according to the order in which they should be stripped, lowered, cladded and hoisted. To make it convenient and visual, they coloured the signs according to the poured sections. The signs were double-sided and assembled on the formwork panels at a 45° angle to make them visible from below and from above. The idea was that the tower crane operator would be able to find the next formwork panel to lower or to lift without needing someone from below to provide directions. If the operator knew the colour of the current pour section for any given day, then the task of sequencing the lifts would be easy – simply follow the numbered sequence for today's colour. In this way, the tower crane operators would be able to predict their lifts, making the crane more efficient, and they could also help to maintain the right flow sequence, providing an extra check.

Unfortunately, sometimes great ideas fail for prosaic reasons. This ambitious idea wasn't realized for two reasons. The first reason was technical: despite their large size, the signs were simply too small for the crane operator to read from their cabin, 60 to 70 m above the ground. The second reason, far more significant, was that despite the clarity of the visual plan sheets and the colour-coded panel numbers, the crew struggled to follow the precise lifting order within the scope of the day. There were a range of small technical problems that arose that prevented particular formwork panels from being lifted in their allotted sequence, but the biggest problem was the culture. The level of resolution of the plan, with single-piece flow, was too high for the crew. They

Figure 18.13 Large signs on top of each of the formwork panels, numbered according to the order in which they should be stripped, lowered, cladded and hoisted

Note: Each cast unit had its own colour based on the colours used in the original plan.

just couldn't think of the process at that level of detail, and over time, couldn't sustain the concentration and control that was needed.

The slab operations defy improvement

The slab operations remained the largest obstacle to fulfilling the plan. The production plan called for starting to erect the slab formwork right after stripping the interior wall formwork panels from the second pour section, i.e. from midday on the fourth day of the floor cycle. In reality, this goal was rarely met because the wall crew prioritized hoisting and placing formwork panels for the third pour section over stripping the previous section. Whenever the demand for the tower crane was high, hoisting was given preference over stripping. It took many cycles for the crew to understand the folly of this approach and to change their behaviour. Until that happened, the start of the slab formwork was delayed until the fifth day of the cycle, which resulted in a nine-day total floor duration instead of the target of eight days.

Worse, even though the overall slab cycle time was planned to be four and a half days, for the most part, the actual time was between five and five and a half days.

Reflecting on the results

What was achieved?

The two towers topped out on 10 November 2015, 40 weeks after the production plan cycle began. Dudi and his crew struggled for months to achieve the target of six floors per month, but unfortunately, they never quite made it. Their average was 5.1 floors per month,[2] with a peak rate of 5.5. As the building grew in height, hoisting and lowering times grew longer, with an adverse effect on the floor cycle time. Drilling down further into the numbers showed that the cycle times varied from eight to nine days per floor and from four to five days per slab.

Despite not quite achieving the ambitious target, the overall result was very satisfying, considering that this was the first time the company had aimed to achieve a 50 per cent improvement in the production rate: 5.5 was a record that Dudi and his crew now proudly held, even if they had not achieved their goal of six floors per month. According to Dudi:

> I have worked in the construction industry for almost 30 years, ever since I was a boy helping my dad. What we did in this project was one of the highlights of my career. I admit that I was sceptical at first, considering the ambitious goals that were set, but along the way I came to trust the people I worked with and the overall system to help me reach our record. The thorough planning process, the new way of looking at things through the Lean perspective of flow, and consistent attention to detail to identify and solve problems on the floor: all of these were big parts of our success.

The flow of the work and the flow stability were measured by one specific parameter: daily concrete pour volume. The experimental project (buildings 11 and 12) had a "twin" project (buildings 16 and 17) with identical geometry, construction methods and crew structure that could be used as an experimental control. This created a great opportunity to check whether there was any real difference between the two production systems – between the high-resolution plan for flow in buildings 11 and 12, and the low resolution "business as usual" approach of buildings 16 and 17. The daily concrete pour volume histograms for the two projects are shown in Figure 18.14. They show that the results supported the theory. The control project, with a low-resolution plan, showed a very unstable rate of concrete pours and the high-resolution plan showed a much more consistent rate of production.

What problems were encountered?

Several issues appear to have prevented Dudi and his crew from reaching the target rate. One of the biggest issues turned out to be the unreliability of the tower crane, which was largely the result of poor maintenance. After both buildings were finished, the crane maintenance report showed approximately 30 breakdowns over the 14-month period. The breakdowns were mostly for a few hours at a time, not full days, but in most cases even a few hours of crane downtime were enough to force postponement of the day's concrete pour to the following day, so that a full-day delay was incurred since the concrete needs to set overnight. The damage to the flow and production rate that resulted from unreliability of equipment was thus amplified. Unlike the effect of stopping a conveyor belt in an assembly plant, where the lost time can be limited to the downtime itself, in the context of construction work the delays are in discrete units of days: the night time is an essential part of the process time because concrete must harden before formwork can be stripped.

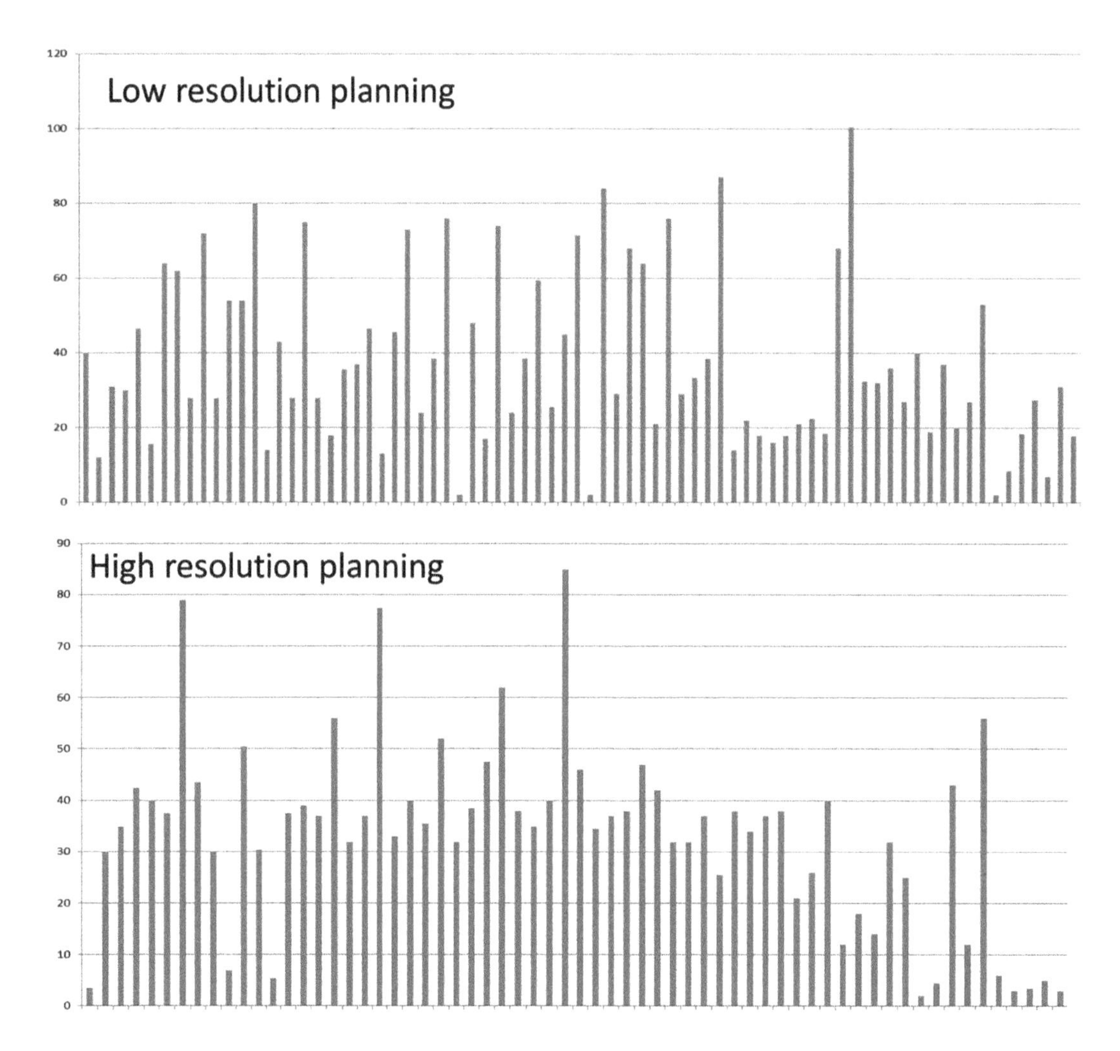

Figure 18.14 Histograms of the daily concrete pour volumes for the two projects: (top) shows the flow of work for buildings 16 and 17 (business as usual); (bottom) shows the flow of work for buildings 11 and 12 (the Underground team's high-resolution plan). The tall peaks are the slab pours; the vertical section pours were far more uniform with high-resolution planning

When working with a low-resolution plan that sets weekly targets and fails to plan or monitor daily or hourly tasks, a few hours of delay are not counted since their effect is not measurable on the scale of a week. A plan with higher resolution is more sensitive to small issues and glitches.

A second problem that was not given sufficient attention by the production planners concerned the motivation of the workers. The crew that worked with Dudi at this time were nearing the end of the maximum period of their work visas, and they were due to return home to China as soon as the project ended. They were highly skilled and professional, but to get the most out of a continuous improvement effort requires long-term commitment and motivation, which they did not have. From their perspective, the fruits of any improvement efforts would be short lived.

A third problem also had its roots in the culture of the construction crews. Construction of buildings 16 and 17 (business as usual) began two months earlier than construction of buildings 11 and 12, and that project was fully two floors ahead of 11 and 12 when the later project began its new production plan from the third floor. As time passed, the gap steadily shrank, as the production rate of buildings 11 and 12 was significantly faster. This did not reflect well on the professional reputation of the construction crew of buildings 16 and 17 – after all, the product, the construction methods, the equipment, the weather and all other parameters were identical. This was a matter of honour for the crews. But unfortunately, instead of working to improve their own methods, they approached the crew of 11 and 12, asking them to slow down. The production planning team members were neither aware of nor sensitive to the issue, and simply grew frustrated from that point on. Perhaps there was also a sense of pride in that their plan was proving itself, and the slower pace of the parallel project was the proof. Although they all worked for the same company, and ostensibly all had a common interest in the economic success of buildings 16 and 17, they had very little, if any, motivation to support the "other" crew.

Epilogue

The effect of these problems became apparent as construction of the third and fourth quarters of Rosh Ha'ayin began (buildings 11, 12, 16, and 17 belonged to the first two quarters of the project which were started earlier). The planning team's high resolution production plan was applied with particular success in buildings 1 and 2, with results that showed that when the problems identified above were addressed, further improvement was indeed possible.

Buildings 1 and 2 were started two months later than buildings 11 and 12. Dima Valdman, the structural works superintendent, decided to take on the challenge and reach the goal of six floors per month. Dima was much younger than Dudi Nasi and very ambitious, but more importantly, he had less ego and was therefore eager to seek and willing to accept the help and advice of others. Almost two months before he was due to start his typical floors he approached Dudi and asked him if he could accompany him and learn the plan, the site and the crew organization.

Dudi became his mentor, and over two months Dima learned everything he could. He understood the plan with all its advantages and challenges and saw the daily problems and the way to solve them. When the start date for the typical floor cycle arrived in buildings 1 and 2, he was ready with a revised plan. His crew was a young group of Chinese labourers, also eager to succeed. Their crew leader was a very experienced form worker who had recently been promoted to the post of crew leader. Everything was in their favour. They started from the outset with four wall pour sections rather than five, because they knew it was doable. Since they already knew that the key to success was in achieving the right flow order by paying careful attention to the order in which formwork panels were stripped from the concrete walls, they put most of their control effort into that aspect. They also paid more attention to the interface between the walls and the slab and concentrated on trying to start the slabs at the right times.

The results were quick to come. Right from the very first cycle they managed to reach the eight-day cycle target. They still struggled with the interface between the walls and the slabs, falling short of the four-day target. Nevertheless, they achieved the highest production rate for this kind of project, setting a new record of a whopping 5.9 floors per month at their best performance. Their overall average was 5.5 floors per month, improving on the 5.1 average that Dudi's team had set on buildings 11 and 12. This was an overall improvement of 8 per cent. More importantly for the production planning team, it was an improvement of 37.5 per cent over the baseline rate of four floors per month that was their starting point. Intelligent production planning, continuous improvement, single-piece flow and careful attention to measuring conformance to plan (rather than measuring quantities against target volumes of work) pays off, and it can pay off big. Focus on process is more effective than management by results.

When asked for his views on the success they achieved, Dima explained:

> It was all about paying attention to details, small details. When I understood that I didn't have to worry about the overall result, I could focus my efforts on the small annoyances that prevent you from working the way you know you should be working, today. The second thing I learned was not to look at the monthly target. It's too far away, and it's too big to be useful. Look at your daily tasks and your single-floor cycle time. If you can control them, the whole process will fall into line.

The Rosh Ha'ayin project became the cornerstone of Tidhar's new standard for structural concrete work. The four-floors-per-month paradigm was broken, and now each project has a minimum target of five floors per month. The target itself led to a new way of thinking. It moved all of the management team – project manager, site engineers and superintendents – out of their comfort zones and forced them to look at the process and not at the target. With the support of the training and experience of lessons learned provided by Ronen's production planning team, the targets led them to think differently, to look for the problems, to search for and to identify the constraints, and find the right solutions. The high-resolution plan required people to pay attention to the smallest details instead of the large picture or the bulk work packages. In doing so, they could deal with smaller problems that were in fact easier to solve.

The traditional, business-as-usual low-resolution plan had another major disadvantage: the number of control points was low, in direct proportion to the level of detail of the plan. When planning a monthly target, there were usually weekly control points, which meant four chances to improve per month. When creating an hourly plan, the control points become hourly, leading to 200 potential opportunities for improvement in every month. In the context of construction, raising the resolution of planning and control is essential for achieving continuous improvement and processes more in line with Lean.

Lesson learned

Lessons learned are:

- The higher the planning resolution, the smaller the work packages, which results in smaller problems with easier solutions.
- Find the right people to collaborate with in making changes.
- Going to Gemba is essential. It is the only way to see the actual process and practice and to discover how people really think about their work.
- Give respect. People on the production floor work hard, and some improvements may not necessarily be perceived to be in their best interest. Giving respect, working with them and genuinely understanding their needs is the best way to engage them in a change process.
- Skilled workers on site are often more willing than managers to innovate because they know their actual capabilities and limits.
- Big leaps are sometimes hard to understand. During development of new processes, visual aids support collaboration by creating common understanding. In implementation, they help everyone responsible to align their expectations, putting all involved on the same page.

Notes

1 *Nemawashi* in Japanese means an informal process of laying the foundation for some proposed change or project, by talking to the people concerned, gathering support and feedback, and building consensus.

2 The number of floors per month is calculated by dividing the total quantity of concrete delivered to the site, in m^3, by the quantity of concrete in a typical floor (slabs, walls and columns).

19 The Fira way

Pioneering companies that have adopted Lean and BIM in an integrated fashion have redefined the level of service in their market, be they real estate developers, general contractors or subcontractors.

Fira Oy is a construction start-up company in Finland. A unique aspect of Fira is that the company's owners, who at the time of writing are 108 of its 248 employees, have come to view it not as a construction business, but as a "people" business.[1] The difference is not just in a word – it is a fundamentally different attitude, which sees the company as a platform for generating value for all concerned in many innovative ways, rather than as a platform for earning profits through constructing buildings. In their own words, Fira is a "Building movement: A phenomenon where people co-create smarter societies and better lives".

This mission statement may sound strange for a construction company, but their behaviour indicates that they mean what they say. In this chapter, we explore their actions: how Fira focuses on creating value for customers, how they promote management by process rather than management by results, how their process focus builds on

Figure 19.1 Construction business versus people business

implementation of BIM and on application of Lean thinking, and how ultimately Fira aims to become a hub of high-value start-up ventures in the construction industry.

Fira's (short) history

Fira was formed in 2002, with the explicit goal of developing from a subcontractor to a general contractor within three years. During the first two years of its existence, Fira built a reputation as a highly skilled concrete subcontractor, capable of building large structures with complex formwork. By 2005, the company's turnover reached €11 million. It began general contracting in 2005, and after initial growth it peaked at €23.7 million in 2007. Following the financial crisis of 2008, it settled back to a turnover of €14.3 million in 2009.

At this point, new leadership brought change to Fira's strategy, introducing its sharp focus on customer value and wholeheartedly adopting BIM. These changes began to take effect, generating an inflection point in its turnover curve: by 2015, turnover had increased sevenfold, surpassing €100 million. During this period of growth, and building on the reputation it gained using BIM and providing excellent customer service, Fira management strongly emphasized development of effective processes that would lead to the right outcomes rather than concentrating only on the results while turning a blind eye to the means of achieving them. This came at some cost – a few projects turned out to be economic failures that hurt the bottom line – but it supported sustained growth and increased profitability in the long run. The key processes that were developed were:

- building production alliances with their trade partners, particularly within the framework of Fira Palvelut, a subsidiary established in 2010 that specializes in bathroom, kitchen and plumbing renovations;
- integrating the entire design process. Fira developed the "Verstas" (workshop) concept for collaborating with clients and designers to produce building designs that had higher value for the client than their original designs (for example, by optimizing the structural layout of a basement floor to increase the number of parking spaces);
- improving procurement processes, by standardizing and removing waste;
- actively managing production and construction information on site, with the introduction of LPS, model-based scheduling and cost controls;
- carrying out detailing and planning in an "intensive big room".

Fira's future

In 2016, Fira's leaders embarked on an ambitious third phase, in which they plan to build on the achievements of the 2009–2015 period by adding innovative start-up nodes within the company. The idea is to create value well beyond what is typical for a construction company, by capitalizing on the knowledge and ability they have gained in process excellence. Three of the start-up nodes are each developing new business models for construction industry processes. SiteDrive, for example, is an application that provides a new transparent approach to scheduling and flow-based workforce management. The idea is to build on and enhance the processes developed within Fira. The software can then be offered as a service to the global construction industry, well beyond the boundaries of Fira's local market. Figure 19.2 shows the new, emerging company structure.

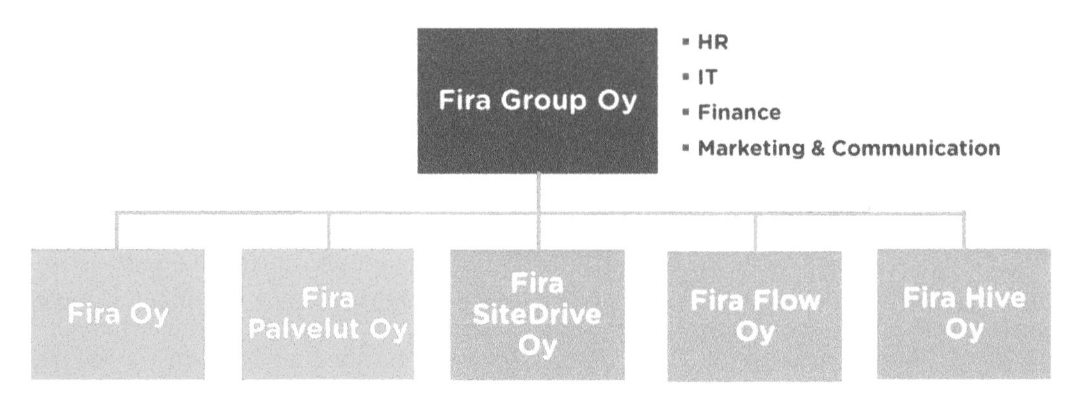

Figure 19.2 Fira corporate structure

Fira's Verstas (Workshop)™ and BIM technology

The first steps that Fira took toward fulfilling their goal of enhancing value for clients were (1) to devise and implement a collaborative, facilitated detailed design, planning, and construction process (which they called "Verstas"), and (2) to adopt BIM technology and integrate it fully with the Verstas process.

Verstas literally means "ironmongery", a place where a blacksmith and workers use a forge. In the Verstas process, Fira project managers, construction planners and engineers engage with clients in a series of intensive, collaborative workshops in which they focus on identifying how the intended building and the construction process can best support the clients' core business processes. The idea is to involve the contractor very early in the project life cycle, before construction costs and the contract price are set. This allows them to reap the benefits of innovative product design and construction method selection, which can significantly reduce the overall cost and lead to improved functional performance that might otherwise be missed. These are usually negotiated projects, in which the client approaches Fira for a construction bid and then works with Fira's people through the Verstas process to refine the project design and finalize the contract price.

Box 19.1 How to deliver real value for Fira's customers

Otto Alhava, Fira's chief technology officer (CTO), explains:

> We are not the best experts on our customer's operations – they are. We should help our customers and users to make correct decisions. During the project development period, we should listen to the users and take their needs into account while jointly converting them into a building design that works as a whole – technically, financially and operationally. Our main tasks are to understand the targets of the customers and help them to reach them, to develop the project in a transparent manner with people participating, to assist the customer in making good decisions, and to generate essential information for users in a format they can use.

The process is also one of building trust between the client and their designers, on the one hand, and the contractor, on the other hand, before the contracts are signed. This develops through the close collaboration in the workshop meetings as the process proceeds. BIM is used to provide continuous visualization of the emerging construction details, and no less importantly, to support frequent preparation of reliable cost and schedule estimates. Short cycle times for developing and evaluating detailed alternatives, often within the time of a workshop (one or two days), mean that multiple alternative solutions can be considered for building systems and for construction methods.

The process is also transparent for the clients, giving them a reliable cost and schedule framework before construction begins. Any proposed design changes can be evaluated for both functional value and for construction cost in short cycle times. The value of

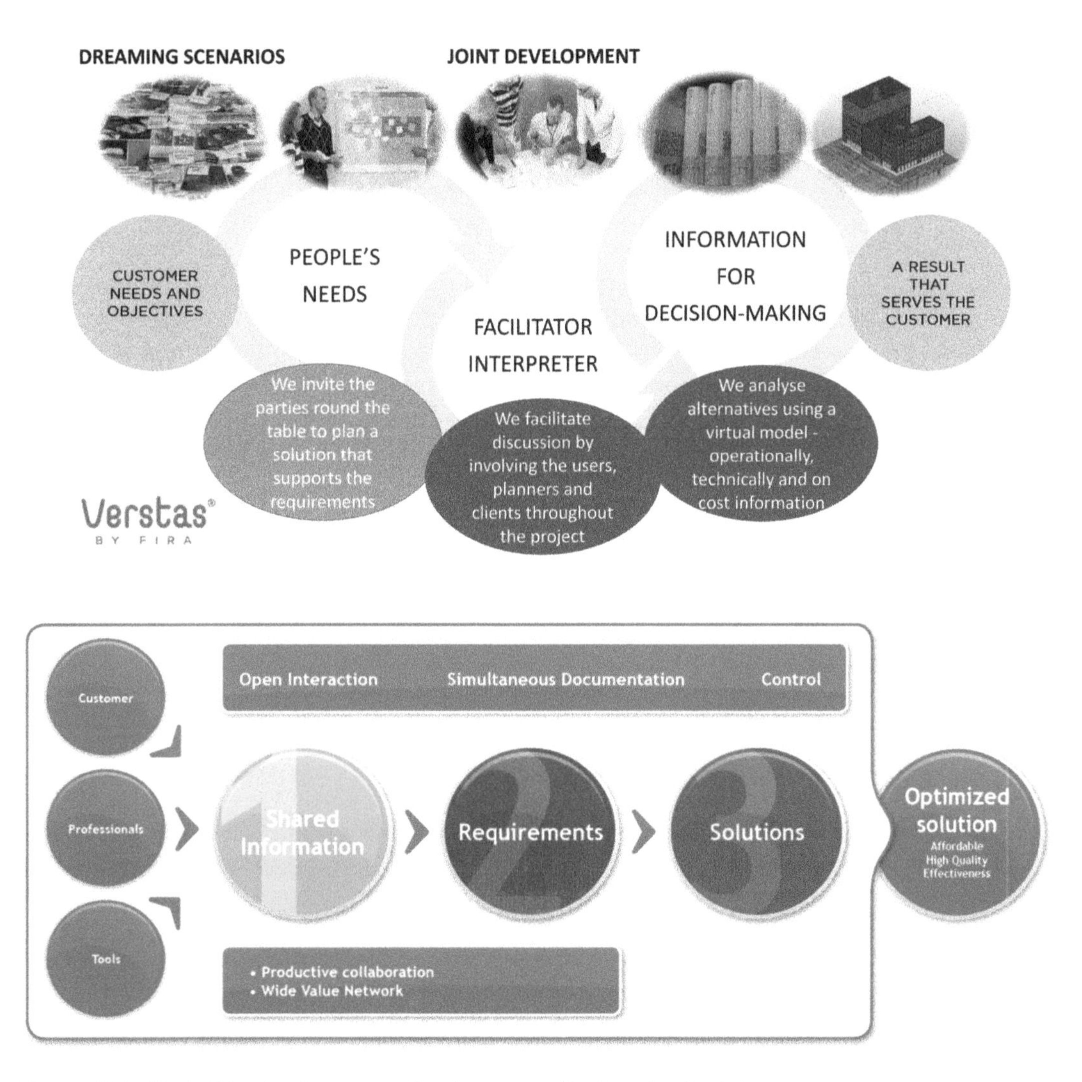

Figure 19.3 Fira's "Verstas" process: (top) process; (bottom) principles

alternate layouts, such as the traffic patterns and the number of spaces in a parking deck, can be quantified and compared; in parallel, accurate cost estimates of the proposed changes can be prepared within days, thanks to the use of the BIM model and automated estimating routines. In this way, the Verstas process reduces risks and achieves schedule and cost advantages. At the same time as they are enhancing value for their client, a secondary goal is to build the knowledge and key competencies of Fira's own people so that they become more customer-oriented, which reduces the risk of making incorrect assumptions about clients' needs in the current and future projects.

The marketing materials that Fira prepared as part of its bid for development and construction of electric power substations for Fortum, a major electricity utility and a leader in clean energy, describe the Fira Verstas process as a "co-creation-based problem solving and development method in every phase of the project". Fira claims that proper implementation of the process leads to the following benefits for the customer: (1) a flexible, agile and reliable service provider; (2) a transparent, modern and efficient project model; and (3) lower risk inherent in the resulting design, thanks to the ability to test the predicted performance of the designs developed in the workshops through BIM simulations and cost analyses. The benefits that Fira aimed to gain from the process were: (1) a new, modern project management model that opens new opportunities for better results in building projects; and (2) an experienced workshop and BIM team who will bring the future tools of the construction management into use immediately.

In practice, implementing Verstas proved to be challenging in the context of the construction industry's traditional culture, which focuses mainly on engineering, risk avoidance and claims. The implementation required intensive training for Fira's own people, as well as a significant investment in "re-educating" clients to understand the potential for win win collaboration so that they could take full advantage of the process by engaging in it thoroughly.

Otto affirms:

> Getting stakeholders and owners to sit down at the same table for value co-creation with architects and designers is not easy. While the benefits of collaborative work are sometimes easy to sell to top management, it is much more difficult to deliver the results with mid-level conservative stakeholders, who tend to stick to the traditional design process of the past, where the parties operated in disconnected "silos". Long distances between the cities of Finland and differences in town plan building provisions between cities also make collaboration difficult. Additionally, Finns are not considered to be among the most collaborative and social people in Europe.
>
> So far, we have learned two key things from successful Verstas implementations. First, team spirit and open dialogue have to be created the beginning. When Fira started to train facilitators, and adopted advanced, third generation methods of facilitation like World Café and Me-We-Us in Verstas projects, the results were significantly improved and customer satisfaction increased. Second, we need an ecosystem of specialists from different business areas. In order to differentiate in the field of construction one cannot rely solely on internal resources. Instead the value of collaboration in the Verstas process depends on our ability to identify the required capabilities and knowledge for solving the customer's business case, and on the ability to connect the identified persons or companies to Verstas in an agile way.

BIM tools were a key component of the Verstas process, for two reasons. First, without fully integrating BIM tools for prototype modelling, estimating and scheduling, Fira could not have achieved the short cycle times for design, analysis and feedback that were necessary for the Verstas service process. Short feedback cycles are key to keeping clients engaged in the process and to helping them build an understanding of what they can reasonably expect, both from the construction process and from the finished building. Second, the construction-specific BIM models prepared in the process enabled the company to move seamlessly from the detailed design and planning phase to the construction phase and at times to overlap the phases by starting construction before design was complete, with minimal loss of information. This overlapping, known as "fast-tracking", is dependent on the availability of accurate, detailed information defining what has already been built so that the design detailing can progress with a minimum of uncertainty, and BIM is the vehicle for carrying the necessary information.

Fira found that a suite of BIM tools was needed to cover all the needs of these processes. A team of BIM champions was formed, each of whom was trained for competency in the whole process. Others were given specialized training to facilitate the Verstas process and indeed to support the whole "E2E" (end-to-end) process over the project life cycle. In addition, a "super-user" was appointed for every software application in use, to provide in-house support. All site personnel have been trained to use the BIM tools that provide access to information, such as model-viewing and model-checking software. Figure 19.4 outlines the different BIM tools in use during this time (2011–2015), the BIM champions and the super-users responsible for each tool.

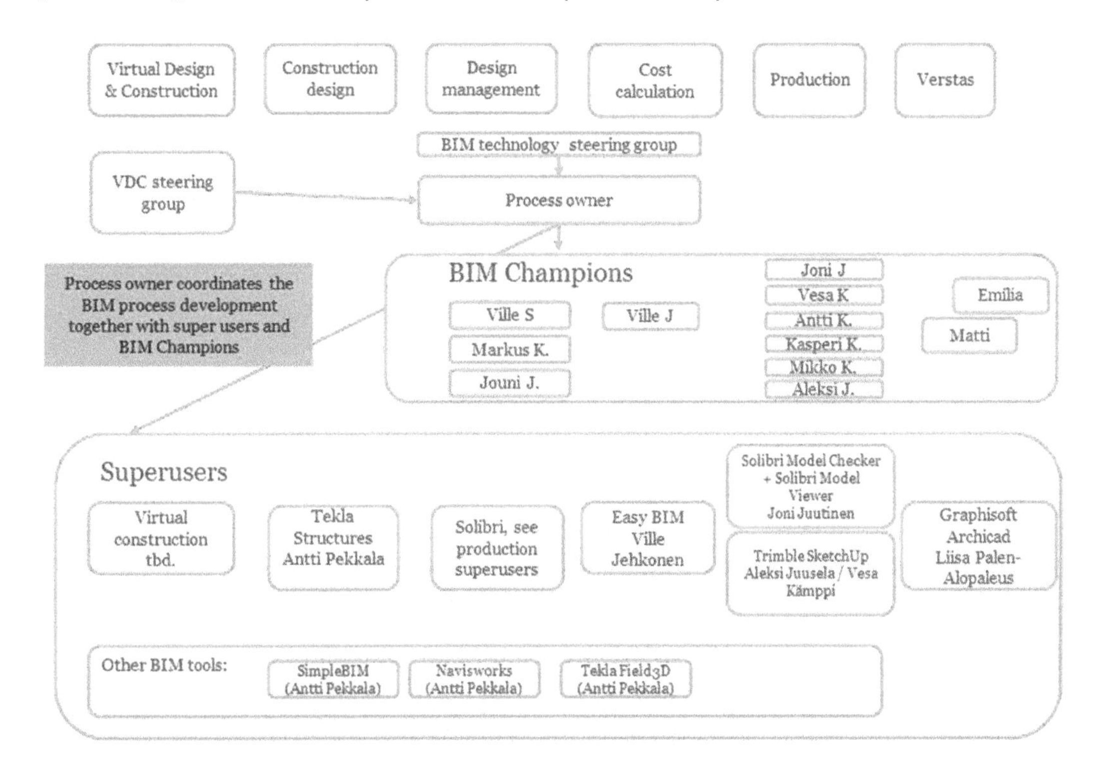

Figure 19.4 Fira BIM processes, BIM champions and BIM tools

Figure 19.5 Fira's depiction of the old way (left) and the VDC way (right)

Note: The image on the left shows a concrete drill used to enlarge an opening in a concrete wall where a design coordination error resulted in an opening that was too small. The image on the right shows coordination of structure and systems to identify such errors in the virtual model, before construction.

Fira's construction engineering and management process was by no means perfect, and there were still many problems. Nevertheless, the process was sufficiently better than their competitors that they could grow quickly. That growth provided a solid financial foundation for new developments, particularly in the areas of Lean Construction and BIM, starting in 2011.

LPS, the "integrated big room" and supply chain management

The two major areas of innovation in Fira Oy between 2011 and 2015 were Lean and BIM. These included adoption of LPS and other Lean tools, better management of suppliers and subcontractors, implementing the "integrated big room" concept and the attempt to run a "virtual big room".

Fira's Last Planner® process owner, Sakari Pesonen, came to Fira after many years of experience with LPS in Skanska Finland and Skanska US. He led a team that developed templates, processes and guidance videos to support deployment of LPS AT Fira. The deployment was divided into small steps and implementation was monitored on a monthly basis at the project managers' production development meetings.

Introduction of LPS was slower and more laborious than expected in terms of subcontractor training and coaching. However, the construction sites at which it was comprehensively introduced were very satisfied with its effectiveness and results. LPS improved the cooperation and interaction between subcontractors, which enabled them to reduce the time buffers between their crews. The improvement was reflected in positive feedback from experienced subcontractors across a variety of trades, who said that LPS set Fira apart from its competitors with respect to the smoothness of work on its sites and the quality of its managers' production planning and control. On sites where LPS was introduced in full, this was also reflected in the priority placed on end customer quality, which resulted in measurably better customer satisfaction. For example, in a project in which Fira refurbished the headquarters of a Finnish mobile operator, Elisa, customer satisfaction was measured periodically using the net promoter score (NPS). Project bonuses were tied to high NPS values to ensure smooth execution of the project and minimal disturbances to Elisa's business. The refurbishment of the office complex was carried out in small sections while the rest of the buildings were in continuous use. The Elisa project team achieved very good results using LPS, but also came to the conclusion that it could have been that much better had it been supported integrally by a BIM model. As a result, the next step of continuous improvement proposed was a pilot of BIM-based LPS, using VisiLean software.[2]

Enni Laine, an experienced construction engineer and BIM manager, was the process owner of the "intensive big room" (IBR) process. Fira's approach to the big room concept is similar to the approach that has been developed in many parts of the world for working with BIM for construction (including at Tidhar): bring representatives of all of the designers, detailers and builders into one open-plan office and organize a work flow in which the functional trade models are developed into a federated, coordinated and verified construction model. In this way, information is generated collaboratively and concurrently.[3]

However, when Fira tried to apply the big room concept in small- to medium-sized projects (from €5 million to €20 million), they found it to be impractical because there wasn't enough work to justify full-time co-location of all of the designers in the big room for an extended period (similar to Tidhar). This led them to begin thinking about replacing an extended period of collaboration in a big room with a series of shorter, more intensive, but intermittent sessions in which the designers and builders would come together to detail a project's design. This series would be called the "intensive big room", or IBR.

Another impetus for the IBR was that Fira placed strong emphasis on customer participation, seeing the big room as a way to co-create value with the client. Clients for smaller scale projects found it impractical and expensive to maintain a continuous presence in the continuous big room. Fira's view of the value of the client's involvement was strengthened by their observation of the problems that arose with regard to customer-defined value in a BIM-enabled design process in the Lahden Sairaalaparkki project, a 5,000 m^2 office building with a car park with 600 spaces. Among the conclusions were:

- Clients' business requirements are always individual and project-specific, and clients' requirements are subject to change during the project due to changes in their business model or their own customers' business needs. Thus the system must be flexible to enable late changes in a cost effective way.
- Clients' abilities to define business requirements and develop them into technical requirements for a construction project are very limited, and clients are unable to understand designed functionalities based on the current standard documentation of construction processes without additional visualization and personalization of information.
- The team responsible for designing and implementing the solution must be brought together as early as possible to create a shared understanding of the objectives, requirements and finally the solution.

- Integrating a team with members from different companies requires changing the prevailing paradigm: participants cannot be selected solely on the basis of the lowest cost bid, and the commercial model must include incentives to integrate the team instead of fragmenting it by forcing participants to minimize their workload.

These conclusions are strikingly similar to the motives that underlie IPD, but IPD was not appropriate for Fira's clients, who were neither sufficiently large nor experienced in construction to initiate a full IPD process. The IBR was devised to meet these challenges: to bring a professional team of client, designers, suppliers and builders together early to co-create customer value, to use BIM for design development and for communication, and to leverage the process to achieve a construction process free of errors and rework.

The business process map in Figure 19.6 defines the IBR process. It consists of a short tendering phase in which workshops are held with subcontractors, after which a contract for services is signed. What follows is a series of weekly IBR workshops, in which all partners come together to define the solution together with the client, all supported by BIM tools.

Enni Laine describes the IBR sessions with the depth of detail that only someone deeply committed to it can:

> Each IBR session has three phases: set-up, integrated concurrent engineering (ICE) and wrap-up. In the set-up phase, participants are briefed and common goals are set (usually centered around identified problems and design tasks). The design status and schedule are reviewed; we encourage teams to use the Last Planner® for this, but

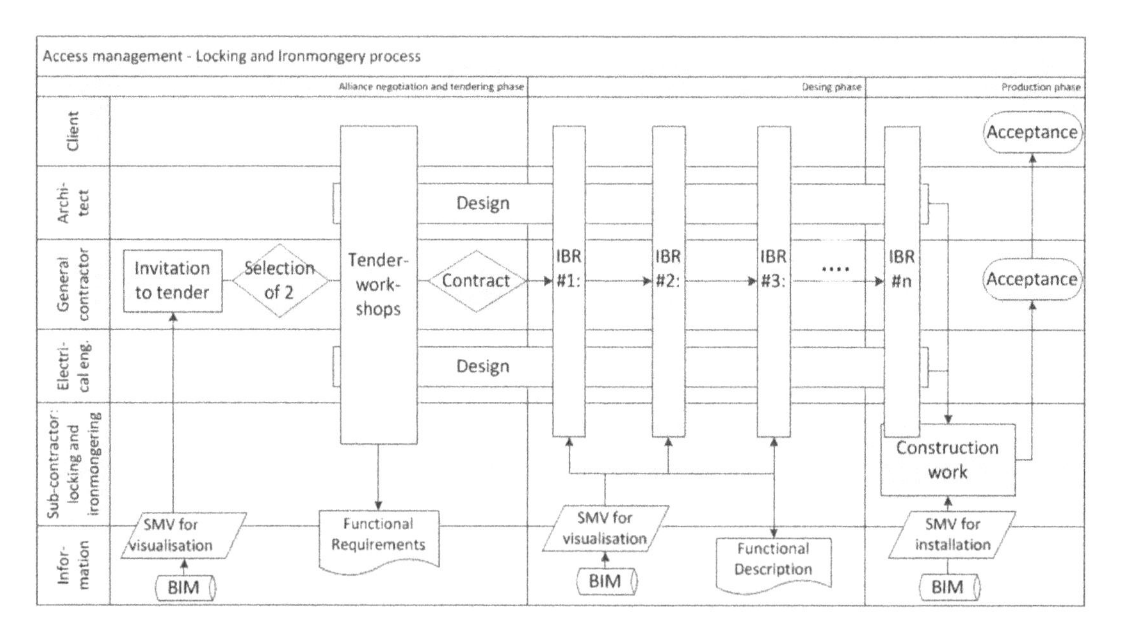

Figure 19.6 Fira's client-centric design management process in which the "intensive big room" and standardized model views are used

Note: SMV stands for 'Solibri Model Viewer'.

not all do. Design issues are classified as green, yellow (an identified problem, which can be solved), and red (a show-stopper problem, which needs interdisciplinary attention and collaboration in the ICE session).

The next phase of IBR is the ICE session, in which there are always two facilitators, the chairman and the secretary. The chairman facilitates, assigning tasks to task owners and helping them form a team for each task. First, the customer's expectations are reviewed and documented, then the systems or machinery requirements are reviewed, and finally the impacts of structural and architectural solutions and their constructability are discussed and decisions are made. Task owners are responsible for documenting the results to the ICE procedure chart for each task. Each task can request additional resources or even interrupt the work of other teams, if the full team is needed for solving the task or to make sure that each and every one is aware of a fundamental change or decision. All this time, the secretary documents the group discussion and especially the decisions. The chairman re-allocates tasks and resources, identifies bottlenecks, refocuses the teams and ensures that all teams achieve their targets by the end of the ICE session.

The IBR sessions are always wrapped up by checking task completion and with self-reflection using Plus/Delta. Each task owner presents the team's conclusions and decisions. The secretary updates the A0 Visual Progress Chart so that the results are visible for all.

Figure 19.7 An IBR meeting at a construction site office

Fira's IBR process went through no fewer than three generations of process improvement (IBR 1.0, IBR 2.0, IBR 3.0) within its first years of implementation. One of the most interesting aspects of this continuous improvement came when Otto Alhava (Fira's CTO), frustrated by a number of less-than-spectacular results, invited an industrial psychologist to consult on the IBR process. Although the psychologist was impressed with the process and the way it fostered collaboration, he raised questions about the motivation of the people involved in the IBR meetings.

What was people's mental state? A check found that people's activity levels dropped significantly after the IBR, with lower motivation than expected between IBR sessions. The participants' professional culture had prepared them to solve problems in personal silos, not in group sessions. One of the conclusions from the observations was clear: facilitators, mostly engineers, had low emotional intelligence (EI) and needed training and support.

Fira continued exploring the avenue of emotional intelligence, working to understand what were the key tasks, skills and desired personal characteristics of a successful IBR facilitator.[4] Otto describes what happened:

> At first, the results caused astonishment among participants, since engineers typically concentrate on tools and technologies rather than on soft skills. After the initial shock, even the engineers felt comfortable with the obvious result: there is a high correlation between an emotionally intelligent transformational leadership style and project quality, as well as with stake-holder satisfaction. The study inspired us to study EI more deeply and continue the efforts to develop emotionally aware co-creation and collaboration.

Implementing one of the key results of this study, Fira introduced training for its IBR facilitators to provide the necessary soft skills.

Management by process

Perhaps the most important lesson from Fira's construction management culture is its strong focus on management by *process* rather than on management by *results*. Management by results is very common in traditional construction because of the tendency to transfer risk. Owners try to push project risks to general contractors, who in turn protect themselves with "back-to-back" contracts with subcontractors, and so on down the line until the risks reach those least able to control them. In this environment, managers demand results – "You need to finish 45 m^2 of flooring by Tuesday afternoon!" – without taking responsibility for the process. Each trade crew is essentially left to determine the details of how the work will be done and continuous improvement is next to impossible because there are no standard procedures.

Fira set out consciously to establish standard processes, with the decidedly Lean goal of improving its work methods through formalized organizational learning. The end-to-end (E2E) process map (see Figure 19.8) encapsulates the strategy. The process is broken down further in hierarchical fashion, so that all steps have documented procedures. Each distinct process is the responsibility of a dedicated "process owner", whose job it is to develop, disseminate and support the front-line employees as they implement the process. For example, LPS has a process owner, who documents the

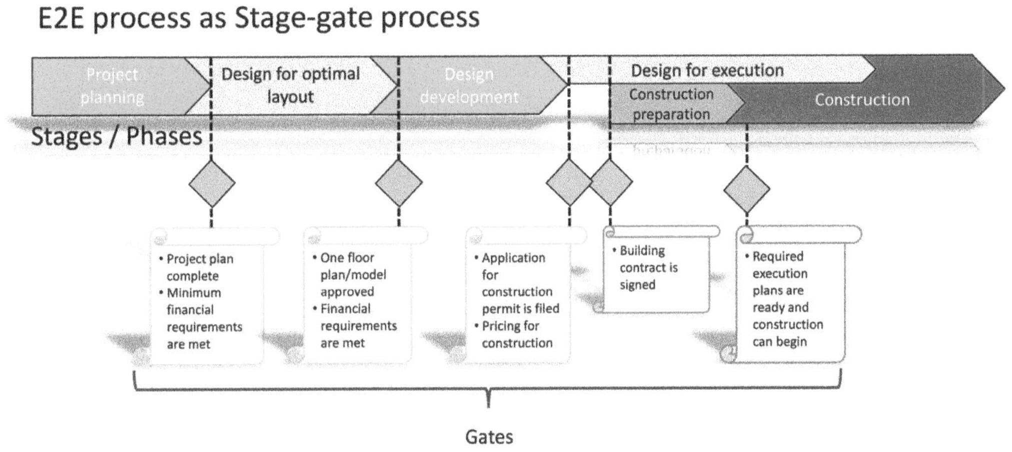

Figure 19.8 Fira's E2E process

procedures, prepares standard forms and materials for running LPS meetings, coaches site engineers and managers, monitors implementation and is responsible for improvement of the process.

Virtual Design and Construction (VDC) is used at all stages of the E2E process. VDC is seen as having three dimensions: information, technology and collaboration. All three dimensions must be integrated appropriately within the process, and the E2E process defines all three explicitly for each stage.

Table 19.1 E2E process stages

E2E process stage	*Collaboration mechanism*	*Technology*	*Information*
Project design	Verstas	Task management	Customer needs
Design for optimal layout	workshops	Tekla, Sketchup Model-based estimation	BIM data -> quantities -> cost estimate
Design development		Multidisciplinary model	Design information Design task management
Design for execution	Big room	Multidisciplinary model Scheduling	Quantities Cost information Initial schedule
Construction preparation		Sketchup site plan	Detailed quantity and location-based schedule
Construction	On-site BIM booth Last Planner® System	Solibri BIM visualization VisiLean	Actuals on site Tekla field 3D

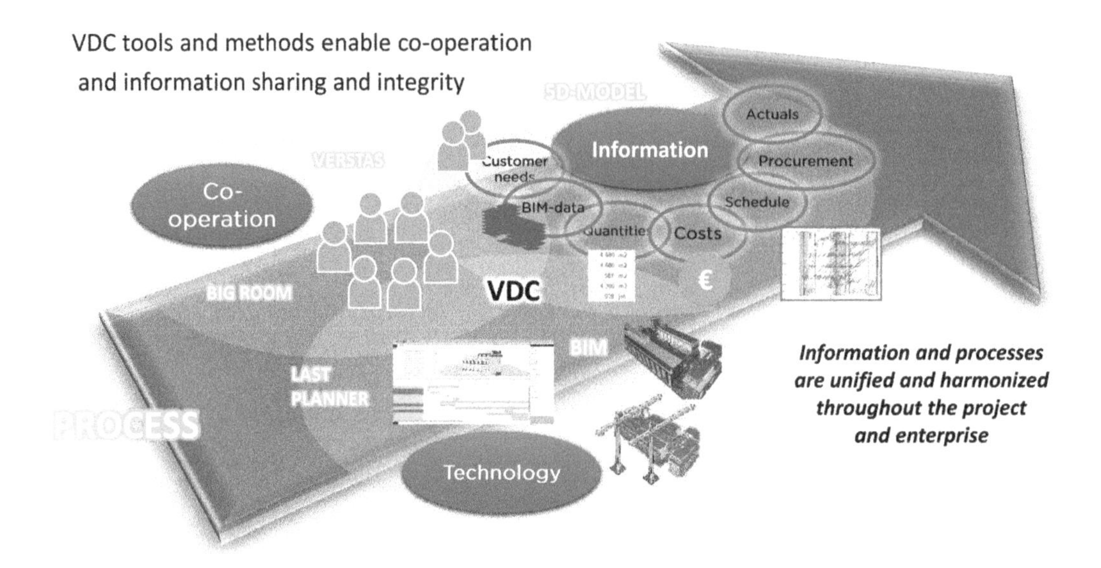

Figure 19.9 Traditionally information and processes are fragmented in construction projects: Fira brings them together in its business and IT processes

Box 19.2 Lean leadership and cultural change

Antti Kauppila leads Fira's Shared Services Center in the head office, a unit with site engineers, service engineers, business controller and invoice coordinators that provides support for the day-to-day operations of business development and production control in Fira's projects. He previously participated in Fira's Verstas process, but from 2013–2015 he was able to devote all of his time to developing a business-process orientation in production, in part thanks to the LCIFIN2 project. LCIFIN2 was an initiative of the Lean Construction Institute of Finland. Within the framework of LCIFIN2, the Finnish Funding Agency for Innovation, TEKES, sponsored a wide range of activities that aimed to instil Lean Construction as part of Finnish construction culture. Eleven companies operating in construction took part in the project and the Industrial Engineering and Management Research Group of the University of Oulu acted as the research partner. Antti Kauppila recalls:

> At Fira, we have been moving towards a process-based organisation and adapting the principles of process-based management. I'm familiar with Lean Construction tools and principles. I coordinated and coached the ten other process owners (POs) and helped them to develop auditing and other measurement tools. I had powerful support in this: Anneli Holmberg, an external consultant with expertise in Lean, helped define the roles and job descriptions and did a great deal of Lean tutoring for the process owners and for project managers. Along the way, I also wrote my Master's degree

dissertation. Not surprisingly, it was titled "A Process-oriented approach to developing operating and management models for production in a construction company".

Within the framework of the LCIFIN2 project, Antti, Anneli and others led a work package dedicated to changing the organizational culture. They held company-wide seminars for production personnel and monthly production process development meetings, they publicized their activity in the *Fira Times* company newsletter, and they rolled out a variety of Lean tools, such as five whys (root cause analysis), visual management with A3 reports, 5S, and reporting and monitoring of the projects on office wall boards. Antti Kauppila goes on to say:

> We discovered that to change the culture of Fira, we needed to introduce scorecards and performance review templates to change process management. The performance-based pay model of traditional construction is strongly tied to the short-term results of a project; there was little or no incentive for attaining company value through long-term development of the company. Compliance with standard practice was not part of the way of doing things. We discussed and developed operational objectives for the project manager scorecards, which included process management and the introduction of Lean manufacturing tools. Coaching was a key way of changing the culture, of changing the management doctrines of Fira's supervisory work and lifting site managers' awareness of production management. Especially in people management (leadership) skills, there is still plenty of room for improvement, which in turn will support Fira's Lean culture.

Antti was not the first Fira employee to integrate the challenges of his day-to-day work with academics, nor the last. Fira sees Master's and PhD students pursuing studies in the broad areas of BIM and Lean Construction as potential leaders of the company's change process, and encourages their employment and their integration in the company. Likewise, it encourages its employees to study part-time. Anneli's contribution to Fira's development was significant. She sees her work with Antti as pivotal in changing the way they work with people:

> When I first joined Fira, in June 2014, they had introduced process ownership and defined the first key processes, but the understanding of why processes are important and how to define and develop them were still unclear to most of the process owners. My first task, together with Antti, was to help Fira clarify the purpose and tasks of process development. We developed a standard for process development in Fira and started to coach the process owners systematically. The target of this, step 1 coaching, was to teach the process owners to define and develop the processes together with their development teams. As the process owners became more independent in their work, my role changed from coaching process owners to coaching managers working in the processes. The target of this, step 2 coaching, was

to teach the managers to coach their team members in developing themselves as well as their work processes. Step 2 coaching aimed to increase managers' understanding that process development and developing people were among their key responsibilities. Later on, the focus changed to a more detailed coaching towards Lean leadership in the daily work of leaders. By this point, Fira Palvelut in particular had already taken the first steps in Lean leadership, where managers plan the work together with their teams at a much more detailed level. They also included the first forms of social subcontracting as a way to improve commitment and cooperation amongst the various subcontractors and employees at the worksite.

Lean innovation in Fira Palvelut

Fira has made great progress in building a general contracting company with innovative processes that integrate Lean Construction with BIM. But perhaps the best example of process development within Fira on the basis of Lean thinking is that of Fira Palvelut (Fira Services).

Fira Palvelut is a subsidiary of Fira Oy, and it is a construction company established to work in the highly specialized, but potentially very large, market of plumbing, bathroom and kitchen renovation projects. This is a big market in Finland, and indeed in many parts of Europe, because the plumbing systems installed in the mass housing projects built in the 1950s and 1960s have reached the end of their service life and must be replaced. Renovations are performed for whole buildings or projects, rather than for individual apartments, so that projects with tens or even hundreds of apartments are common.

Fira Palvelut was separated from Fira itself in recognition of the fact that the type of project on which it would specialize required processes that were significantly different from the processes that were in place for new construction. The company is not only very different from most other construction contractors in its approach, its processes are also quite different from those of other renovation contractors. Two key words guide the "Fira Palvelut way" of doing things: *flow* and *value*.

Flow

During the first two years following Fira Palvelut's establishment in 2010, the company pursued a number of projects using a traditional approach to construction, in order to learn. Its leaders realized that they could not affect the cultural change that they wanted to until the company had grown to at least 15 employees.

With a background in Lean production and in the Theory of Constraints, Fira Palvelut's production manager, Jaakko Viitanen, and his team, identified five key wastes in the traditional production process that was used in those early projects:

1 Poor location flow. The team installed webcams in a few apartments in a typical project and observed the movement of workers. They found that, on average, apartments in which work had begun were empty for 82 per cent of the working hours during the process. This translates into an amount of work in progress that is

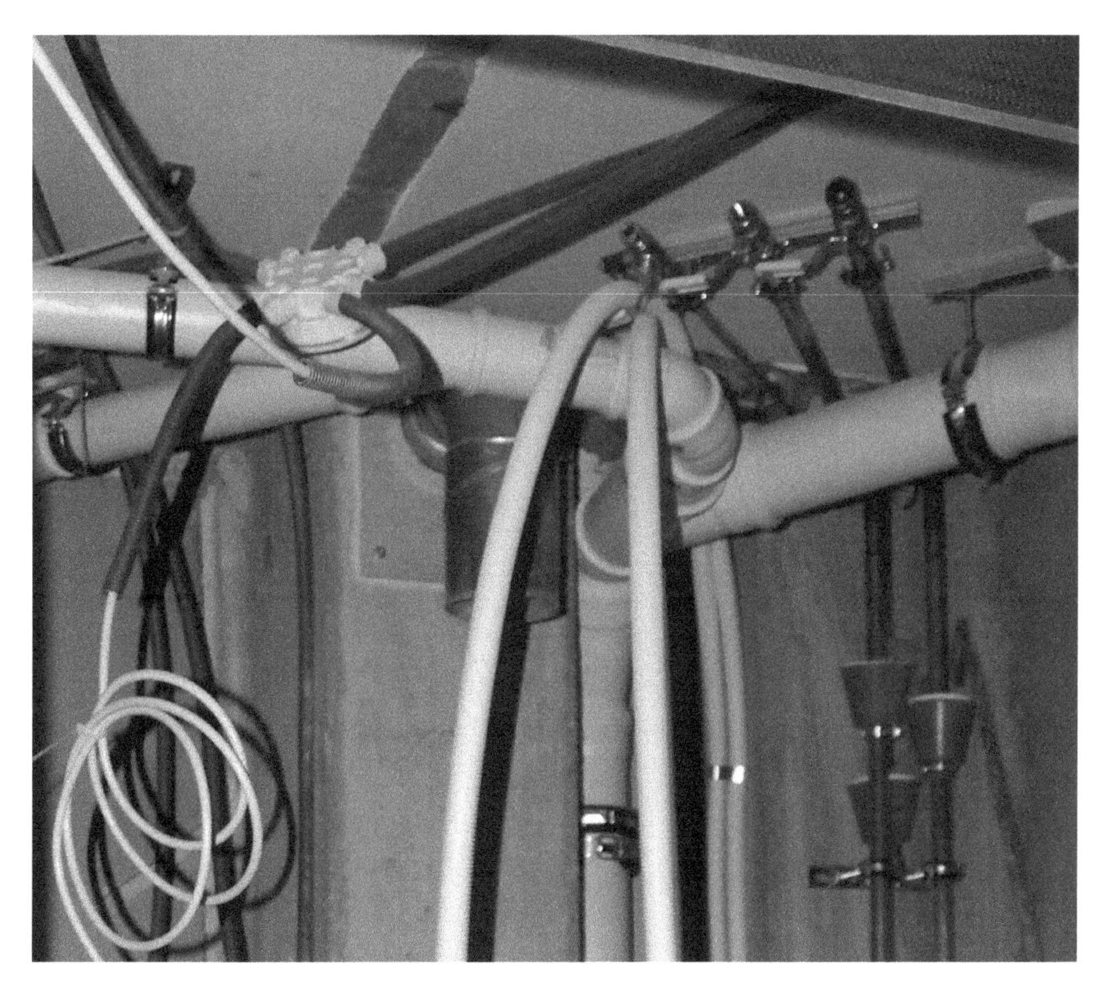

Figure 19.10 New heating, electrical, water supply and sewage piping on the ceiling of a bathroom

five times as large as it needs to be in theory because at any point in time, work is only being performed in one apartment in five. Conversely, it means that the overall work duration in an apartment is approximately five times as long as it could be.

2 Inefficient work practices. There was a great deal of unnecessary movement of workers between apartments.
3 Wasted time for subcontractors. In addition to waiting during projects for materials, information, equipment or preceding tasks to be completed and inspected, subcontractors had to wait between projects.
4 Loss of the opportunity for learning and real productivity improvement. The lack of continuity between projects also meant that the composition of the crews changed from project to project, as they were mobilized ad hoc for each new project but then moved to other jobs before the next project begins, which in turn meant that workers' knowledge of the process, the habits of collaboration they

had developed within each project as they became familiar with their fellow workers, and their technical know-how of process and methods were diffused after every project.

5 Poor information, both design information and process flow information. Inaccurate or flawed design information was a root cause of much waste, but lack of process information could be a more insidious problem due to the fact that renovation projects are particularly prone to generating unwelcome surprises. As the cycle time is shortened, the rate of deviations increases. As a result, the foremen needed real-time knowledge about the progress of each worker, the ability to see instantly the conditions and situation on each work site, real-time access to all related information for making timely and correct decisions, and the ability to communicate directly with everyone. Instead, in traditional conditions, the workers suffered from disruptions in work and waiting while foremen spent more time walking around the site or chasing missing information. Poor process information led to reactive actions instead of proactive management and leadership.

The average cycle time for completing individual apartments in projects with 100 to 200 apartments was some 12 to 16 weeks (depending on the number of floors and the number of apartments in each entrance). Residents had to vacate their homes for extended periods and were, on the whole, dissatisfied with the process.

How could these wastes and the long cycle times be avoided? The solution lay in creating a stable flow of work for the contractors by aligning projects to be performed in sequence. In this way, each subcontractor could move from one project directly to the next, without discontinuity and without disbanding and reforming their work crews for every new project. Perhaps equally importantly, if all the trade subcontractors working

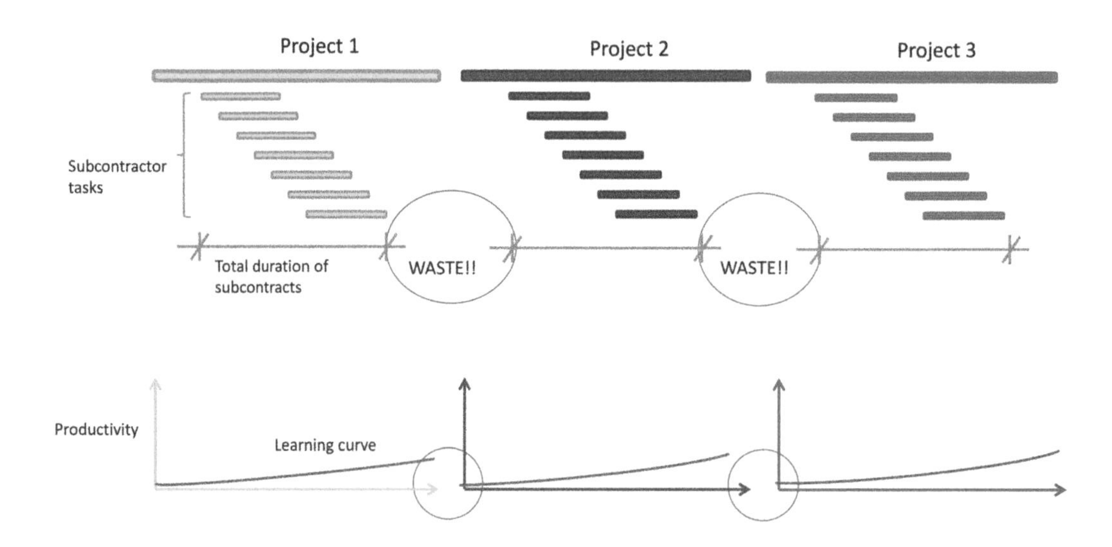

Figure 19.11 The waste of waiting between consecutive projects not only means unproductive time for subcontractors, it also means that every project has a new team and there is no opportunity for long-term learning from project to project

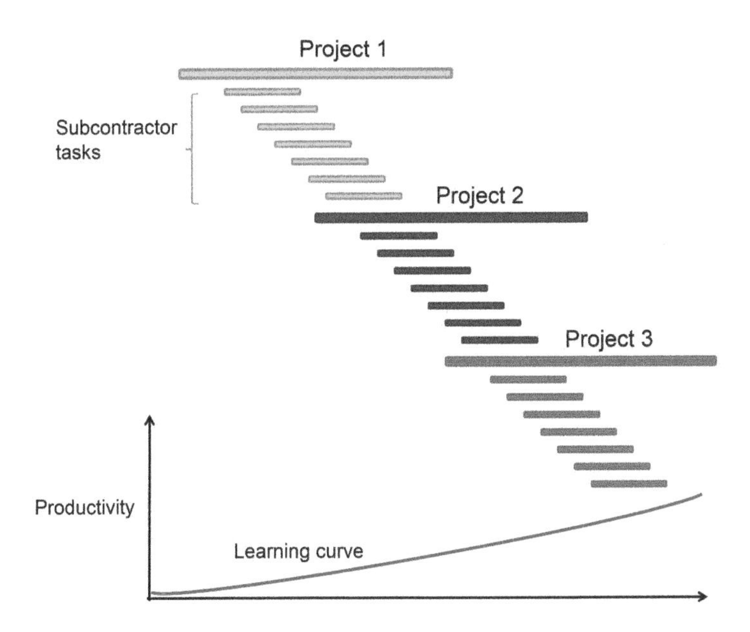

Figure 19.12 Continuous flow of projects, continuous flow of trade crews from project to project and continuous flow of apartments within the process

Note: The gains of the learning curve are maintained as most of the workers in the alliance continue working with one another from project to project.

on a project moved in sequence to the next project, then they would maintain the same working relationships between crews. In turn, the shared learning of how to hand over from one crew to another would also be maintained. In this way, a group of independent subcontractors could function, over time, as a more cohesive workforce that could sustain productivity gains through a learning curve that continues from project to project.

Achieving this "ideal" flow of work requires the general contractor to maintain a sufficiently stable flow of work so that subcontractors can be assured of continuous flow of work from project to project. This is dependent on three factors:

1 A sufficiently large volume of work in order to provide a continuous work stream.
2 A high degree of commitment to and compliance with the pull production plan on the part of the subcontractor trade crews.
3 Stable work durations, so that the company can both have the flexibility to set project start dates (allowing them to link the work between projects to maintain continuity of work for the subs) while also meeting the time windows promised to each project's customers.

Fira Palvelut achieved these conditions by offering excellent customer service and competitive prices, which built demand; by employing novel contract forms to align subcontractors' interests and behaviours with their needs as production manager; and by managing the production for flow and minimal waste, thereby significantly shortening cycle times and making schedules stable so that customers could have confidence in the completion dates promised. Turnover increased steadily over the five years up to 2015, amounting to €6 million, €8.6 million, €14.6 million, €21.9 million and €28.9 million annually. Of course, an almost continuous flow system like this is more vulnerable to interruptions than is a system that is replete with buffers. Fira Palvelut deals with disruptions by using capacity buffers; where necessary, overtime or additional workers are used to prevent one crew's slowdown from slowing the rate of work for the rest of the crews.

By the summer of 2016, Fira Palvelut was operating with four "trains" of subcontractors, each producing approximately 250 apartment renovations per year. In three of the trains, subcontractors worked under cooperative subcontracts which offered continuous work over consecutive projects in return for reliable allocations of labour to the project and the ability to influence the selection of individual workers. The fourth train worked within a "production alliance" model.

Box 19.3 Production alliance

The production alliance in use in the fourth "train" is comprised of Fira Palvelut and five key subcontractor firms (demolition, drywall framing, plumbing, electrical, flooring and tiling). The alliance is governed by a board on which Fira and the trades are represented. The board must approve each new project that it will take on, and the board has the authority to decide which individual workers will continue from project to project. The alliance agreement is open-ended, so that it can continue moving from project to project as long as Fira maintains the supply and the board members agree.

The concept for the production alliance was based on the assumption that in order to manage and to build effectively, three components were essential: (1) common organization; (2) a commercial framework; and (3) a Lean production system. These are the same three essential components of IPD. In traditional construction, where each subcontractor works in a silo – with no relationship with the other subcontractors – none of these three components exist. The production alliance framework offers all three, as follows:

A common organization: the alliance production contract sets up a governing structure with a governing board and a day-to-day management team with a culture of trust, mutual commitment and intense cooperation.

A commercial framework: a joint contract, with common goals and an equitable gain and pain sharing mechanism. The alliance as a whole commits to a fixed price for the customers, with any savings over the course of the project distributed among the alliance partners. Similarly, the alliance partners run the risk of carrying any cost overruns that may occur. Figure 19.13 illustrates the gain/pain sharing mechanism.

A Lean production system: based on TFV (transformation-flow-value) theory and emphasizing value for money for customers. The production system strives for one-piece flow by reducing the number of apartments in each batch as much as possible. The mechanisms used include flow-line scheduling, big room design management, extensive use of BIM and LPS, close cooperation with the material supply chains, built-in quality, and kaizen (continuous improvement).

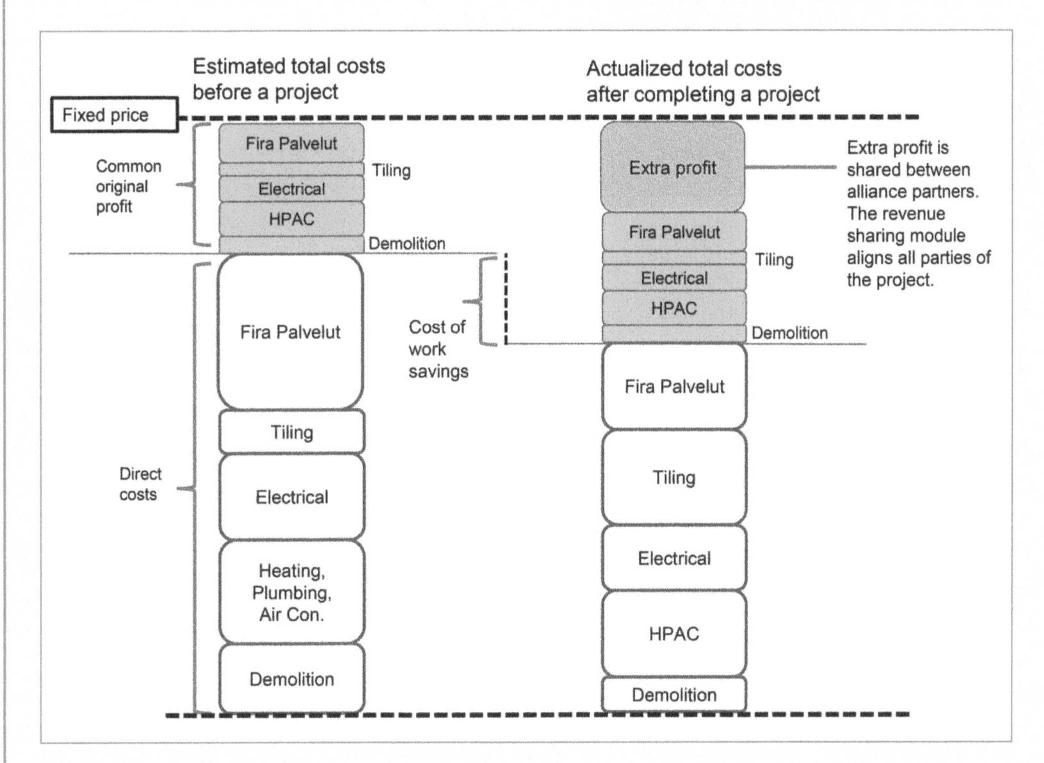

Figure 19.13 Gain-sharing in the event that the project costs are less than the original budget

Note: In an early project built by the production alliance, the total estimated costs were €1,287,000. The actual costs were €1,185,000 and therefore €102,000 was distributed among the alliance partners.

Using the flow system, Fira made some remarkable gains. Figure 19.14 explains how reducing batch sizes can reduce apartment renovation cycle times. In the top image in Figure 19.14, each batch has four units and the total cycle time is 12 weeks. In the bottom image, units are handed over one by one and cycle times are reduced to five weeks. Fira calculated that if crews worked three shifts, the cycle time could be reduced to two weeks – 144 hours of work and 24 hours of setting/drying.

The overall annual throughput for the company as a whole increased significantly, although perhaps not as consistently as theory would predict. The cycle times in the better projects reached five weeks, with a WIP count of 45 apartments, corresponding

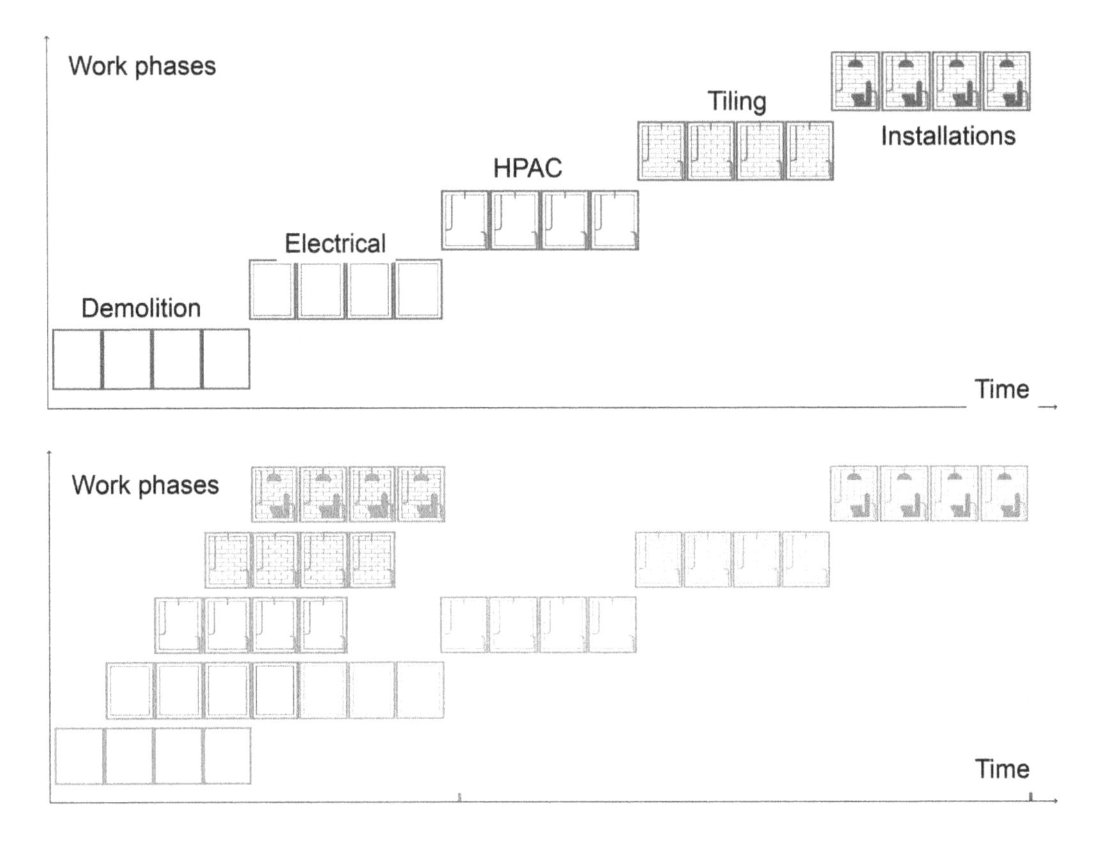

Figure 19.14 Fira Palvelut work phases for bathroom renovation (top); reduction of batch sizes reduces the cycle time for individual units without the need to work any faster in each unit (bottom)

to a throughput of nine apartments per week,[5] or 450 per train per annum, if the flow could be maintained. In reality, each train averaged 250 apartments per year. This indicates that there is still room for improvement, particularly in the transitions from project to project.

Some of the flow improvements that were achieved resulted directly from the new understanding of the value of flow and from the collaborative culture of continuous improvement, which were able to thrive thanks to the stability and the continuity of demand that the process as a whole provided. The case of the demolition crews is a good example of this. Traditionally, the demolition crew subcontracted diamond drilling work to a specialized service provider who owned the diamond-drilling equipment. Naturally, from his point of view, the driller was only willing to come to the site when a few holes had accumulated, so that he could maintain high productivity and avoid excessive set-up times and travel. This meant that often workers would have to wait before they could install piping because the holes had not been drilled.

With collaboration and a focus on flow for the train now key priorities under the new paradigm, the crews and production managers realized that this "fast" trade was working on large batch sizes and thus significantly extending the overall cycle time for each unit.[6] The solution? The demolition crew purchased a diamond drilling machine for concrete (about €7,000 and €1,000 a week for bits) so that they could drill holes as and when needed in single-piece batches. This removed the fast but unreliable drilling specialist from the train of trades. With the new, much higher throughput, this was a profitable investment for the demolition crew.

But there were also some surprises. In one of the production alliance projects, electrical work originally estimated to cost €186,000 ended up costing €199,000. The root cause was found to be a lack of commitment and motivation on the part of the electrical workers to the goals of the overall system. The problem was dealt with through training, discussions and coaching, but also, workers who found it difficult to adjust to Fira's way

Figure 19.15 Webcam images show a record apartment plumbing renovation cycle time, from demolition to handover, of just 12 days

of working were replaced by their superiors. On the following project the electrical package estimated at €232,000 was completed at an actual cost of €192,000.

Value

Fira Palvelut's second guiding Lean principle is *"value"*. Fira engaged Maria Snäkin, a customer service consultant, to help the team that was tasked with preparation of the Fira Palvelut business development meeting protocols. When Maria began working with the company, a project manager was assigned to teach her the ropes, to explain all he could about the apartment pipe renovation business. He told her that the client in every case was the building management company, with whom the contracts were signed, and that the residents, whose apartments were being renovated, were therefore not the customers: "We cannot treat them as customers, because then they would come with all sorts of special requests and demands that we cannot manage and could not fulfil without losing money on the job!"

Later, when meeting with the management team developing the new customer service-oriented protocols, one of her first questions was "Guys, what background music are you going to play when you present the project to the clients?" "Music?!" they responded, incredulously. "Yes, music!" she insisted. "You need to become aware that for a customer the experience includes every facet of the interaction with you, including the environment in your offices, the nuances, the body language, etc. It's not just your costs and your schedules and your nice BIM models. And by the way, the residents are absolutely your customers!"

In the new process, called "Fira AGILE", a customer service representative is assigned to each family. These are civil engineers, but with a strong customer service orientation, and they are the single point of contact for each family. Customers are provided with a detailed information pack that clearly lays out a process map showing all points of contact with the company through the life of the project.

Among the services provided under Fira AGILE are: design workshops, an online information service, an online 3D bathroom modelling software,[7] and a construction trailer fitted out with all of the bathroom design options – wall and floor tiles, sanitary fittings, cupboards, mirrors, light fittings, etc. – that is parked at the job site for the convenience of the customers. Each customer is invited to view the options and make their selections, in consultation with their customer service representative.

Box 19.4 Fira AGILE

Our primary task is to build a sense of well-being for our customers. The following customer service components help us to do so:

STARTER PACKAGE

Thorough and clear informational materials help residents (apartment owners) understand what happens in renovation projects and the practical issues they must deal with.

HOPES AND WISHES QUESTIONNAIRE

Online questionnaires determine our customers' opinions, hopes and wishes.

FIRA WORKSHOP

A workshop where we plan the renovation project together, share information and survey hopes and wishes.

WORK SITE VISITS

During the planning, the possibility is arranged for residents to visit another Fira Services work site to get acquainted with the everyday routine of a plumbing repair project.

KOTIO NOTIFICATION CHANNEL

A developed housing company notification channel in real time, allowing everyone the chance to personally choose the types of communication they prefer.

THE HOUSING COMPANY'S OWN SERVICE ENGINEER

Fira's unique way to serve each shareholder on a personal level.

SHOWROOM

With Fira AGILE, a showroom container is brought to the site of the housing company, where the apartment owners (shareholders) can discuss their ideas of renovation with regard to their own bathrooms as well as get more familiar with the fittings and surface materials.

COUNSELLING SESSIONS

If the shareholders wish to do so, they can reserve a counselling session with the service engineer to help them make the choices affecting their own homes.

BATHROOM DESIGNS IN 3D

A 3D image is drawn of each bathroom so it is easy to understand the planned solutions.

INFO SESSIONS

During the project, two info sessions are organized for all customers. In these events, the project personnel introduce themselves, the project stages are reviewed and time is allotted for questions and discussion.

Summing up and lessons learned

Fira has experienced remarkable growth since its founding in 2002. Clearly its youth and lack of traditions are part of the reason that it has been able to develop a company culture that is quite different from that of more venerable construction companies. It looks and feels more like a start-up company whose equity is in its knowledge, people and processes, rather than a company whose value derives solely from the profits it earns from each separate construction project.

Fira places a very strong emphasis on management by process, rather than on management by results. It has numerous carefully documented and standardized processes. Each process is assigned to a process owner, whose job it is to focus the efforts of structured experimentation and continuous improvement for his or her process in line with Fira's *Process Owner's Manual*. In fact, Fira employs a dedicated process engineer who supports the process owners. According to Otto:

> The problem with processes that are not owned is there is a vacuum. For example, we experimented with a quality control-monitoring IT solution based on a tablet computer – some people adopted the tools ad hoc, which led to information islands. For any process to succeed, we need to have a process owner who can evaluate, select and institute the process and take responsibility for standardizing it. Once standardized, they become responsible for continuous improvement. There are three steps: develop – standardize – continuously improve. Process owners need to have the knowledge, authority and conviction to be able to tell people: "Your way is good, it is based on your experience, but it is not the Fira Way, and we need you to adhere to the Fira Way."

The key to Fira Palvelut's approach to achieving steady flow is to manage all of the three aspects of the PPO model[8] – project portfolio, processes and operations – simultaneously, ensuring continuity for subcontractors within projects and between projects:

- They manage the flow of projects in each company's portfolio, ensuring that the projects start consecutively such that all of the subcontractor crews working on a project can move together in sequence to the following project.
- They manage the flow of processes in each location carefully, identifying the bottleneck operation in each process and adjusting crew sizes accordingly to align all with the predetermined Takt time.
- They manage the operations themselves, striving to continually improve each task of the process.

The move from design-bid-build to collaborative design-build with the Fira Verstas process has allowed Fira to leverage their expertise in BIM and Lean Construction in order to evaluate what brings value to the client, and then to work together with the client to provide that value. Together with its process efficiencies, this has led to a high degree of success in bidding for construction projects; while the average success rate for construction contracting companies in the Helsinki area to win bids was about 10 per cent, Fira achieved a 44 per cent overall success rate after the introduction of DB and the

Verstas process, which climbed further, to around 80 per cent, for projects with tender bids prepared in detail using BIM.

Even with the Verstas and IBR processes, many questions remain open even after work begins on site, and projects still run into issues like being over-budget, having poor quality and not coming in on time. Fira has succeeded in achieving deeper engagement with clients in defining the product, but their leadership is not strong enough during the early parts of the design. Designers tend to only show fairly complete work, which means bigger batches of information and longer iterations after presentation to clients. Anneli Holmberg identified a particular difficulty with site engineers, who tend to have weak understanding of Verstas and IBR, and who still find it difficult to adapt and engage with the new processes:

> There is not yet a set ideal picture of a successful project manager or site superintendent. We need coaches and mentors at every level. We do not have a tradition of servant leadership in construction, and we do not have management awareness of the need for cultural change. We need to develop ideal profiles and standards for the project leadership roles, and then coach people about how to achieve them.

Fira has strong leadership in encouraging participation but less in forcing people to work in a particular way. Site managers are independent, and management fears forcing them to comply because then they might leave. According to Anneli:

> I'm idealistic: I want things to be done "the Toyota way". For example, in quality, I would expect each worker to judge the work they receive, and reject it if it is not good. But they don't, because of time pressure, lack of competence, and so they "build-in" poor quality. No flags are raised. Therefore, we need leadership to inculcate the attitude of personal responsibility for quality, for smooth production flow and for continuous improvement. Site supervisors are more committed to the project than they are to the product. Company leadership is not yet strong enough. Some project managers are good, but they are not changing the culture. I feel that the leadership is aware of the importance of their role in driving cultural change, but they remain reluctant or fearful of pushing too hard.

One of the ways in which Fira tries to inculcate working according to its processes is to support those processes with appropriate IT systems. In pipe renovation projects, Fira Palvelut uses a number of different software packages and mobile apps. Initially, Fira started using WhatsApp for real-time collaboration between workers and foremen on site. Good results inspired Fira's start-up, SiteDrive, to develop a mobile app for workers, and not just foremen, to receive and send different types of process information. Simultaneously, the Finnish software company Congrid released a mobile app for their software, which enabled users to add mark-ups, punch lists, RFIs and pictures from site to plans drawings online. As the development of software has been very rapid, internal processes are commonly adjusted during a project and not only between projects. Therefore, Fira provides on-the-job training on site as new applications and versions are published and working procedures and processes are changed. It sounds like an oxymoron, but standardization needs to be dynamic.

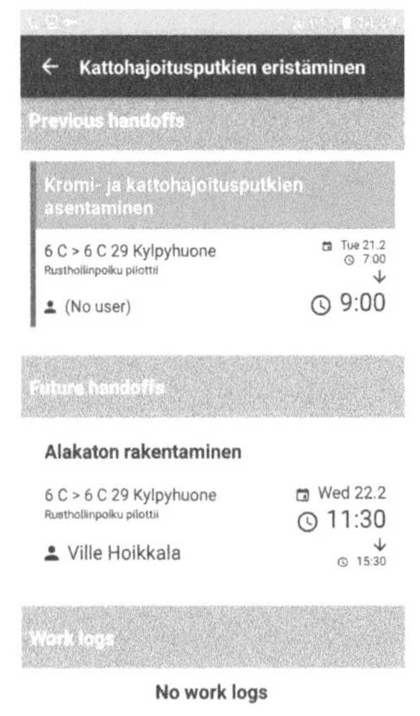

Figure 19.16 (Left) The foreman updates the process information based on changes that were decided in a meeting at the site office; (middle) Mobile applications allow the whole team on site to see the progress and problems of others transparently, which improves the ability of the team to react wisely to changes and even to react proactively to prevent problems; (right) A construction worker notices the change in allocation on the mobile app and can replan his own schedule for the rest of the day

Otto adds:

> In the software business, you have to learn all the time, otherwise you are out of a job. In construction, requirements develop all the time, but there is no culture of learning or change. Stagnation is built in. But this is changing in Fira. We are changing the business model and that enables new behaviours – the alliance contract, BIM and Verstas, digital tools that are carefully designed to promote workflow and collaboration, are all features that arise with the new business model. We are slowly building a new construction management culture.

Notes

1 According to the *Oxford Living Dictionary*, a "people business" is (a) a business in which contact with customers, or customer satisfaction, is (portrayed as) the most important element; a business in a service industry, and/or (b) a business whose most important asset or element is the (skill or talent of) the people it employs. Both of these definitions apply in the case of Fira Oy.

2 VisiLean is a "software-as-a-service" application that enables teams to use BIM models as the backbone for their implementation of LPS. For details, see www.visilean.com.

3 The description of Fira's process is based in part on a research paper co-authored by Otto Alhava, Enni Laine (a former Fira construction engineer) and Professor Arto Kiviniemi of the University of Liverpool: Alhava et al. (2015b).
4 The results of this effort were reported in a paper presented to the European Conference on Product and Process Modelling. See Alhava et al. (2015a).
5 Using Little's Law, TH = WIP / CT = 45 / 5 = 9 apartment renovations per week.
6 This phenomenon is explained in Chapter 15, "What flows in construction?"
7 Fira provides access to an online 3D bathroom and kitchen modelling application called InnoPlus Web, a product of Compusoft GB Ltd. See www.compusoftgroup.com/en-gb/bathroom-design/innoplus-web.
8 The PPO model is described in Chapter 15, "What flows in construction?", and more details can be found in Sacks (2016).

20 Pulling it all together

Every once in a while, the stars align and there is a confluence of positive factors. This was the opportunity of the Rosh Ha'ayin project – a project of perfect size, a perfect team and perfect timing. Within four years from start to finish, Tidhar built 1,036 apartments, using all the tools they had acquired since the "Growing the Margin" initiative began: a full BIM modelling process, co-located design in a big room, production planning and control based on Lean principles, and more.

Introduction

Gil Geva, Tidhar's co-founder and group CEO, saw in the Rosh Ha'ayin project (see Chapter 13, "Tidhar on the Park, Rosh Ha'ayin") a great opportunity to establish the Lean and BIM learning across the company. When the project was acquired, Tidhar was two years into the "Growing the Margin" strategic plan initiative. This seemed to be an ideal large-scale project in which all the knowledge and the tools that the company had worked so hard to develop could be applied. The idea was to use the project as a lever to spread the knowledge across the company and to establish new standards for design and construction management. Falling back on his experience as an operations officer for a military commando unit, Gil described it as an opportunity to achieve a breakthrough:

> We had two "big guns", BIM and Lean, but only a small number of our people knew how to use them. We estimated that some 80 or more of our core staff would take part in the project and be exposed to those wonderful tools. That was one fifth of the overall number of employees in the company at the time. Once the project was over they would take the knowledge and the processes they learned at Rosh Ha'ayin to other projects, creating a wave that would change the face of the company.

The Rosh Ha'ayin project had similar characteristics (in terms of scale, market, construction methods, etc.) as the earlier Yavne project (see Chapter 7, " Tidhar on the Park, Yavne"). Yavne started three years earlier and its duration was planned for 5.5 years. The goal for the duration of the Rosh Ha'ayin project was set to four years, fully one and a half years less. Gil explained:

> It was part of the deal. We wanted to market the project to the "Hever" consumer club, just as we did in Yavne. When we negotiated the deal, one of their demands

> was that all of the apartments had to be occupied by the first quarter of 2017. This was potentially a deal breaker, but I had confidence both in the capabilities and the experience of our people, as well as in the new tools and methods, and I knew that it was doable.

Time proved him right – the last apartments were delivered to their owners in March 2017.

In Rosh Ha'ayin, Tidhar applied many of the BIM and Lean tools it had developed in pilot applications in earlier projects. Among them: co-located design in a big room, intensive use of BIM for design, for coordination and for detailing of building systems, LPS and the social subcontract, BIM modelling for production planning and control, a dedicated production planning and control team, a production "situation room" at the site, BIM modelling to support client change management and others. Most of these applications are discussed in earlier chapters of this book. The goal of this chapter is to show how all of these different aspects came together as part of a whole. Each of the following sections highlights a different aspect of Lean Construction or BIM implementation in the project. Where the details of each innovation are provided elsewhere, we refer readers to the relevant chapters.

Lean thinking and BIM in design

Design began as soon as the partnership deal was finalized. The first major target was to obtain the building permits, and it became apparent immediately that time was of the essence, not only because of the overall project schedule but also due to the vagaries of local politics and bureaucracy.

The town plan for the whole neighbourhood accommodated 14,000 apartments. The regional planning commission was issuing the building permits in batches, but it had decided to limit the total number issued to just 2,200 of the total until two preconditions were met: construction of a large interchange to connect the neighbourhood to the local highway and completion of a new sewage pipeline from the neighbourhood to the Dan metropolitan wastewater treatment facility. In March 2013, when Tidhar's deal was signed, the interchange was well under way, but political issues had stalled the start of the pipeline construction. Many developers were in the process of applying for building permits; it was clear that the first to the post would get their permits and the others would be forced to wait for some indefinite time.

At the time, Revit was the most commonly used BIM tool among design consultants in Israel, but overall penetration remained low. For Tidhar, the pilot project in Yavne had used Revit, so it was clear that Rosh Ha'ayin would have an architectural BIM model and that all of the architectural drawings would be produced directly from it. Since the engineering consultants were only starting their BIM adoptions, Tidhar assigned one of its modellers, Eli Renzer, to model all of the MEP systems for them. Eli was the interface between the project building model and the engineering consultants, who continued to prepare CAD drawings but benefited from the coordination of their systems through the model.

Given the severe time constraint and the need for close communication among the architects, Tidhar's modeller (Eli) and the engineering consultants, synchronous collaboration of the kind that was needed was possible only through co-location in a big room. Collaborative design using BIM in the big room was no longer simply a learning opportunity – it was a strategic imperative. Tidhar invested the necessary resources to set up a 166 m^2 suite of offices to accommodate the team, equipped it with BIM workstations, conference rooms, etc., and added contract terms to its agreements with the engineering consultants to ensure the presence of their designers in the big room.

The decision to establish the first big room in Israel was made very early in Tidhar's Lean and BIM adoption process. It came only six months after a group of leading managers returned from a study tour to Finland and Denmark, where they saw working examples of the concept. With the extremely tight schedule and potential communication issues, the CEO (Gil Geva), the vice president for design (Zohar Raz) and the R&D lead (Ronen Barak) believed that it was the obvious solution, but they faced resistance – the design consultants, and indeed some of Tidhar's own design management staff, did not share their conviction. Gil felt that it was important to build consensus for the move. It took hours of meeting and personal conversations to convince the heads of the design offices that it was in their best interest to work in this environment. In May 2013, the big room was first occupied and Maia Davidson, the design manager, Eli Renzer the model manager and four architects were the first to work there. Two structural engineers and then the plumbing and electrical engineers joined them in quick succession.

The effort paid off. It took the team just three months to complete the design, prepare the drawings and submit permit requests for the first two quarters, and by December 2013 the first permits were secured. What was normally a 12- to 18-month process for a project of this size was accomplished, with the help of BIM tools, the right processes and good communication, within only eight months. The full story is told in Chapter 14, "BIM in the big room".

During the four months that it took for permit approval, the team progressed to the detailed design stage to prepare the project for marketing. The major task was to prepare 1,036 sales plans, specifications and apartment purchase contracts. This was Tidhar's first attempt to extract most of the data for those documents directly from the model. At first, the data were exported to Excel sheets and then embedded in text files by document automation routines in a two-step process, but later, the process was automated to work directly from the model. This process that was established became standard practice across Tidhar.

On Friday 22 November 2013, barely eight months since acquiring the rights to the project, Tidhar held a large marketing event. Within five hours, no fewer than 800 of the 1,036 apartments were sold. This was quite an accomplishment: two of the critical milestones (building permits and sales) were met according to plan, ensuring that construction could start as planned. The permits for the third quarter were secured in March 2014, but those for the fourth quarter were delayed because the regional planning commission's quota of 2,200 had been filled in the interim. In the end, they were released in July 2014, three months behind schedule, forcing a change in the planned sequence of construction of the quarters.

Production planning

While Rosh Ha'ayin was similar in many respects to the earlier Yavne project, it differed in two important ways that increased the expectations of Tidhar's leaders that it could

Figure 20.1 Tidhar on the Park, Rosh Ha'ayin sales event, November 2013

Note: As they wait for their names to be called according to a lottery, potential apartment buyers watch a status board displaying which apartments are still available.

be an opportunity to plant and nurture the seeds of new practices in the company. First, the project was more uniform – the variety of apartment types and building types was small compared with Yavne. The high volume with low variability meant that improved practices developed and implemented at the start would have long periods in which to incubate and become accepted standard practice within the company. Second, the project would be built over a short duration, which meant that there would be a much higher concentration of Tidhar's engineering and management people on site than any previous project. At the peak, there were some 70 Tidhar staff employed full time at the project, between one quarter and one third of the company's site staff at the time.

Production system design

Beyond the need to provide the management teams, construction equipment, logistic systems and subcontractors needed to build the project, the goals set for the production system designers called for broad adoption of Lean Construction tools, such as a collaborative big room for construction planning and control (the production situation room), visual management, LPS and continuous improvement. The challenges were to formalize standard work procedures, implement them across all the project teams, and follow a rigorous PDCA cycle to ensure improvement and organizational learning.

Toward the end of the design phase, four months before construction began on site, a team of senior managers began meeting once a week in the design team's big room to design the production system. In these meetings, the team decided on the main principles for selecting construction methods, planned the manpower needed, set-up an organizational structure, planned the site layout, began to mobilize client services and more.

The production plan called for division of the overall project into five smaller projects. Each quarter was organized as a separate project, and the public park was the fifth project. A central headquarters office was formed to manage the four quarters as a single cohesive project despite their independent organizational hierarchies. This HQ provided centralized services for all five projects. Full details of the staff allocation are provided in Chapter 13, "Tidhar on the Park, Rosh Ha'ayin", and illustrated in Figure 13.4.

Master planning and phase planning

The first step of the production planning, in line with LPS, was to create a master plan. Generated using Excel, it showed each of the buildings in a single row, with the series of operations to be performed shown as Gantt bars in a continuous sequence of work from excavation through to handover to clients. For each building, the plan showed the milestones for completion of these main phases: excavation, cast-in-place reinforced concrete (RC) pile foundations, the RC underground structure, the reinforced concrete structural works for the residential building, interior finishing works, handovers to clients and the final contract completion milestone.

The construction plan was deliberately prepared at a high level of abstraction and low resolution. Although it only established target dates for work to begin and end in each zone, it helped everyone involved to understand the timeframes. It was only updated twice during the 40 months of production, once when the permit for the fourth quarter was delayed, and a second time when instead of the expected soft soil, the excavation of the first quarter (455) revealed hard rock, causing a two-month delay. The master plan was hung in every site office, making it a visually accessible reference for everyone involved.

The next production planning step was to create phase schedules. The tools used for phase planning were selected to suit the complexity of the processes in each phase:

- The relatively straightforward excavation and underground basement structure phases were scheduled using standard critical path scheduling software. This was sufficient to capture the simple "finish–start" relationships that govern these phases.
- For the reinforced concrete works for the building's structural frames, the team used the BIM model to plan the pour breaks and to extract quantities for both the concrete and the formwork equipment. With all the information in hand, a 4D phase plan was created using Autodesk Navisworks software. Daily plans were produced from the phase plans (see Chapter 18, "Raising planning resolution – The four-day floors").
- The interior finishing work phase was planned as three "waves" that would flow through the building in succession, following the standards developed in earlier

projects (see Chapter 16, "Production planning and control in construction"). The first wave included the bulk work: partitions, plumbing and electrical systems, waterproofing and insulation, flooring and painting. These all involved in situ fabrication. In the second wave, windows, doors, kitchen cabinets and surfaces, air-conditioning systems and sanitary hardware were all installed. These were "made-to-order" products fabricated off site. The third and last wave prepared individual apartments for handover to clients. The team used two tools to plan this phase: a business process model notation (BPMN) chart with swim-lanes for each subcontractor trade crew to plan the logical task sequence and location-based scheduling with VICO Control software to compute and communicate the plan using flowline charts. The details of this planning step are provided in Chapter 16, "Production planning and control in construction", with examples of the flowline charts that represented the interior work phase plans (see Figure 16.3).

Production control

Production situation room

With 70 Tidhar employees and some 550 subcontracted construction workers on site on an average workday, formal communication of production planning and control decisions and status via traditional channels would not be enough. The production team proposed to turn their workspace itself into a "visual management" tool, a place where anyone and everyone could – ideally at a glance – get a clear idea of the production status and the production plan for the next few days.

A "production situation room" was established in a 50 m^2 open space in the site offices. In it worked Ortal Yona, Tidhar's first production engineer, Eli Renzer, the BIM model manager, Ronen Barak, Tidhar's R&D manager and BIM and Lean lead for the project, and Vitaliy Priven, the recently hired company process engineer (Vitaliy had previously introduced LPS to Tidhar at the Yavne project while conducting his PhD research. That story is told in Chapter 17, "The Last Planner® System"). This team was responsible for planning and controlling production.

The situation room was the nerve centre of the site office compound and as such, its location was critical. The situation room's walls displayed all of the weekly work plans and the phase plan flowline charts for all four quarters. In addition, they created a special section of the wall in which they could show PDCA items. At first, the situation room was located on the second floor, above the offices of the five sub-projects, but strategically located – between the entrance and the dining room. All site personnel passed through it and were exposed to the status information at least once every day. When the office cluster was moved to make way for landscaping of the central park, the situation room was transferred to occupy the main entrance space of the cluster, which had continuous traffic. Ortal, the production engineer, was responsible for operating the situation room status boards. She prepared the detailed production plans that drove the make-ready process, participated in all Last Planner® weekly work planning meetings, created control reports and updated the plans according to production performance.

Quarter & Building #	# Stories	01/14	02/14	03/14	04/14	05/14	06/14	07/14	08/14	09/14	10/14	11/14	12/14	01/15	02/15	03/15	04/15	05/15	06/15	07/15	08/15	09/15	10/15	11/15	12/15	01/16	02/16	03/16	04/16	05/16	06/16	07/16	08/16	09/16	10/16	11/16	12/16	01/17	02/17	03/17	04/17
		1	2	3	4	5	6	7	8	9	10	11	12	13	14	15	16	17	18	19	20	21	22	23	24	25	26	27	28	29	30	31	32	33	34	35	36	37	38	39	40
455-11	21			Excavation		Foundations		U.G Parking				G.FL	FL 1-19		4	5	6	7	8	9	FL 20	FL 21	Roof	Finishing		3	4	5	6	7	8	Certifi	Hando	Contract completion							
455-12	21			Excavation		Foundations		U.G Parking				G.FL	FL 1-19		4	5	6	7	8	FL 20	FL 21	Roof	Finishing									Certifi	Hando	Contract completion							
455-13	9	Excavation				Foundations		U.G Parking					1	2	3	4	5	6	Finishing		3	4	5	6	7	8	Certifi	Handovers				Contract completion									
455-14	9	Excavation				Foundations		U.G Parking										1	2	3	4	5	Finishing		3	4	Certifi	6	7	Handovers		Contract completion									
455-15	9	Excavation				Foundations		U.G Parking					1	2	3	4	5	6	Finishing								Certifi	Handovers				Contract completion									
454-16	21		Excavation		Foundations		U.G Parking				G.FL	FL 1-19		4	5	6	7	8	9	10	FL 20	FL 21	Roof	Finishing		3	4	5	6	7	8	Certifi	Handovers			Contract completion					
454-17	21		Excavation			Foundations		U.G Parking			G.FL	FL 1-19		4	5	6	7	8	9	FL 20	FL 21	Roof	Finishing									Certifi	Handovers			Contract completion					
454-18	9	Excavation				Found	U.G Parking					1	2	3	4	5	6	7	Finishing		3	4	5	6	7	8	Certifi	Handovers				Contract completion									
454-19	9	Excavation					Found	U.G Parking										1	2	3	4	5	Finishing		3	4	Certifi	6	7	Handovers		Contract completion									
454-20	9	Excavation					Found	U.G Parking				1	2	3	4	5	6	7	Finishing								Certifi	Handovers				Contract completion									
453-01	21					Excavation			Foundations		U.G Parking				G.FL	FL 1-19		4	5	6	7	8	9	FL 20	FL 21	Roof	Finishing		3	4	5	6	7	Certifi	Handovers		Contract completion				
453-02	21					Excavation			Foundations		U.G Parking				G.FL	FL 1-19		4	5	6	7	8	9	FL 20	FL 21	Roof	Finishing							Certifi	Handovers		Contract completion				
453-03	9					Excavation				Found	U.G Parking				1	2	3	4	5	6	7	Finishing		3	4	5	6	7	8	Certifi	Handovers				Contract completion						
453-04	9					Excavation				Found	U.G Parking										1	2	3	4	5	Finishing		3	4	Certifi	6	7	Handovers		Contract completion						
453-05	9					Excavation				Found	U.G Parking				1	2	3	4	5	6	7	Finishing								Certifi	Handovers				Contract completion						
456-06	21										Excavation			Foundations		U.G Parking				G.FL	FL 1-19		4	5	6	7	8	9	FL 20	FL 21	Roof	1	2	3	4	5	6	7	Certifi	Handovers	
456-07	21										Excavation			Foundations		U.G Parking					G.FL	FL 1-19		4	5	6	7	8	FL 20	FL 21	Roof								Certifi	Handovers	
456-08	9										Excavation				Foundations		U.G Parking				1	2	3	4	5	6	7	1	2	3	4	5	6	7	Certifi	Handovers					
456-09	9										Excavation					Found	U.G Parking				1	2	3	4	5	6	7	8	1	2	3	4	5	6	Certifi	Handovers					
456-10	9										Excavation					Found	U.G Parking				1	2	3	4	5	6	7	8							Certifi	Handovers					

Phases: Excavation | Foundations | Underground Parking | Structural Work | Finishing work | Certificate of occupancy | Handovers | Contract completion

Figure 20.2 Master plan setting milestones for the main construction phases for all 20 buildings in the four quarters of the Rosh Ha'ayin project

Production engineer: a new role

Traditionally, responsibility for production methods and engineering in Tidhar lies with the site construction engineer. However, in practice, the site engineer is usually involved mostly in the structure/framing – ordering the rebar, checking the rebar before pouring concrete and so on. The interior finishing works often had no engineering support, since they would begin well before the structural work was completed. But when planning for Rosh Ha'ayin began, the decision was made to provide more support for the finishing works in particular, given that the project had very high volume with relatively few basic design variations ("low mix, high volume" in manufacturing parlance). Since the project was divided into sub-projects, each with its own complement of support staff and site engineer, there was a need to standardize processes across the four quarters. A dedicated employee was needed to provide cohesion and coherence to the master plan and to coordinate between the project phase plans.

Ortal Yona was chosen for the job. She had graduated as a civil engineer with a specialization in construction management, and she had worked in Tidhar's control department where she had mainly prepared budgets and dealt with project adminis-

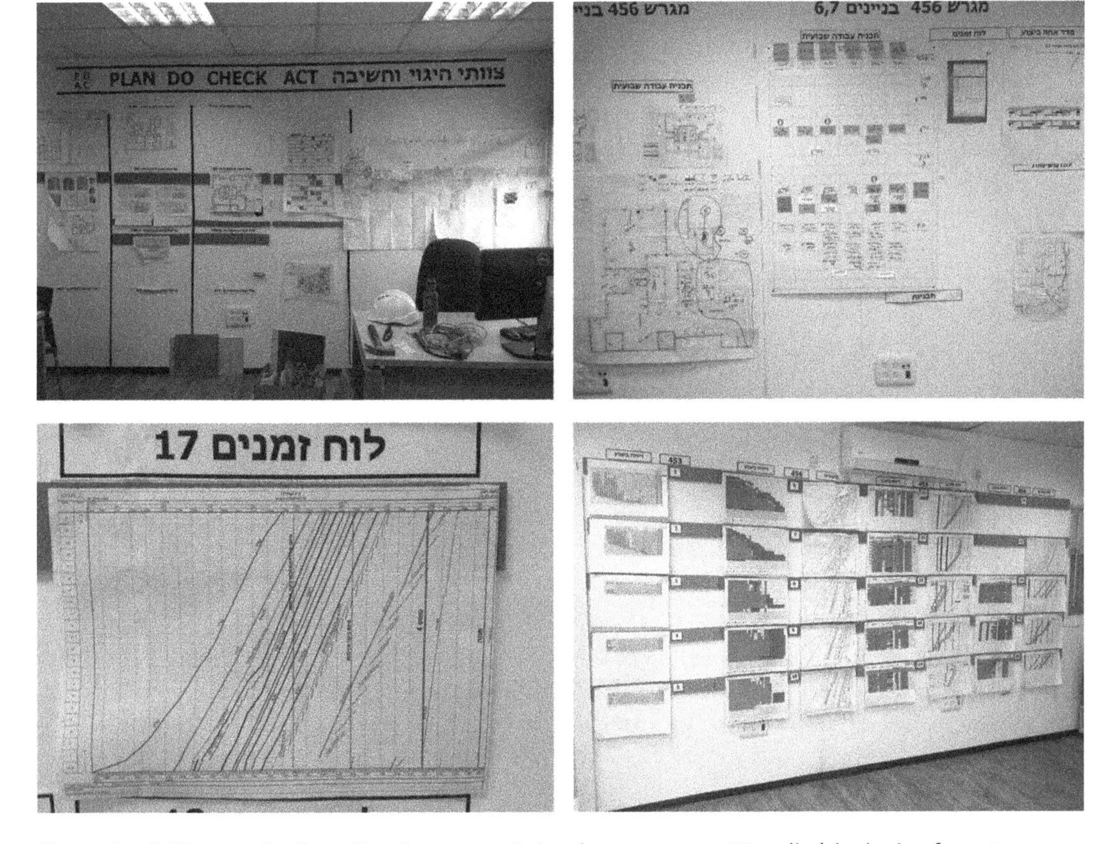

Figure 20.3 The production situation room "visual management" walls (clockwise from top left): PDCA board; weekly work plans and measures; the status information boards; flowline phase plan for building 17

tration. She was given the title of "production engineer". She was the first person at Tidhar to hold the title of production engineer, and perhaps the only one at the time in similar companies in the market. She was responsible for the entire production system, tasked with addressing questions that weren't being answered by anyone else in the company: how should the BIM model be prepared to support production planning and control? What was the optimal sequence of the detailed production tasks? This was the first time production would be managed at such a high resolution.

Ortal began learning by interviewing five experienced finishing works superintendents. She asked each one to sketch out the processes they would use to build apartments like those in the project. She learned how they thought and how they approached planning. She found that despite the company's relative sophistication with regard to documenting work methods, there were different ways of doing things. Some superintendents prioritized a clean work space, while others were more pedantic about getting to done-done.

Her first goal was to first extract the underlying principles and then work together with the superintendents to design an "optimum" sequence of trades. This harks back to the principle of creating standard work: by meeting with them all and allowing each to contribute, she could capture the best way the company collectively could come up with to build the project. Some guiding principles came to light:

- Minimize re-entrance (i.e. design work so that trade crews visit each apartment as few times as possible). Each time a crew has to return to an apartment in which they have already worked, the work flow is interrupted and the process made more complex. Controlling re-entrance adds to the workload of the superintendents.
- "Done-done": every activity in the process should be 100 per cent complete before a trade crew was allowed to progress, since leaving "loose ends" open cost a lot more time and effort in terms of rework later. In the interim, incomplete work is more susceptible to damage. For example, if floor tiles are not grouted immediately after they are laid, debris will build up quickly in the gaps between the tiles, resulting in the need for deep cleaning before finally installing the grout. A good production plan allows each trade to get to done-done and ensures that the product is not vulnerable to damage by subsequent trades.
- Create a plan that is acceptable to all. To increase the chances of success, get buy-in from all involved. Ask parties to commit to a plan, in order. Subcontractors who feel that they are getting the short end of the stick are not conducive to long-term success.
- Dirty works first: minimize the amount of cleaning work, as trades progress from dirtier to cleaner tasks. Plaster before flooring is better than the opposite, otherwise plaster debris must be cleaned off the tiles.
- Fragile works last. Minimize unintended collateral damage from workers and tools.

Next, Ortal worked with the same superintendents and with some of the subcontractors to get better estimates of work rates for each task. She had to push beyond general rules of thumb (such as "painting takes four days per floor") to get to average daily production rates in units of square metres or cubic metres per day. With these in hand, she could get the quantities to be built from the BIM manager and compute reliable durations for each task, since they would reflect the variations in quantities between buildings and floors.

Based on the sequence of tasks and durations, Ortal put together a detailed production schedule. From there, it was possible to apply industrial engineering tools, like identifying bottlenecks and deciding where to work continuously versus where to split up the work into non-continuous segments. She decided on batch sizes and crew sizes that would be required to support the plan. Also, since each building of the 20 in the project had its own start date, Ortal integrated each of the plans for each subcontractor across the different buildings. "Fast" trades could be scheduled to jump back and forth between buildings, preserving their work continuity but not delaying any given building.

The Takt time target was about a week for each floor per task. This allowed her to adjust the number of workers in each crew to aim for a week's duration for each. The final plan was modelled in VICO Schedule Planner, a location-based management software tool that creates line-of-balance charts (see Figures 16.3 and 20.3). The VICO output was a preliminary project timeline and a work plan for each building. The next step was to check whether the schedule's end date was in line with the contracted finish date. If not, compromises would be needed, such as relaxing continuity constraints for some trades or shrinking batch size. The LOB charts helped identify waiting times for apartments, since these show up as empty space on the chart where no trade is working in an apartment. Production planning of this kind, *par excellence*, had never really been done at this level of detail – it had been intentionally relegated to the trades to solve for themselves, or to just work itself out in the doing. The goal was to develop a plan that was as close to reality as possible, one that could facilitate coordination, not just a rehash of tired rules of thumb thrown into fairly arbitrary boxes on a Gantt chart. Ortal's plan was a feasible plan, one that could be acted upon and used for production control.

Another important task Ortal performed as the production engineer was to define the work content of each task, and in particular, what "done-done" meant for each operation. This entailed specifying clear requirements for completion. For example: cut floor tiles adjacent to door frames were typically installed only once the door frame was placed. Was this the floorer's responsibility or that of the carpenter? Did the mason have to create opening penetrations for air-conditioning ducts in interior partition walls or was this the resonsbility of the HVAC contractor?

Once work began, Ortal was tasked with controlling execution of the plan. She attended all of the WWP meetings for all the buildings, ensuring that each building was progressing according to plan. She could identify deviations, discover problems and either work to resolve issues by making work ready in consecutive weekly cycles or adjusting the plan itself. Deviations arose from incorrect production rates, inaccurate quantities, less-experienced workers or delayed learning curves. She knew exactly what was planned for each upcoming week and saw the non-completion reports from the prior week. In parallel, she received weekly updates from the superintendents and site engineers. Finally, to further triangulate the accuracy of the reports, she toured every building at least once a week. With all of this information in hand, Ortal reported project status to the project managers and the regional manager weekly. They assessed risks and decided on countermeasures, such as mobilizing additional resources, as needed.

All told, establishing the role of production engineer proved to be a major boon to the project. Tidhar policy was changed to require the role in all projects of sufficient scale or to allocate more of site engineers' time to production planning and control on smaller projects. Sadly, however, the company's traditional head office control department did not engage with the role of production engineer on site. They remained set in their perception of the task of project control as one of monthly measuring and reporting. Even within a dynamic company, some people remain set in their ways and management may find it difficult to relinquish or substantially change functions that had previously been perceived to be an essential part of the company's way of working.

Weekly work planning and control

The Last Planner® System WWP meetings provided the most detailed level of production planning. Until the Rosh Ha'ayin project, WWP meetings had only been used for interior finishing works – here it was applied to all phases. During the excavation and foundation phases, all WWP meetings were held in the production situation room, but as soon as the ground floor of each building was built and could be made safe, the meetings were moved to specially prepared rooms within the buildings. By then, Tidhar had learned from previous projects that not only did trade crews feel more comfortable meeting in the buildings in which they were working, but also that the completed weekly work plan boards should be readily accessible to all crews on a day-to-day basis.

After every WWP meeting, Ortal Yona photographed the boards, printed them in colour on an A3 sheet, posted them in the situation room and shared them with the crews via their specific project's WhatsApp group.[1] Simple, low-cost technology can be very effective.

Managing the second wave – Identifying bottlenecks and reducing cycle times

The interior finishing works for an apartment were reconceived as consisting of three waves. One of the key lessons from Vitaliy Priven's research at the Yavne project was that whereas the LPS WWP meetings had a profound effect on the first wave, once the second wave was reached, the effect was small. In some of the Yavne buildings, all of the gains achieved in significantly reducing durations for the first wave were lost during the second wave due to constraints beyond the control of the site team.

The reason was not hard to decipher. Whereas the WWP meetings had been thoroughly applied, no one had tried to implement the LPS's make-ready process. For the long lead-time prefabricated parts, a persistent make-ready process was essential to bringing the installation activities to maturity before their planned execution dates. Although the net duration of each activity was short (in some cases just a few hours per apartment) the entire process could take months.

Two steps were taken to try to correct the situation. A new role, called the "installation manager" was created. The installation manager's responsibilities were to coordinate, monitor and facilitate the logistics of the suppliers and installers of aluminium framed windows, custom kitchens, sanitary fittings, doors, air-conditioning systems, etc. Two dedicated installation managers began work on site once the tall buildings in the first quarter were ready to begin their second wave of interior finishing works. The long lead-time items, the windows and kitchens, required from 60–90 days from order to installation. The managers compiled a control system using an Excel spreadsheet, with 1,036 rows, one for each apartment. The make-ready steps were tracked against the production plan, and whenever an off site component passed a milestone preparation date without reaching the required status, one of the installation managers began following it up.

Box 20.1 Why, why, why ...

In one instance, on buildings 10 and 11 in the Yavne project, Vitaliy ran into a brick wall when he reached the second wave. Having achieved excellent flow in the first wave through the LPS WWP meetings, he observed that, all of a sudden, WIP grew to encompass four to six floors of the building:

> The aluminium window crew would come to site and install all of the windows on a floor, but not the sliding doors for the apartment balconies. Without these doors, we could not close the apartments to the weather, and so we could not proceed with the other installations. Apartments without doors accumulated, joining the WIP pile, and the downstream flow of work essentially ground to a halt. I had to get to the bottom of it, so I began asking "Why?" repeatedly, until I finally understood the problem.
>
> "Why do you install everything except the balcony doors?" I asked the windows fabrication workshop manager. "The sheets of glass for the sliding doors are in our workshop, but unlike the window panes, the building code requires that they be tempered", they answered.
>
> "So why has it not been tempered?" I asked. "Because we need to send it to a kiln in Jerusalem for the treatment, and we can only do that once we fill a truck. It's too expensive to send one at a time".
>
> "So why can't you send a batch in advance?" I continued probing. "Oh no, we can't do that, because each door has to be cut to size before being tempered, and the size must be measured on site, because the way you guys build, each apartment has a slightly different floor to ceiling height".
>
> "Why can't you measure on site and cut them to size in advance, then, and not delay the construction flow for so long?" came my next question. "We can only measure a door once you complete the flooring in the apartments, and so we can't send any glass for tempering until the flooring is complete in enough apartments for us to measure and cut a whole batch", was the final reply.
>
> So it turned out that the root cause was that the inconsistent flooring height in the apartments we built, combined with the supplier pursuing a local optimum by batching sheets of glass on a truck for a 100 km round trip, was delaying the entire construction process! Four to six floors, some 16 to 24 apartments, were forced into a batch for reasons that no one had properly understood. And just as Little's Law predicts, as the WIP went up, so did the cycle time for the apartments, which meant that the overall project duration grew longer as a result.

This was a Sisyphean task, and it soon became clear that real improvement in the second wave would require delving into the processes of the off site fabrication suppliers. Vitaliy, fresh out of the Technion after completing his PhD, was employed and assigned the task of investigating and improving the two installation processes that had the most profoundly deleterious effect on the process: the aluminium window frames and the custom kitchens.

As described in Chapter 16, "Production planning and control in construction", he began by forming a small team and visiting several kitchen and aluminium window manufacturers. The visits to the kitchen factories revealed that orders had to pass through the customization and design departments, through the fabrication departments, to the assembly line, and hence to packaging, shipping and finally to installation on site. There were two bottlenecks. The first was that suppliers were required to have the apartment owner approve the design, and this was difficult for them to control; the problem was solved by transferring this aspect to Tidhar's own customization process managers, who had a closer relationship with the clients and could focus their attention in a timely fashion.

The second was that the kitchen fabricators would take their own "as-built" measurements on site before beginning fabrication, to protect themselves from the variability in construction tolerances, similar to the bottleneck Vitaliy had discovered with the balcony glass in Yavne (see Box 20.1). Their employee preferred to measure multiple kitchens in a building in a single visit, and so, for the sake of local optimization, they waited to let multiple kitchens accumulate as a batch before waited to the site. This naturally increased the cycle times. After some deliberation, Tidhar assumed the responsibility for the measurements for all of the standard kitchens (those with no or minimal client changes). The site superintendent would make the measurements, mark them up on the drawing and sign off on them. These shifts in the process had real impact. By the end of the project, 95 per cent of the kitchens were being installed within a week of their planned date.

Raising planning resolution

The four-day floor

Chapter 18, "Raising planning resolution – The four-day floors", tells the story of the production team's efforts to break the company record for the standard production rate for cast-in-place structural concrete framing work for residential buildings. For many years, the record stood somewhere between three and four floors per month, depending on the actual layout of the site (one or two buildings) and the equipment allocation (cranes, formwork etc.). A typical building storey is usually between 600 m^2 and 700 m^2 in area and has four or five apartments. The quantity of concrete per storey is around 350 m^3, with nearly equal portions in the walls and in the slabs.

The target set for Rosh Ha'ayin was to improve this rate by 50 per cent, to achieve a production rate of six floors per month (equivalent to casting an entire floor every four days). The team used BIM tools to plan the concrete casting cycles and Lean methods to reduce waste in the process. The main idea behind the solution was to increase the planning resolution. Traditionally, the structural schedule planning considered each wall casting section to be a two-day task, and each storey was commonly decomposed into four to six sections. The slab was a six-day task. Construction planners had always viewed both the wall section and slab tasks as "black boxes" without any production details. The new production plan delved deeply into the details for each wall casting section, depicting the daily work packages for every team in the crew (formwork, reinforcement, concreting and formwork stripping). With a detailed production plan, it was much easier to control the production rate, but more importantly, it was also easier to identify the problems that prevented smooth process flow. The planning team and the construction team worked together and spent many hours in the Gemba to find solutions. Most of the problems and the waste were found in the organization of the tasks and in the interfaces between them. For example, if a specific exterior wall formwork panel was not stripped at the right time, it also could not be hoisted at the right time for the next section. In this case, the superintendent shifted his attention to the stripping process, making sure that the formwork was stripped in the correct sequence, which then ensured that the next section could be prepared correctly.

The first reinforced construction crew to use the new plan (working on buildings 11 and 12 in tandem) broke the record of four floors per month, reaching a peak production rate of 5.5 floors per month (the average was 5.1 floors per month). The second crew to use the plan learned from the first, and on buildings 1 and 2 they set a new peak record of 5.9 floors per month, while averaging 5.5 floors per month through the whole building.

Detailed planning and control of façade finishing work

All of the buildings in the Rosh Ha'ayin project were built using an external wall system called the "Baranovich" cladding system. In this system, stone cladding pieces are arranged in rows and secured on wall-size steel formwork shutters together with the wall's reinforcement on the ground, and the panels are then hoisted by crane into position for casting. The shutters are closed with an additional steel form to make the inside of the wall, and concrete is poured to form a pre-finished wall. The details of the system are provided in Chapter 18, "Raising planning resolution – The four-day floors". Its major advantage is that no fixed scaffolding is needed for fixing the cladding, but it does require some cleaning, which is done using hanging work platforms.

The façade finishing is done by a subcontractor who performs several tasks on each façade: replacing any broken cladding stones, cleaning concrete residues, closing formwork tie-holes and waterproofing the façade. The subcontractor supplies the crew and the hanging scaffolds according to the required production. Their work is generally planned ad hoc, with the tasks generally running late and interfering with other tasks, such as window installation and landscaping.

Ortal Yona, the production engineer, decided to improve the process by using the BIM model to calculate the amount of work and the exact number of scaffolds needed to get the job done on time and without interferences. Ortal and Eli Renzer (the model manager) modelled the hanging scaffolds and separated the façades into vertical wall strips that could each be done with a single scaffold drop. They then extracted the surface area quantities for each drop and working with the subcontractor, Ortal prepared a detailed schedule for each drop and for each building, making sure that the overlaps with other tasks were avoided. Since there were two façade finishing subcontractors and 20 buildings, a detailed overall production plan was essential for achieving the right flow of equipment and crews.

With the model and the plan, the work could be managed by a single Tidhar superintendent. As the work progressed from building to building, he improved the process by adding more visual aids derived from the BIM model to show the crews the amount of work remaining and the schedule. It was quite a success. The façade finishing work ran

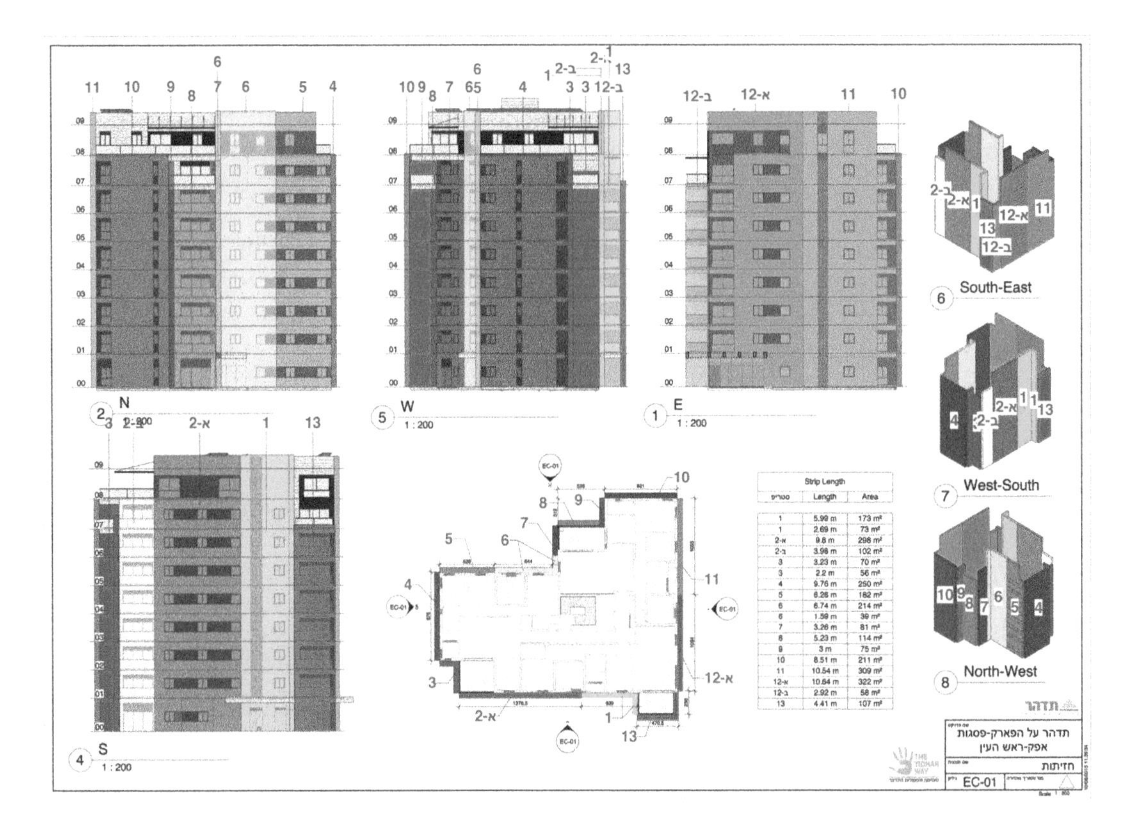

Figure 20.4 Façade finishing production plan drawing, showing the strips order and quantities made directly with the building BIM model

with almost no deviation from the plan and was essentially transparent to the other crews on site (no time–space conflicts were reported).

BIM for customer design change management

When the management team for Rosh Ha'ayin began planning the project, they assumed that the detailed architectural BIM model of the project could be leveraged to greatly improve the customer design change process. With 1,036 apartments to manage and the expectation that there would be many changes, this was a major concern due to the disruptive effect of customer changes on the flow of work, particularly in the interior finishing stages. They assigned a team of two to come up with a process for customer design change management in which the company's customer coordinators, mostly architectural technicians by training, would use BIM models to manage the design change process and provide model outputs directly to the subcontractor trade crews during construction.

Gal Salomon, Tidhar's first and most experienced modeller at the time, and Hadas Zaig, an architect who led the customer design change team at Yavne, set about developing and testing a process for customer changes in the Rosh Ha'ayin project. They explored both the technical aspects of using Revit in the process (model objects, level of detailing, separation of those parts of the apartment that could be changed from those that were fixed) and the information flow aspects (what inputs the customer would need to provide and when, how the coordinator and the customer would work with the model during their meetings, what content the construction team would need and in what format). The process that was developed appears in Chapter 11, "Virtual design and construction".

With the Yavne project (Chapter 7, "Tidhar on the Park, Yavne") still ongoing, Gal and Hadas ran a pilot of the process on the last building of that project, essentially as a first-run study for Rosh Ha'ayin. Once they had refined their plan, it was presented to management and approved. Perhaps the most critical decision in implementing the process was when to extract and separate individual apartment models from the central project model (this was necessary due to technological constraints). On one hand, since Rosh Ha'ayin was a large-scale project, with more than 1,000 apartments, preparation needed to begin as early as possible. On the other hand, once the models were split, any updates to the overall building would need to be applied "manually" to the apartment models, exposing the models to human error and creating rework. Making the split as late as possible – at the last responsible moment – was the right approach, and the process was designed to accommodate it.

Eight customer change coordinators were trained to work on the project. Most of them had some experience with Revit, but none of them knew the full process. In all, 55 different apartment type models were generated, resulting at the end of the process in 1,036 individual apartment model files. A major advantage was that all the information for an apartment was available in one model file: architectural, plumbing, electrical, HVAC, etc. All of the detailed construction drawings were issued from that model. From the customer's standpoint, the ability to review their apartment in 3D and to better understand the actual product proved very helpful. 3D PDF files were generated and provided to customers before the coordination meetings so that they could come better prepared.

Building on the idea of enhancing customer's cognition of their apartments during the design phase, Amit Dayan, an MSc student at the Technion, explored the potential of virtual reality (VR) to enhance the customer experience. He used one of the Rosh Ha'ayin apartment types and ran experiments with customers and coordinators at the site offices and in the Virtual Construction Lab at the Technion (see Chapter 21, "Reflective practice and practical research"). However, while the technological aspects proved feasible, the majority of customers in the test group at Rosh Ha'ayin were reluctant to invest the extra time and suffer the discomfort needed to view the VR models. It seems that viewing 3D models on screen was sufficient and convenient and did not involve the discomfort and extra time required to use the Oculus Rift VR goggles. This is discussed in more depth in the next chapter.

Continuous improvement

The scale of the Rosh Ha'ayin project and its timing in the overall scope of the "Growing the Margin" effort made it fertile ground for implementing and testing a range of process improvements that had been proposed in earlier projects. One of those was to drastically reduce the number of paper prints of construction drawings delivered to the site by instituting a pull process in which construction engineers could use the project collaboration platform to call for specific drawings to be printed and delivered only when needed. This replaced the traditional push process, in which copies of every new version of every drawing were printed and delivered to site.

However, some of the more significant improvements proposed required changes to the designs of the buildings' structures, which meant that they could only be tried on new projects where they could be initiated from the start of design. One of those was the proposal to make far-reaching changes to the way in which interior partition walls were built using gypsum blocks. The wastes in the existing process included multiple handlings of the blocks, non-value-adding process steps such as measuring and cutting many blocks and high levels of material waste. Chapter 8, "The waste of non-value adding work", discusses this in detail.

Some aspects of the solutions, such as avoiding stacking pallets of blocks on top of one another, were relatively simple to implement (although they did require the support of the production team to provide detailed layouts of the floors from the BIM models, clearly marked with the correct locations for pallets to be delivered) and had become common practice on all of the company's construction sites. Other aspects, such as the proposal to design the net internal storey height to avoid the need to cut blocks for the top rows of partitions, required intervention from day one of the building design, because the floor slab-to-ceiling clearance is one of the first decisions the design team makes. As Figure 8.9 shows, the dimension was set to 290 cm to accommodate the block dimensions.

With "designed-in" waste reduced to a minimum (some blocks still had to be cut at the ends of partitions), and with the detailed BIM model of the partitions available for accurate quantity take-off, it now also became reasonable to consider supplying exactly the right amount of materials, without adding any buffer to the orders. The various kinds of block needed – standard white gypsum blocks for standard partitions, water-resistant green gypsum blocks for bathrooms and kitchens, door lintel blocks, acoustic insulation strips – were all modelled and quantities were extracted for each delivery location. The production team issued two distinct drawings for each location. First, a floor plan and quantity sheet for the structure superintendent, showing where each pallet was to be placed by the tower crane prior to closing up formwork for the next slab. Second, an A0-size shop drawing for the builders, printed in colour, showing the partition material classification and its proper construction details. The effort of increased planning resolution paid off handsomely. Material waste was reduced by 20 per cent when compared with the last buildings of the Yavne project and the number of hours worked for transporting blocks within and between the floors was halved.

Lessons learned

The Rosh Ha'ayin project showed that significant improvements in construction processes can be achieved when Lean and BIM are applied together with forethought, particularly during the planning of the production system. Not all of the initiatives were successfully implemented, nor were all of the goals realized. Success must be measured not only by waste reduction or improvements to workflow, but equally by the degree to which viable process changes became company standards. Lean and BIM processes must not only be worthwhile; they must be sustainable.

Figure 20.5 Pipes routed around a square "void" space in a parking basement reveal the location during construction of the tower of a construction crane

Note: The crane's former position is marked with a yellow square on the floor and ceiling. Detailed system routing and coordination using BIM needs to account for constraints as they exist during installation of the systems.

Among the major successes that have subsequently become standard practice in Tidhar are: the BIM in the big room collaborative design process, reductions of material waste in most of the interior finishing trades of the first wave, better make-ready processes for installations in the second wave, greater detail in product modelling and process planning and customer design change management using BIM. While the cycle times for the reinforced concrete structural work to build a typical floor have been reduced, the detailed practices of careful sequencing of the Baranovich perimeter wall, including colour-coding and numbering of the formwork panels, proved difficult to sustain.

The dedicated production management team and the production situation room proved their value. In addition to their direct day-to-day contribution to smooth flow of construction work, they play a major role in continuous improvement. In contrast to factory settings, workers are relatively permanent, whereas in construction, the front-line production workers are almost all subcontracted and therefore transient, moving from company to company as they move from project to project. The production team are long-term employees of the general contractor, and they can be process owners within the company across multiple projects. Their learning and their improvements are more likely to be sustained.

A key lesson of the Rosh Ha'ayin project is that integration of design and construction, of product and process, are vital if the benefits of Lean Construction and BIM are to be realized. Figure 20.5 illustrates the point vividly. It shows a section of the ceiling of a parking basement with an array of pipes that are routed around what appears to be an empty space. The MEP work in the project was fully coordinated using BIM, but the models do not show this detour of the pipes. The change was made during construction by the trade crews in ad hoc fashion, because when they reached this segment, they found their path blocked by the tower of the construction crane that was installed in that spot.

Some readers may consider it strange that the BIM model of the project did not consider the presence of the tower crane and lead to MEP design coordination that would have taken it into account. After all, space time conflicts of this kind are common in many projects, and indeed, the ability to identify them and account for them in design and planning using BIM has been clear since Akinci's work in the late 1990s.[2] However, the failure points to the need to build integrated teams of people, not to deficiencies of technology. It was not enough for the engineers who designed the site layout to use BIM and to place it in the overall model of the project. This conflict, and others like it, might have been identified if the sequence of modelling the construction product using BIM had been aligned with the actual construction sequence, in a purist VDC fashion. Yet design sequence and construction sequence cannot be perfectly aligned because the scopes of effort needed over time are different. Carefully designed coordination and review procedures are needed, and by and large, these are developed through experience rather than forethought. Perhaps errors of this kind will be avoided in future Tidhar projects – and perhaps not.

Notes

1 Vitaliy Priven's PhD thesis (Priven, 2016) had shown that social cohesion amongst the various trade crews in a project could enhance workflow coordination. In a direct implementation of the findings, Tidhar Construction CEO Tal Hershkovitz instructed all site supervisors' teams to set-up WhatsApp (a cross-platform messaging application) groups to engender informal communication.

2 Akinci et al. (2002a; 2002b).

21 Reflective practice and practical research

"A business man once stated that there is nothing so practical as a good theory."
(Kurt Lewin, 1951)

Throughout Tidhar's process of Lean and BIM adoption, researchers from the Technion – Israel Institute of Technology, followed the change process. They measured production rates and aspects of waste, they interviewed managers and workers, they helped train people at all levels, they developed prototypes and ran experiments and they learned a great deal. Their work led to new insights, notably about the social aspects of production planning on site and the impact of social networks on planning and on the Last Planner® System, about the nature of flow in construction and about possible ways of measuring flow, and about the impact of customization on production flow. Through interaction with the researchers, Tidhar's people learned, developed, experimented and improved the company's work methods, management practices and technology systems. The collaboration was a win win endeavour.

This chapter describes the ways in which the Technion researchers engaged with Tidhar's efforts. It begins by describing the roots of the collaboration and summarizes some of the results of their work. It concludes with a brief discussion on the nature of the industry–academia partnership from the specific point of view of Lean and BIM adoption, and it attempts to answer key questions about industry–academia cooperation in construction management research and practice.

Tidhar–Technion collaboration

The collaboration between Tidhar and the Technion, between industry and academia, began, as many successful professional collaborations do, on a personal level. Ronen Barak, R&D manager at Tidhar at the time of writing and one of the authors of this book, first worked for Tidhar from 1999 to 2002, developing information systems. Ronen then completed his MSc degree in 2007 at the Technion under the supervision of Professor Rafael Sacks, another author of this book, and stayed on at the university as a researcher. He returned to work at Tidhar in 2011. His knowledge of BIM and Lean spread within the company, and captured the attention and imagination of Gil Geva, Tidhar's CEO. When Gil decided to launch Lean and BIM throughout the company later that year, he invited Rafael to discuss his plans. Rafael's response to Gil's long explanation about how he was going to transform Tidhar by adopting Lean and BIM was simple: "Gil, I have been researching these topics for ten years. I can certainly help you. In fact, if you think you're going to do this without me and my students both leading and following every step of the way, you're wrong!" This book is one of the many fruits of the collaboration that ensued.

Using the terms defined by Rogers' bell curve figure of the innovation adoption cycle,[1] Tidhar was an "early adopter" of BIM and Lean Construction in an international context, but for all intents and purposes, it was still an "innovator" in its local market. Apart from a failed Lean Construction initiative in a larger local construction company in 2005–2006 (discussed in Chapter 2, "False starts"), Tidhar's effort was the first on the part of a sizable construction company in Israel. Given the lack of understanding of Lean and BIM among its suppliers, subcontractors and external designers, it was a bold step. The knowledge accumulated through years of study and research at the Technion was therefore a valuable resource. Many of Tidhar's construction engineers and project managers were Technion graduates. They had learned BIM in courses such as Graphical Engineering Information and they had been introduced to basic production concepts in the Planning and Control of Construction Projects course. A few had also taken the Technion graduate courses in Lean Construction and Advanced BIM while studying for Master's degrees.

At the same time, Tidhar's construction projects provided an ideal "living laboratory" in which the Technion researchers could explore new ideas about production in construction and the synergistic interactions of BIM and Lean Construction. Construction management research is often severely restricted by limited access to the field. Constraints related to construction safety, commercial interests, intellectual property and especially to limited research budgets, typically lead researchers to make do with surveys, interviews or short-term observations of construction practice on site. Longitudinal studies are rare and researchers are almost always barred from significant interventions in live projects (such as changing management methods and schedules or introducing new information technologies) for the sake of experimentation. It was exactly the opportunity to apply such interventions, with the trust and participation of the people on site, that made the collaboration attractive for research.

The social subcontractor

In Chapter 17, "The Last Planner® System", we introduced the role played by Vitaliy Priven, then a Technion PhD student, in setting up the first implementation of the LPS at the Tidhar on the Park project in Yavne. As a researcher, Vitaliy's goal was to create the conditions for exploring the impact of LPS on the social networks that develop among the people in a construction project. He focused on the people at the construction site during the interior building system and finishing work, because he wanted to find out whether, and how much, the effect of LPS might be traced to improving personal relationships and communication. If this were the case, then the rigour with which LPS's steps was applied was less important. It might also explain why LPS implementations that only used weekly work meetings but neglected the phase planning and the "make-ready" constraint removal process nonetheless had been found to have fairly good results.[2]

Figure 21.1 lays out the thinking behind this research. The conventional thinking shown with solid arrows #1 and #2, was that LPS improves the project coordination, reduces variation, ensures that work packages are ready so that they can be done right the first time and thus improves production flow, in line with the mechanisms described by Ballard[3] and explained by the flow aspect of Koskela's TFV theory.[4] The dashed arrows (#3, #4 and #5) represent a different explanation, which was the hypothesis to be tested: the WWP meetings engender communication and strengthen the subcontractors' centrality (reducing the relative centrality of the superintendent), as defined in social network theory. In turn, according to coordination theory, better communication leads to improved coordination and thus to better flow.

Vitaliy worked closely with a number of Tidhar's superintendents, coaching them in LPS and measuring their progress. The experimental design called for a straightforward comparison across two groups: of the 32 buildings in the Yavne project, roughly half were to be built with LPS, while the remainder were to serve as a control group. In a refinement to the experimental set-up, the "social subcontract" – an experimental device inspired by Jean-Jacques Rousseau's "social contract" and designed to enhance the social relationships and communication on a construction site – was to be applied to the teams at eight of the buildings in each group. This would have led to four different experimental groups with different interventions in each group allowing the possibility of comparing results in two directions (two groups each with and without LPS, and two groups each with and without the social subcontract).

Unfortunately, construction management research of this kind is messy. Projects and the teams building them cannot be insulated from outside influences that affect outcomes and render data "impure". The major hurdle encountered in this research was ironically of its own making: LPS was so successful in the initial pilot that Tidhar's

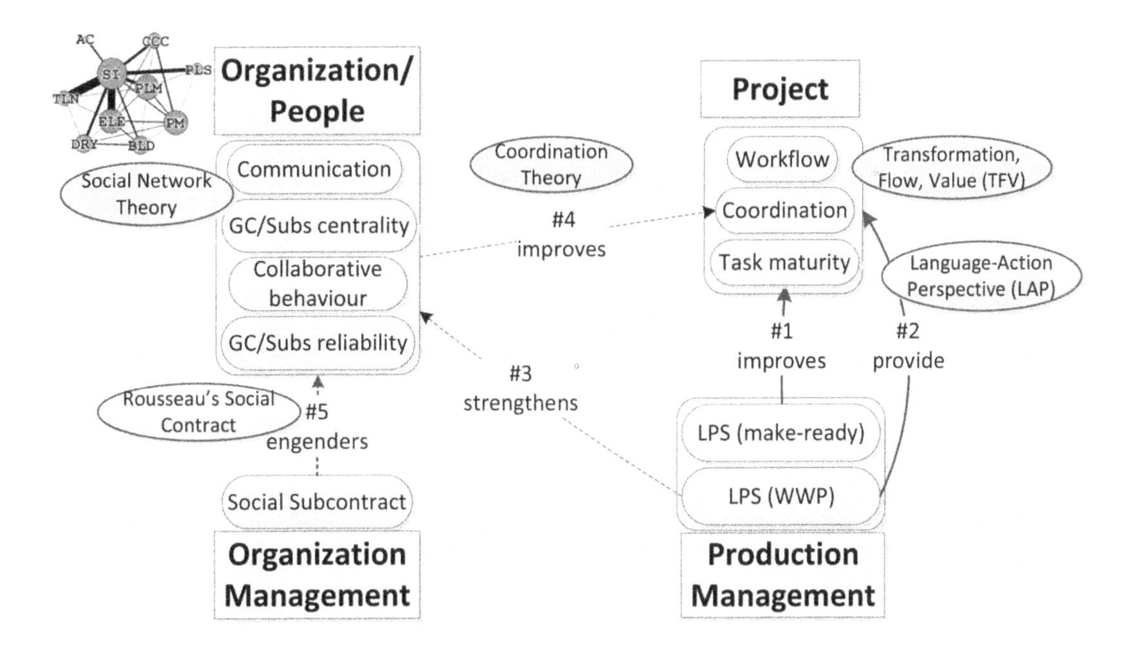

Figure 21.1 Hypothetical cause and effect diagram for the impact of LPS and social subcontract on workflow in construction projects

Source: Priven and Sacks (2016b), with permission from ASCE.

management instructed that it should be rolled out to all of the buildings at Yavne! This reduced Vitaliy's "no LPS" group to a size that was much smaller than anticipated, too small to use as an experimental control:

> Rethinking the problem led us to adopt a method to measure the extent of LPS implementation and the degree of change in the social networks as a way of differentiating among the projects to enable comparison of their results. An important question that arose was how "success" or "failure" of each of the projects was to be measured. Budget, quality and schedule measures were considered imperfect measures of success, because factors outside of the control of the project teams influenced them. This in turn led directly to a realization of the need for a direct measure of the quality of production flow in the projects.
>
> To measure the extent of LPS implementation, we used the "Planning Best Practice" (PBP) index, which was developed by Professor Carlos Formoso's team at UFRGS. The PBP index consists of a checklist of 15 Last Planner® System best practices measured with a simple full (1), partial (1/2) or none (0) score for implementation of each practice for any given construction project. The table [Table 21.1] lists the best practices and the scores for four of the 12 Tidhar projects studied. Two of these were among the best of Tidhar's LPS implementations (the highest score among the projects was a full 15) and the other two did not use LPS at all, but did have formal weekly planning meetings of Tidhar staff.

To measure the strength of the social networks, Vitaliy measured the intensity of coordination information exchanges among the various actors on site. Over a period of some three months per project, he measured the intensity of communication between the people on site at the 12 projects. The roles of the actors on the sites are listed in Table 21.2.

The first conclusions from these measurements concerned the growth of the social networks on the sites. The sociograms for these four projects, shown in Figure 21.2, illustrate the intensity of communication for coordination (represented by the width of the lines between the actor nodes) and the degree of centrality of each actor (represented by the size of the circle at each node). The differences between the two projects with low levels of LPS implementation and the two projects with high levels of LPS are stark. The superintendent is the hub of almost all coordination in the "no LPS" projects; in the projects with strong LPS implementation, there is much more direct coordination among the project's participants (and much more communication in general).

The research results for the 12 projects confirmed that LPS implementations do indeed strengthen the social networks and improve communication, even across language and ethnic groups. Even partial implementation of LPS proved to be sufficient to strengthen the social network because the WWP meetings appear to be the main source of this effect. The more thoroughly LPS was implemented, the more central the trade crew leaders became.[5]

The social subcontract was implemented as a series of discussions held in addition to the standard LPS WWP meetings, with the same participants. Details of the work plan and task coordination were excluded from the social subcontract discussions. At the first meeting, the team formulated a "work relationship covenant", in which each partner proposed how they wanted the others to behave, and the group was asked to reach

Table 21.1 PBP index scores for four Tidhar projects

	Practice	*Yavne 30* 17 storeys 63 apartments	*See Unik* 22 storeys 80 apts.	*Ir Yamim 2* 22 storeys 72 apts.	*Ir Yamim 3* 24 storeys 75 apts.
1	Formalization of the planning and control process	✔	✔	½	½
2	Standardization of short-term planning meetings	✔	✔	✘	✘
3	Use of visual devices to disseminate information in the construction site	✔	✔	✘	✘
4	Use of PPC and corrective actions based on the causes non-completions of plans	✔	✔	✘	✘
5	Critical analysis of data	½	✔	✘	✘
6	Correct definition of work packages	½	✔	✘	✘
7	Systematic update of the master plan, when necessary	✔	✔	✘	✘
8	Standardization of the medium-term planning	✘	½	✘	✘
9	Inclusion of only work packages without constraints in short-term plans	✘	½	✘	✘
10	Participation of crew representatives in decision-making in short-term planning meetings	✔	✔	✘	✘
11	Planning and controlling physical flows	✘	½	✘	✘
12	Use of indicators to assess schedule accomplishment	✔	✔	✘	✘
13	Systematic removal of constraints	½	½	✘	✘
14	Use of an easy to understand, transparent master plan (e.g. LOB)	✔	✔	✘	✘
15	Scheduling a backlog of tasks	✘	✘	✘	✘
	Total scores	9.5	12	0.5	0.5

consensus on the ten most important behaviours they wanted everyone to exhibit. Later, as the project progressed, they met every two weeks to review and discuss how they were getting along, to introduce new members to the covenant and to continuously improve their working relationships. Figures 21.3 to 21.5 show an example of a typical covenant, a photograph of a social subcontract meeting and a close-up view of the scorecard that was displayed alongside the covenant on the meeting room wall and filled in at the start of every review.

In the final analysis, there were too few sites at which LPS was implemented sufficiently weakly to allow clear-cut conclusions about the effect of the social subcontract when implemented without LPS. The results provided anecdotal evidence that the social subcontract activity enhanced the effects of LPS, resulting in better coordination and work flow than would have been the case with application of LPS alone. It was also

Table 21.2 Key actors at the interior finishing works stage

ID	*Crew/Role*	*Number of workers per crew*	*Organization*
BLD	Builder (masonry walls)	1–2	Subcontractor
ELE	Electrician	2–5	Subcontractor
PLM	Plumber	2–4	Subcontractor
AC	Air-conditioning installer	2–3	Subcontractor
TLN	Tiling crew	2–5	Subcontractor
PLS	Plastering crew	2–3	Subcontractor
DRY	Drywall crew	2–3	Subcontractor
CCC	Client changes coordinator	1	Tidhar employee
SI	Site superintendent	1	Tidhar employee
PM	Project manager	1	Tidhar employee

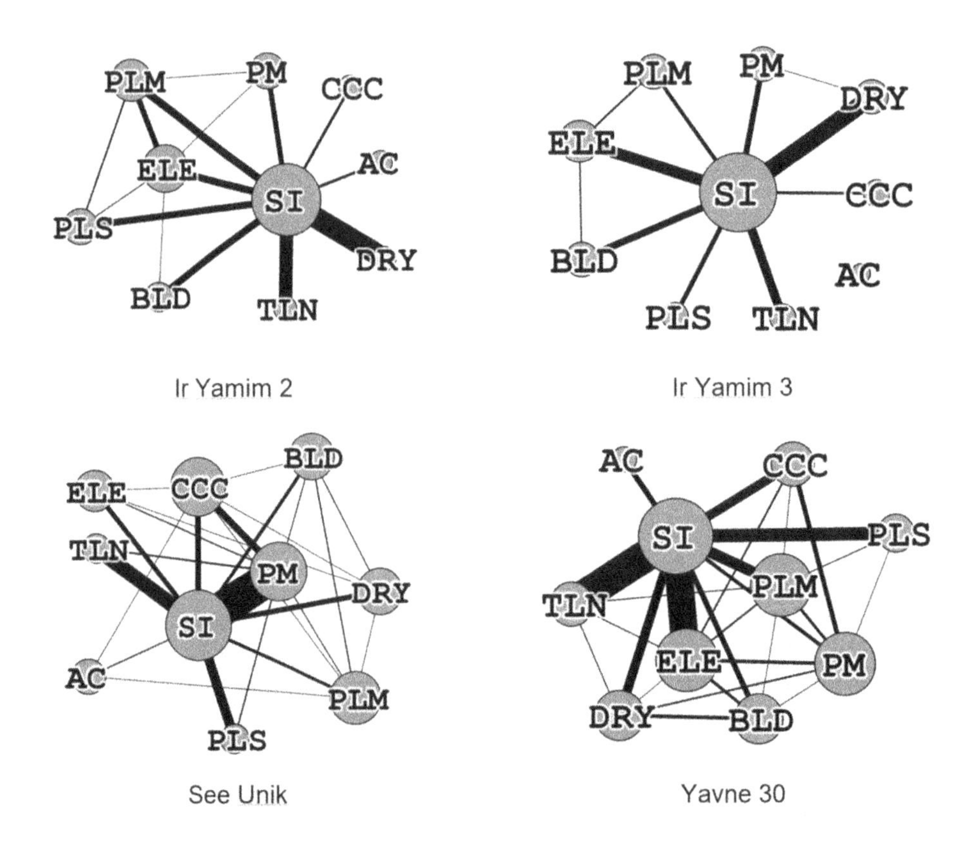

Figure 21.2 Social network analysis "sociograms" showing communication density (line thickness) and centrality (node size) for four Tidhar projects

Source: Priven (2016).

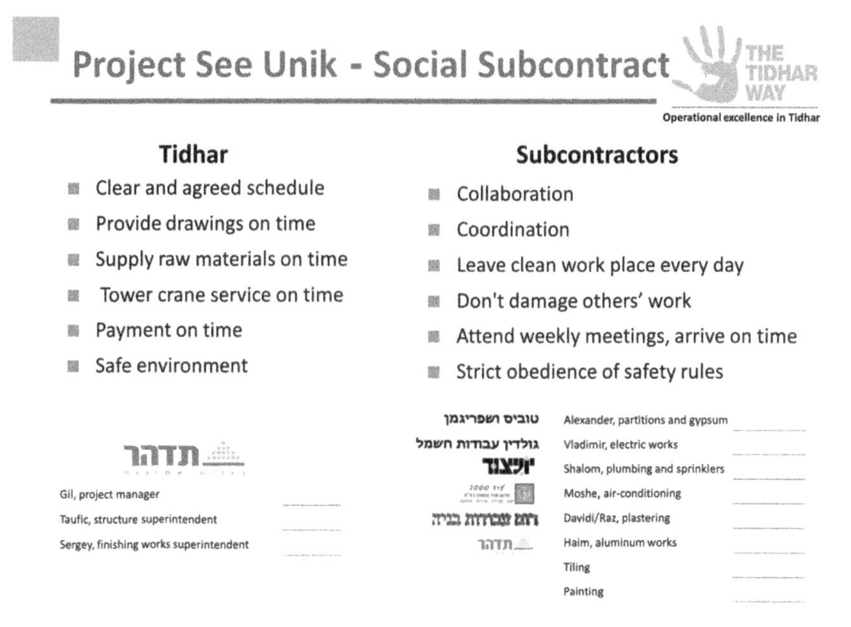

Figure 21.3 An example of a "social subcontract" covenant at the See Unik Project, with the explicit responsibilities of the general contractor (Tidhar) on the left and the responsibilities of the subcontractors on the right. All participants committed to the covenant by signing their names at the bottom.

Source: Priven and Sacks (2016b), reprinted with permission of the ASCE.

interesting to observe the effect of the social pressure on Tidhar's project managers. When they failed to provide reliable work, when materials were delivered late or when equipment was not available when promised, the subcontractors quickly used the review boards to shame them, knowing that the boards were shown to Tidhar's senior management from time to time. The social subcontract encouraged all participants to improve the quality of their work and respect their commitments.

Construction flow index

Earlier in this chapter, we saw that the need to measure the quality of the work flow in a construction project arose in the context of research concerning the impact of LPS and the social subcontract. All of the progress measures in common use in construction measure the results but not the quality of the process. Measures of "earned-value" and its derivatives (such as the schedule performance index [SPI] and the cost performance index [CPI][6]) describe what has been achieved over time, but reveal nothing about the quality of the workflow. The Percentage Plan Complete (PPC) from LPS, monitors the quality of commitments and planning, but not the quality of flow per se.

What was needed was a process measure rooted in the flow view of production, rather than the measures of results from the transformation view. Perhaps the main practical

Figure 21.4 An LPS and social subcontract meeting on the Lagoon project site

Note: The scorecard and the signed covenant can be seen on the top row of display boards. The bottom row of display boards has the LPS weekly work planning and PPC review boards.

Source: Priven and Sacks (2016b), reprinted with permission of the ASCE.

significance of the flow view is that it highlights the waste of time from the perspectives of both constructed spaces and construction crews. Constructed spaces where no productive activity is taking place are considered analogous to products in a production line that accumulate in front of a machine before they can be processed. Wasted time for crews is time spent working on non-value-adding activities or waiting for materials, information, equipment, space, completion of preceding activities, etc.

In factories, flow can be monitored using a variety of metrics, many of which are closely related to the parameters of Little's Law.[8] Little's Law states that throughput (TH) is equal to work in progress (WIP) divided by cycle time (CT): TH = WIP/CT. WIP and throughput are easily measured: WIP by counting the number of products present in a production line, and throughput by measuring the rate of output of the last machine in the line. Other common metrics measure aspects that impede flow (e.g. set-up times, waiting and inspection times, batch sizes).

The research team proposed a composite "construction flow index" (CFI) that would encapsulate properties of flow that can be measured and observed in flowline charts. "As-made" flowline charts can be compiled easily by recording the actual start and end dates of each trade in each constructed space. The idea grew from the observation that if an experienced Lean Construction planner looks at a flowline chart, he or she can

Project See Unik

Social Subcontract Monitoring Board

THE TIDHAR WAY

	Commitments	Partitions	Electrical	Plastering	Air Con	Plumbing	Tiling	Drywall	Windows
GC	Clear and agreed schedule	☹	☹		☹	☹		☹	
	Provided drawings on time	☹	☺		☹	☹		☹	
	Supplied materials on time	☹	☺						
	Tower crane service on time	☹	☺			☺			
	Payments on time	☺	☺			☹			
	Safe environment	☺	☺	😐	☺	☺			
Subcontractor	Collaborate and coordinate								
	Clean work place every day	☹		☹		☹			
	Respect others' work								
	Attend weekly meetings on time	☺	☺	☺	☺	☺		☺	
	Strict obedience of safety rules			☹					

Figure 21.5 An example of a social subcontract behaviour review board

Note: The GC is evaluated by each of the subcontractor trades, who place emojis in the GC section (upper rows). The trades are evaluated by the GC's representatives (project manager and superintendent), who place the emojis in the subcontractor section (lower rows).

Source: Priven and Sacks (2016b), reprinted with permission of the ASCE.

immediately see what the quality of the work flow has been up to the current time, and whether the trend is toward better or worse production flow.

Consider the flowline charts in Figure 21.6. The top left chart is an idealized project plan that has the best possible flow pattern. Production rates are stable, WIP is at a minimum, constructed spaces are performed in single-piece flow, constructed spaces experience continuous work, as do trade crews (crews never need to wait for work) and throughput is optimal. The other three charts are "as-made" records of the production performance of interior systems and finishing trades in construction projects. The horizontal axis is time and the rows each represent a distinct location (a constructed space) in the building. It is apparent that the top left project has the best flow, that the bottom left project has a bottleneck operation which caused a large amount of WIP to accumulate, and that the upper right project suffered significant variation in the production rates of many trades.

In practical terms, for optimal flow, production planners in construction should try to achieve the following nine conditions:

1 near equal Takt times across all trades (i.e. the duration of work in each constructed space is similar. This is achieved by balancing production rates and work quantities across spaces);
2 stable production rates for each trade crew (low variation);

3 small batch sizes;
4 minimized waiting/time buffers between operations;
5 minimized non-value-adding time within operations (minimized layout/set-up, waiting and inspection times);
6 datisfactory quality to avoid delays for rework;
7 minimized number of operations;
8 just-in-time delivery of off-site assemblies (windows, doors, kitchens, etc.);
9 just-in-time delivery of bulk raw materials.

Requirements 1–4 relate to the process (flow of constructed spaces), 5–7 to the operations (flow of crews) and 8–9 to the supply chain flows. Of these, numbers 1–4 and 7 can be measured directly from flowline charts such as those shown In Figure 21.6.

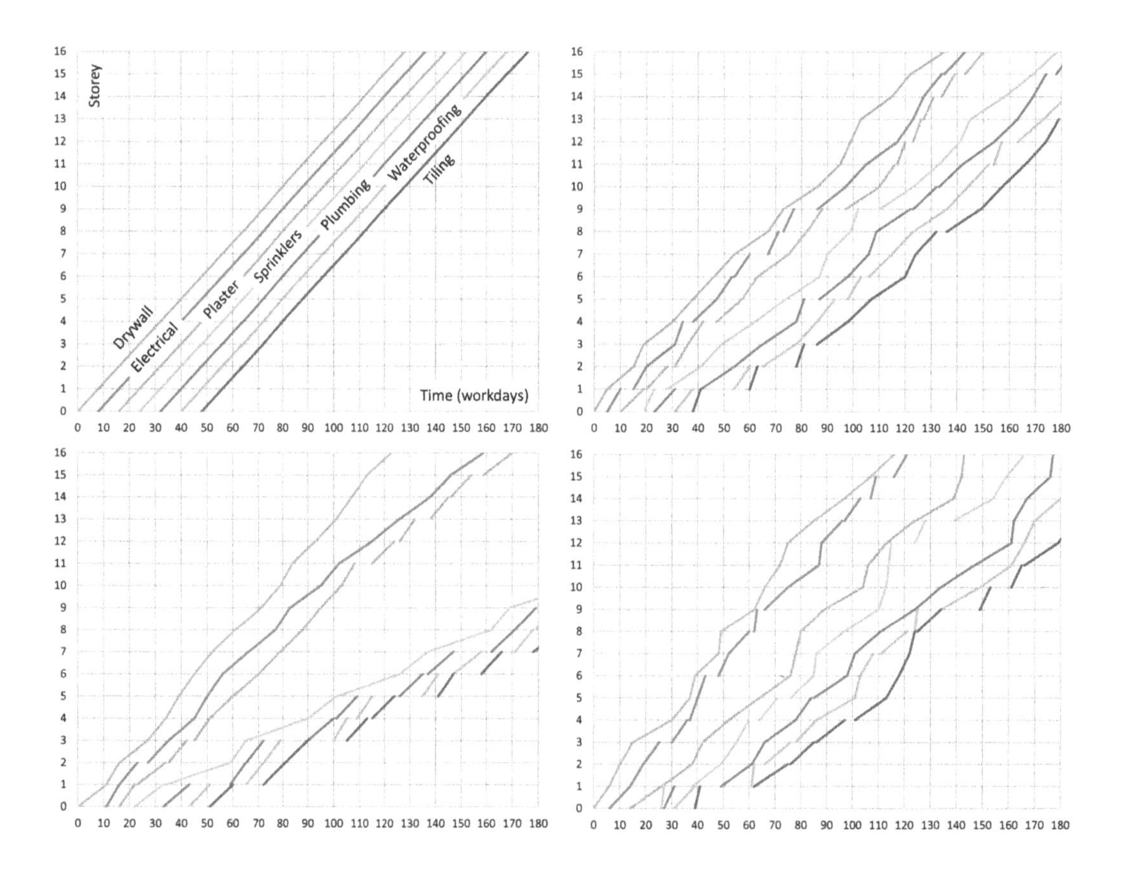

Figure 21.6 Flowline charts of production in construction projects

Note: The chart on the top left is an ideal project production plan and serves as the legend for the remaining three charts, which reflect "as-made" records of the construction of interior systems and finishing trades in three construction projects.

To test the idea of a composite CFI, the researchers obtained a set of 12 project records and then interviewed a number of experienced Lean Construction practitioners to solicit their opinions as to the quality of the workflow in each of the 12. They used the results to calibrate the weights in the proposed CFI formula so that for each of the test projects the CFI would give a score that was as close as possible to the scores given by the expert.[9]

The basic CFI formula was: $CFI(t) = 10\sum_{i=1}^{7} W_i P_i^{xi}$ *where* $X_i \in [1, 2]$

and the resulting calibrated CFI formula was:

$$CFI(t) = 10[0.30P_2^{\ 2} + 0.32P_3^{\ 2} + 0.06P_4^{\ 2} + 0.24P_5^{\ 2} + 0.03P_6 + 0.05P_7^{\ 2}]$$

The period t is the time span over which the CFI is calculated. Table 21.3 lists the parameters of the formula, and Table 21.4 lists the set of production flow measures that are used to compute the parameter values.

Table 21.3 Parameters used to compute the CFI

Parameter	*Description*
$P_2 = \frac{\bar{P}}{\bar{P} + \bar{P}\,\overline{STD}}$	Standard deviation of the duration *P* normalized using the average of *P*
$P_3 = 1 - \frac{NB}{TFP}$	Proportion of transfers from constructed space to constructed space for a crew that are continuous (i.e. percentage of the spaces after which a crew will not have a break after finishing the space)
$P_4 = \frac{\bar{P}}{\frac{ND}{TFP} + \bar{P}}$	Normalized proportion of average actual working time to total time spent on site for all trades
$P_5 = \frac{NT}{\sum_{i=1}^{P} NT_i}$	Proportion of spaces worked on in a given period to the total number of spaces with work in progress over the same period
$P_6 = {}^{1}/_{e(10*BP/TFP)}$	Indicator of the proportion of work performed out of sequence
$P_7 = \frac{TFP-BF}{TFP}$	Normalized proportion of spaces performed out of sequence (a trade crew performing spaces out of order or in parallel – not according to the plan)

Source: Priven et al. (2014).

The CFI formula was validated with the help of Professor Olli Seppänen, Aalto University in Espoo, Finland. Professor Seppänen had earlier measured detailed records of the production progress in three large projects in Finland, and he calculated the moving average of the CFI over time using the project records. The validation led to identification of various issues with the original formulation. These were corrected for and the weights and definitions shown above are the final result.

Table 21.4 Proposed production flow measures

Symbol	*Description*	*Unit of measure*
P_i	Duration per space for each trade *i*	Days/space
STD	Standard deviation of duration per space	Days/space
ND	Days of break for all tasks	Days
NB	Number of breaks for all tasks	–
TFP	Sum of all spaces produced by all trades	Count of spaces
NT	Number of tasks considered (active tasks)	Tasks
BP	Number of times a task is performed before its predecessor in a space (work out of sequence)	Tasks
BF	Number of times a crew works on space X before space X-1 (space out of sequence)	Tasks
WIP	Work in progress (WIP) – number of spaces with work in progress	Count of spaces
NTL	Number of active task spaces for a trade	Count of spaces

Note: The symbols for the flow measures are defined in Table 21.4.
Source: Priven et al. (2014).

One of the more interesting outcomes of computing the CFI was that there was no correlation between CFI and the PPC values for the projects. This reinforced the understanding that the PPC is not necessarily a good measure of work flow. The CFI has potential applications in both practice and research. When measured during the life of a project, CFI can give advance warning when a project's production flow begins to deteriorate. It can guide action to help the site team to improve production flow and avoid chaos. In research, an objective and quantitative measure of work flow quality is an essential tool for investigators whose research methods require measurement in the field or evaluation of simulation outcomes.

VR for client customization of apartments

In Tidhar's market, most buyers are offered the opportunity to customize their apartments. Customization can be as simple as choosing the floor tiles, bathroom fixtures, kitchen cabinets and paint colours; but more often than not, it extends to changing locations of interior partitions, addition of air-conditioning systems and ducts with false ceilings, addition of electrical outlets, etc. In extreme cases, buyers employ interior designers and completely redesign their apartments. The process, the problems it causes and the way Tidhar now manages it using BIM are described in some detail in Chapter 11, "Virtual design and construction", in the section entitled "VDC for apartment buyers' design customization".

In general, developers and general contractors dedicate significant managerial and technological effort to manage the customization process with their customers, but the process remains inefficient. Often customers are unable to interpret construction drawings, as is the case for most customers who don't have a background in AEC, or lack the knowledge or competence to deliver the decisions and information that is required from them. This leads to significant rework and/or frustration once customers see the physical manifestation of the decisions they made during consultations with the apartment change representatives. Decisions made "on paper" turn out to be unreliable where the customers' perceptions of the eventual reality were weak. For customers who are not construction professionals, decision-making is often a challenging and unpleasant task , in large part due to insufficient product cognition.

As the client change process got underway at Tidhar's Rosh Ha'ayin project, the Technion research team proposed an experiment: could the quality of the clients' customization decisions be improved by showing them their apartments using immersive VR as a supplement to the standard 2D drawings? Would better cognition lead to shorter meetings with the company's representatives, to shorter cycle times for decisions and to less rework? Would these in turn lead to less waste and more satisfied clients?

Tidhar's marketing and client customization managers were enthusiastic about the idea, and gave their full support to the experiment. With the help of the staff of the Seskin Virtual Construction Lab at the Technion, a graduate student built a VR model of a typical standard apartment. The source information was Tidhar's Revit model of the apartment, and it was exported through Autodesk 3D Studio into the Unity game engine for display using an Oculus Rift 2.0 display. The display and the model were set-up in the client change coordination offices at the construction site, and representatives were trained to show the model to clients. Figure 21.7 shows a view of the stereo display of the virtual reality model of the apartment.

The experimental design called for measurement and comparison of the cognition of the apartment design, layout and dimensions for two groups of Tidhar customers – a control group who were shown only the 2D drawings, reflecting standard practice, and

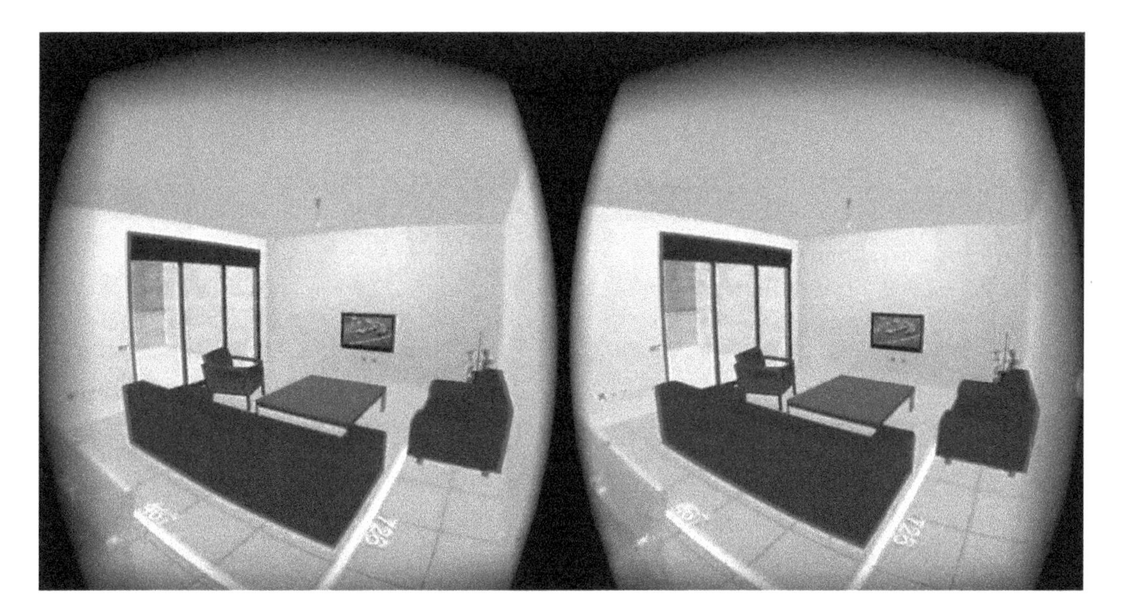

Figure 21.7 Stereo image of the virtual reality model of the apartment as shown in the Oculus Rift headset

Note: The original Revit model was enhanced with extra detail and visualization of technical information. Among the objects added to the model were furniture, room labels (stamped on doors) and significant dimension lines (on the floors and walls), as can be seen in the figure.

an experimental group who were allowed to tour the apartment using the VR model before beginning their change coordination meetings. A questionnaire was prepared, in consultation with experts in cognition, to test subjects' understanding of the apartment. Questions that probed a person's perception of dimensions, of walking times from one point to another in the apartment and so on, were accompanied by questions that asked them to express how confident they were in each of their responses.

However, despite the simplicity of the experimental design, the sophistication of the technology and the care taken to prepare the questionnaire, the researchers encountered an unexpected hurdle: not enough of the clients who came to the site office were willing to participate. Understandably, those intended for the control group were reluctant to spend time on the questionnaire. More surprisingly, many of those in the experimental group were unwilling to take the extra time needed to tour the virtual model and to fill out the questionnaire. Despite careful planning, researchers, like builders, sometimes face surprises.

In order to complete the research, university students were employed as subjects. Students who had families, and who had no prior experience with construction documents or drawings, were selected (students of architecture and civil engineering were excluded). Table 21.5 presents a small sample of the data describing the responses to those questions which showed the most significant difference between the control and the experiment groups.

Table 21.5 Selected results for questions concerning cognitive differences between the use of drawings and immersive virtual reality for cognition of an apartment's spaces and their proportions.[10]

Question	*Control group (drawings)* Mean (range 0–6)	*Experimental group (VR)* Mean (range 0–6)	*Confidence of difference* (P value)
How easy was the task of evaluating the suitability of the bedrooms for use by two children?	4.5	5.4	0.00
How clear are the room's precise areas?	3.8	4.8	0.06
How sure are you of the accuracy of your answer to the last question?	3.0	4.3	0.01
How easily and precisely can you estimate the distance from the living room sofa to the TV screen?	4.1	5.1	0.00
How easily and precisely can you estimate the available storage volume in the walk-in closet?	2.4	3.6	0.00
To what extent did you feel that the VR/drawing (accordingly) helped you form a clear understanding of the apartment's spaces, areas and dimensions?	4.1	5.8	0.00
To what extent did you feel that the VR/drawing (accordingly) helped you form a clear understanding of the apartment's functionalities and possible everyday uses?	4.4	5.7	0.01

Overall, the research measured four aspects of the subjects' experience:

1. Perception of spaces and their dimensions. All the measures in this area had higher mean scores for the VR group, with P values for significance analysis from 0.1 to 5 per cent.
2. Perception of the functionality of the spaces and of the apartment as a whole. Here, some of the measures showed a clear advantage for the VR group, while other measures were inconclusive.
3. For questions that required estimation and recall of quantities (such as the number of electrical outlets in the kitchen), the spread of the answers was far lower for the VR group than for the control group, which implies that they had a clearer understanding of the apartment.
4. The last group of questions tested the extent to which people felt that the tools helped them understand the apartment. The VR group found that the tools they used supported their understanding significantly more than the group that used drawings did.

This final story of research collaboration between industry and academia, between Tidhar and the Technion, is a sobering one. Perhaps the most important lesson learned is not that VR is useful, but that engineers and researchers can be enamoured with the novelty of high-tech tools to an extent that prevents them seeing that clients may not be quite as impressed as they are. When offered the opportunity to experience their apartments in virtual reality, many clients did not consider the opportunity to be sufficiently enticing or valuable for them to sacrifice significant extra time beyond their regular customization meetings.

Reflective practice

A great deal of the practice of managing construction, the core "business" of a general contractor, is performed intuitively, by engineers and managers who act more on the basis of experience and learned behaviour than on the basis of technical rationality. In university schools of civil engineering future construction managers are taught to apply CPM to solve scheduling problems rigorously, yet every experienced project manager knows that the resulting construction schedules are of marginal relevance to the actual management of production on site. Construction managers are forced to rely more on experience and intuition.

Many of Tidhar's managers and engineers, and those of Skanska, Lease Crutcher Lewis and Fira, have struggled with Lean Construction and with BIM. Both paradigms challenge the accepted modes of working and require them to change behaviours, redefine their roles in the construction process and try new methods. Change of this kind is difficult not only because it is perceived to be risky, but also because highly experienced construction managers find it difficult to explain to themselves, let alone to others, why it is that they do things the way they do.

In many of the earlier chapters of this book, we described situations in which conflicts arose in meetings where a proposed innovation was being discussed. Two examples were the arguments over how strictly to impose the requirement that site staff should be required to adhere to the BIM model in Chapter 11, and the heated arguments about whether or not Tidhar should concern themselves with the productivity of partition-building masons in Chapter 8. In most of these situations some people wholeheartedly support the innovation while others object. Once the initial rational arguments for and against are exhausted, with neither side able to prove their case, people become more passionate about their opinions, in part because the nature of their work has led them to trust their managerial or engineering instincts more than the economic or production models. This can stymie even the best-intentioned initiatives.

Therefore, perhaps one of the key advantages for a construction company in partnering with an academic research team is the opportunity to step back and reflect on everyday practice. This can be done through "learning-by-doing" in controlled experiments. If the interaction with researchers and or Lean consultants can sidestep criticism of the accepted way of doing things, the experience and the conversations that ensue allow practitioners to see their work in a new light.

In his book, *The Reflective Practitioner*, Donald Schön[11] described the ways in which professionals rely on improvisations that are learned through practice, rather than working solely with the knowledge gained from their formal education. Experimentation is a key aspect of reflective practice – trying new processes, often conceived intuitively, to see whether they are better or worse than the current best practice. Experimentation in the real world enables professionals to improve their performance, even if sometimes they are unable to formally explain or to model the mechanisms that underlie the behaviours they observe. This process of trial and error is formalized in Lean thinking as "continuous improvement". Continuous improvement requires rigorous measurement of the outcomes of both the baseline process and the proposed alternative process. It also relies on standardization of a process, making it both more consistent than professional trial and error and more applicable to the work of teams and organizations.

Tidhar has adopted and applied the process of continuous improvement extensively, albeit not always with sufficient rigor. Chapter 6, "Continuous improvement and respect for people", tells the story of Tidhar's attempts to solve specific problems as well as their much more difficult mission to instil a culture of continuous improvement. Perhaps the most significant controlled experiment that Tidhar has undertaken is the attempt to significantly improve the construction process of their typical apartments, which required redesign of the product and the work methods. The more recent Rom Tidhar project, with only 124 apartments, has been a laboratory for this experimentation. The experiment began at the earliest stages of architectural design, with the decision to go from a standard system of masonry block partitions and MEP systems below floors laid on filler material, to drywall partitions with MEP systems in the ceilings and no flooring filler material at all. At Rom Tidhar, Tidhar performed first run studies not on one apartment but on four, testing different variations of the changed process in each apartment. The lessons learned were then rolled out as a new standard which was repeated over the whole project.

Lessons learned

The lessons learned from the collaboration between Tidhar and the Technion reinforce the understanding that this was a win-win collaboration. Without the ability to test interventions, to experiment and to measure production outcomes, it would not have been possible to explore the ideas of the social network impacts of LPS, the social subcontract, or to develop the CFI measure. The CFI is a good example of an idea that was developed for the needs of the research but that has potential for application in the construction industry at large.

The flip side of the coin is that Tidhar's projects and people benefited greatly too. The researchers initiated adoption of LPS at the Yavne projects, and this was taken up throughout the company. The quantitative results of the research that examined the waste in building partitions from gypsum blocks (see Chapter 8, "The waste of non-value-adding work") provided a strong and direct impetus for Tidhar's managers to seek ways to reduce the waste in this and other operations. The numbers from the research papers were often a first line of defence in internal arguments for Lean Construction initiatives.

One last aspect of the benefit that Tidhar gained from the industry–academia collaboration is perhaps the most significant in the long term. With very few exceptions, most of the graduate students and undergraduate interns who participated in the joint research projects were recruited to work at Tidhar soon after their graduation. They hold positions as BIM modellers, project managers, planning and control staff. In one extreme and exemplary case, a Technion PhD graduate, who was hired by Tidhar to apply Lean Construction tools to support efforts to improve construction methods, specifically requested to be assigned the job of site superintendent for the duration of a project so that he could learn first-hand how the work was done. In this real case of dedicated "going to Gemba", he became the most highly qualified construction superintendent Tidhar – and perhaps any other construction company – has ever had.

To sum up the argument in support of industry–academia collaboration in the areas of Lean Construction and BIM research specifically, and perhaps in construction management research in general, here is an imaginary Q&A session:

Why do construction companies need theories of construction and production management?

The way that people think about a problem dictates the solutions they find and the way they behave. To start with, having the right vocabulary to describe the concepts is a prerequisite to discussing them and devising appropriate work practices. Specifically, for the case of BIM and Lean Construction, some of the essential terms – like *simulation of building behaviour, variation, virtual design and construction, construction flow, big room collaboration, work in progress* and *cycle time* – are new to the vocabulary of most traditional designers, construction managers and trade crew leaders.

Once the terms are understood, they can be used to build theories that predict the ways in which construction production systems will respond to different management actions.

Why do academics and researchers need access to construction projects?

The short answer is that useful research is therefore dependent on construction sites as "living laboratories" in the best case, and where not feasible, as the source of the behavioural data that is needed to set-up and calibrate laboratory or computer simulations.

The long answer is that construction management is a specialization within the field of management science and, as such, for research to be relevant, it must be grounded in the field. Management science has been challenged from time to time as being disconnected from reality because many academics have assumed simplified systems to make their research tractable. This has led to research of cases in which highly sophisticated mathematical and/or computational models have been derived for highly specialized problem formulations that cannot be applied in real systems.

Professor Lauri Koskela has been the most eloquent critic of management science from the area of construction management. In a paper published in the journal *Construction Management and Economics* in 2017 entitled "Why is management research irrelevant?", Koskela argued that "the rejection of production as an integral part of organizations and management has been perhaps the most damaging feature" of a pair of reports on management education in the US that were published in 1959 and had an overarching influence on the nature of management education, research and practice in the decades since.[12] The main idea is that management research came to view production as a "black box" that could be abstracted out of the other aspects of management, such as finance, human resources, etc. Lean Construction calls this view into question, and Koskela's own work on the theory of production fundamentally rejects this abstraction by highlighting the flow and value views as complementary to the transformation (black box) view.

Thus, if researchers want construction management research to be relevant, they must delve into the nitty gritty of production, to begin to understand what flows and how it flows in construction. To do that, researchers must have access to, and must engage with, construction projects firsthand.

What do construction companies gain from engaging with academic research?

The benefits range from obvious short-term benefits to subtler long-term benefits. Since researchers strive to be at the forefront of the state-of-the-art, they are often faced with the challenge of finding the right platform to test proposed innovations or theories. One way to achieve the conditions needed for meaningful testing is to work with a company or project team to implement whatever innovation they would like to experiment with, as the Technion team did with the Last Planner® System at Tidhar. These experiences can benefit the collaborating construction company greatly, if – and only if – the company dedicates the right staff to work with the researchers, people who have both the propensity to learn and the authority to establish the innovations, whether technical or managerial, in their project and later throughout the company.

In the mid-term, companies gain from the access to high-quality human resources. Research students who go on to careers in industry often find their exposure to the forward-thinking people in the kinds of construction companies that fund and engage in research to be meaningful and even inspiring, leading them to seek employment with those companies after graduation.

The long-term benefits are improvements of the productivity and profitability of construction companies. Productivity in the construction industry has long stagnated, as highlighted by Professor Paul Teicholz[13] of Stanford University. Technological and process changes like BIM and Lean Construction are the best hope for improving the quality of product information in design and construction and for improving process management. Further improvements can be expected when software tools that leverage the information in BIM models to support Lean Construction, specifically by providing real-time process information, become broadly available. Almost all of these efforts germinate in research, and the research cannot be completed without experimentation in the field.

What is the difference between controlled experimentation in the context of Lean Construction, on the one hand, and experimentation for theory building, on the other hand?

In Chapter 18, "Raising planning resolution – The four-day floors", we describe how the Rosh Ha'ayin Underground, Tidhar's ad hoc production planning team for the reinforced concrete work at the Rosh Ha'ayin project, pursued a series of controlled experiments to determine the best plan (work structuring and sequencing) for building the wall sections. This is a good example of how incremental experimentation can be used to improve processes.

However, the broader questions raised about the ways in which subcontractors form social networks and how the strengths of those networks cause or correlate with improved work flows, described earlier in this chapter, is an example of experimentation for theory building. Their goal is to enable people to build a conceptual perception of how production actually works in construction. The ideas expressed in Chapter 15, "What flows in construction?", are largely based on observations gleaned over time and through multiple research projects.

To some extent, this practical-academic distinction can also be applied to understand the very different roles of Lean consultants and academic researchers. Cheni Kerem, Tidhar's consultant, observed that "consultants primarily aim to solve existing problems, with less focus on preparing the organization and the workforce to deal with future problems any more successfully than they have in the past." Consultants analyse specific production processes in light of their experience in other companies or in other industries and apply the Lean tools in which they specialize to solve specific problems with specific solutions. Structured experimentation, based on careful before-and-after measurement and standardization to entrench the solutions reached, is one of the formal Lean tools they can apply. As professionals who gain experience through practice, i.e. as *reflective practitioners*, they often reach important conclusions about the ways in which things work. In the best cases, they are able to frame that understanding in conceptual terms and convey it to others.

The approach of academic researchers is different. More often than not, they will start by inducing a hypothetical explanation of some phenomenon based on observation of a number of occurrences. For example, they may observe that some projects have less waste and higher productivity than others when trade crews work continuously rather than with interruptions. They would then design an experiment or gather available empirical evidence to try to deduce whether the hypothesis is valid or not. This generally leads to a more formal learning that, in ideal circumstances, can be applied broadly, across construction projects and organizations.

The goal of an academic researcher is primarily to learn; she may improve processes incidentally through pursuit of her research. The goal of a consultant is to improve processes; he may learn incidentally through pursuit of his practice. They both contribute to advancing construction and construction companies.

Notes

1 See Rogers (2003).
2 Many researchers have observed that even when LPS is only partially implemented, workflow is still improved. See for example Alarcon et al. (2005), Hamzeh (2009) and Viana et al. (2010).
3 Ballard (2000b).
4 Koskela (2000).
5 More information on this research can be found in Priven and Sacks (2016a). This paper won the Thomas Fitch Rowland Prize of the ASCE Construction Institute.
6 Hendrickson and Au (1998).
7 Hopp and Spearman (1996).
8 Little andGraves (2008).
9 The calibration procedure used a goal-seeking algorithm written by Jonathan Savosnick, a research intern in the team. For more details, and the full CFI formula, see Sacks et al. (2017).
10 See Dayan and Sacks (2017).
11 Donald Schön (1984).
12 Koskela was referring to the 1959 reports on American business education, written by Pierson, Gordon and Howell. See Koskela (2017).
13 Paul Teicholz (2001).

22 Conclusion

Staying the course

Both BIM and Lean are endothermic endeavours. You need to keep applying energy to keep the initiatives alive. BIM processes seem to become exothermic after a short time – once you get going, you'll never go back. Lean Construction, however, seems to take a much longer time to become established in a company. You have to keep people interested and motivated, even after new practices become standard practices.

Skanska Finland, Lease Crutcher Lewis, Tidhar and Fira all benefitted from strong leaders who had the foresight to see that BIM and Lean were valuable, the confidence and strength to apply the resources needed and the skilled people to make it happen. They certainly did not succeed in everything they did, but they now build buildings quite differently than they used to as a result of their efforts. Productivity is higher, customers get better value and the companies are more profitable than they were when they began. What lessons can we draw from their experience that might help others stay the course and succeed?

Motivation – Why change?

Indeed, why change? An existential threat to a company, such as pending insolvency as a direct result of the inability to compete, is a great motivator for change, but this was not the case for Tidhar. They were already in fine form when Gil Geva launched the "Growing the Margin" initiative. Tidhar was considered a progressive, well-run and reliable developer and construction contractor – one of the best in its market. So why change?

In short, because if they didn't, one of their competitors would.[1] The construction industry in most countries is highly fragmented, with many companies competing for the same resource (paying customers). If a company has the necessary capacity and it can find a way to gain more customers at the expense of its competitors, it will do so. An increased profit margin for a construction contractor is the equivalent of extra height for a tree in a forest, so that companies that have the wherewithal and the capacity to invest in improvement have an incentive to increase their productivity, their margin, their market size and their profits. BIM and Lean Construction both require significant investment of resources to become effective, but once they do, they can give a company a competitive advantage.[2]

Process changes in construction companies are not random, they are the result of purposeful leadership and decision-making. Both BIM and Lean are step-wise improvements (as shown, for example, by the ladder of progressively sophisticated BIM adoption in Figure 11.2). At each step of the way, company managers ask simple questions: is the marginal cost of investment in the next step justifiable? Is it necessary? If we have already invested enough to get ahead of the competition, why invest more?

Some of the "cost" of investment in process change is the cost of overcoming the inertia of people accustomed to doing things in a particular way. Firms tend to adopt the aspects of BIM and of Lean Construction with the lowest marginal cost, not the greatest benefit–cost ratio. The low-hanging fruit are not usually the ripest and juiciest, they are simply the easiest to pick. It is not surprising then, for example, that Tidhar has fully adopted the weekly work planning meetings of LPS, but not the make-ready process; Skanska Finland no longer uses the RFID tag-enabled supply chain monitoring system for precast concrete components; Lease Crutcher Lewis uses an informal rather than a "purist" form of IPD; and Fira has not persuaded all of its Fira Palvelut subcontractors to join its production alliance.

Most companies begin adopting BIM with the parts that require least effort and are most visible to their clients – clash detection, renderings, 4D movies, colourful FEM or CFD or energy results. Some stop with the renderings, until competition forces more – collaboration on models with other disciplines, big room co-location and so on. Most companies begin adopting Lean Construction with the parts that require least effort and bring the most immediate results – VSM to identify and remove the most obvious wastes, WWP to stabilize production and so on. Where leadership is lacking, even these efforts may fail if the investment of time and effort for the individuals involved is not maintained.

Education as a motivator

Is the ability to "see the waste" a motivator for change? Tal Hershkovitz's Lean study tour to Japan definitely helped motivate his efforts to introduce incremental process improvement in Tidhar. The management group tour to Finland and Denmark in late 2012 provided Tidhar's senior managers with a vision of what they could aim for. The Lean Construction boot camp devised by Tidhar's Lean consultants (Lean Israel) was a key factor in managers' learning to see the wastes and the opportunities, largely through their experience working at the Gemba.

Education is not only a facilitator of skills, it is also a strong motivator, and should be at the heart of any company's efforts at all levels. Personal excellence, the desire to do as good a job as one can, is also a motivator, along the with the profit incentive. Tal Hershkovitz asked in a conversation sometime in 2014:

> Do you know what I would really like to do? What I would really like to do is leave this office and work as a site superintendent for a year at least. I can think of so many things I would change, that I would do differently. The amount of process waste I could get rid of, if I could just control things and do it all my way!

This might sound strange coming from the CEO of a construction division, but it reflects the deep – almost moral – sense of frustration that one feels when the waste is obvious but changing the system requires change by multiple people across multiple organizations.

Tal carried the responsibilities of the CEO, and so, like many proponents of Lean and BIM, he had to be patient and work methodically to bring about the deep process change that was – and largely still is – needed.

Motivation and education go hand in hand in the long run. Company culture is changed by policies and standards, by spreading training to all employees, but also by subtle cues in the workplace. Visual production control boards, BIM models displayed electronically and in print, and even posters such as the 4 × 4 flows and preconditions poster, shown in Figure 14.4, all play a role.

Two steps forward, one step back

The stories of Skanska Finland, Tidhar, Lease Crutcher Lewis and Fira are the stories of real companies, and the real world is messy and complex. Readers should not finish reading this book with the misconception that all of their steps to adopt BIM and Lean Construction in their operations succeeded just as their change agents intended or as theory predicted. On the contrary – change of this magnitude is a mix of achievements and disappointments.

Reviewing which BIM and Lean process changes became rooted in a company's everyday way of working is an effective way to measure success or failure. In Tidhar's case, the following key practices became standard:

- weekly work planning as part of the Last Planner® System;
- BIM in the big room design and coordination for all projects in which Tidhar is the developer;
- allocation of a BIM modeller to construction site offices to serve as the information hub for product and process information on site;
- detailed production planning and monitoring of reinforced concrete work, based on BIM models and 4D simulation;
- proactive make-ready management of the second wave of interior finishing work;
- BIM modelling of customer changes to apartment interiors and use of the models to drive production directly;
- Involvement with building design at the earliest stages to enable changes to construction methods.

Process ownership has proved to be a differentiator in terms of the depth of adoption of standardization of new processes. BIM plays a far more significant role in Tidhar-owned projects than in projects in which Tidhar is the general contractor. To some extent, a similar pattern emerges at the practical level. Processes being "owned" by specific employees, who carry personal responsibility for development and implementation of business processes, is key to their longevity.[3] The role of Tidhar's VP for design and of its big room design managers is essential for aligning the practices of diverse designers. Ortal Yona's role as production engineer at Rosh Ha'ayin was essential to standardizing the phase planning and WWP practices in the long term. Fira is keenly aware of the essential role of formal process ownership and employs a dedicated process engineer who supports the process owners. As Otto Alhava explained in Chapter 19, "Process owners need to have the knowledge, authority and conviction to be able to tell people:

'Your way is good, it is based on your experience, but it is not the Fira Way, and we need you to adhere to the Fira Way'."

Among the processes that have not succeeded are:

- Standardization of a Tidhar apartment with a fixed set of alternative modular bathroom, bedroom and security room layouts that could be compiled into any new apartment design with the goal of standardizing parts, work methods and procurement. Libraries of BIM families for standardized walls and other details are used and enforced, but modular room-size families proved to be a step too far. This hampers off site fabrication.
- Fully fledged quantity take-off from BIM models. For a small number of the typical line items in traditional bills of quantities, they have not been able to achieve conformance between BIM model quantities and the traditional methods of measurement.
- Abandoning reactive production status measurement and control using the "work backlog" method, and its replacement with proactive progress monitoring using BIM and reporting by trades in LPS meetings. The latter is done, but the continued work backlog practice adversely affects project managers and site supervisor's production control behaviours, incentivizing them to "push" production forward.
- In projects where Tidhar has the role of general contractor, the dictum "build only from the model" is not faithfully followed. Lack of control over the client's design team and local legal precedent whereby designers retain full responsibility for detailed design has weakened Tidhar's resolve to model everything before building it. Contrast this with Lease Crutcher Lewis and with Fira, both of which apply early engagement with public and commercial construction clients, often well before the contract is awarded, to ensure value for the client through design team collaboration using BIM.

One of the opportunities missed at Tidhar is that "learning to see" was not institutionalized. Despite the strong impressions and influence of going to Gemba on its senior managers, the act of spending significant time to observe processes did not become a habit. When engineers and supervisors see quality defects on site, they demand and ensure repair through rework, without taking the opportunity to delve into root causes with a view to fixing processes. Urgent problems take precedence over important ones. This is similar to the lack of follow-up after the Lean boot camp. Participants were assigned personal projects but there was no point at which they were asked about the results. Naturally, the day-to-day matters of management took precedence over the improvement projects. It also stems from a continuing inability to measure work flow – and what cannot be measured is difficult to manage.

Perhaps the main disappointment in Tidhar's journey is that the "Tidhar way" imagined by the proponents of Lean and BIM adoption embodied a vision of a Lean company, one that achieved better margins through greater productivity of a relatively small number of people applying Lean thinking and BIM thoroughly. The reality is that the company grew quickly, too quickly for the change agents to embed new processes deeply enough. Growing fast by taking on new projects and employing more people can retard BIM and Lean adoption. New people bring with them the dysfunctional culture of the generic construction industry and act as a brake to change. Education of the original 250 direct employees had really only just begun during 2012, let alone the many more hundreds of subcontracted workers, and the influx of 340 new employees by 2017 made broad and thorough cultural change that much more difficult.

Tidhar pushed people to do everything (LPS, VSM, 5S, BIM, VDC, 4D, customer design change management, VR and more) at every phase (design, structural work, off site fabrication and the three waves of interior systems and finishing), but paid insufficient attention to measurement, follow-up and standardization. The company supported employees who showed initiative, but that often depended on individuals pulling what was needed. Had the Lean and BIM culture been better established earlier, perhaps new people would have adapted to it more quickly. Fira's leaders are keenly aware of the challenge, and despite their rapid growth, their efforts to prioritize emotional intelligence represents a conscious effort to change the way people think about their work.

Necessary conditions

Are there any desirable, or even essential, conditions or characteristics for a construction company to successfully adopt Lean Construction and/or BIM? If we assume that a company has highly skilled and educated people and access to technology, what else is needed?

Integration

One of the very few things that Tidhar has in common with Toyota is that, like Toyota, Tidhar's business model extends from product design through to production. Toyota designs cars for ease of production. Likewise, to the extent that Tidhar has control over design of its apartments (at least those for which it is the developer) it can, over time, tailor its products to suit its production processes. Its role as employer of its subcontractors, designers and builders makes it possible to coordinate and optimize the end-to-end production system.

Fira's Verstas process aims to optimize the full project life cycle by bringing the owners and designers of candidate projects into their orbit in the very early stages. This is an innovative way to aim for overall optima despite fulfilling the role of general contractor with limited scope. Similarly, Fira's production alliance attempts to extend production planning and control to its subcontractors, also for the sake of global optimization. Lease Crutcher Lewis, too, strives to support clients early in the process, well before Lewis has been formally contracted to build a project, and also aims to engage its supply chain in its BIM initiative.

Lana Gochenauer of Lease Crutcher Lewis gave a pertinent, practical example: "In BIM an architect may model a single column rising 23 floors. But for estimating, for scheduling, and for creating work packages, Lewis must break the column down into parts for each floor level. Unless we can work with them to maintain integrity across these two representations, adding all of the detail needed for construction in the field would be very difficult."

Ronen Barak responded: "In Tidhar's case, when we engage the design team directly, modelling instructions of this kind are part of our BIM execution plan, so we get a model that fits our needs as a GC."

The common thread is that these companies recognize that the extreme fragmentation common in the construction industry is a major barrier to Lean and BIM adoption. They recognize the value of integrated management of the full life cycle of product design and production. Integration of the entire value stream, from conception to project completion, is a key ingredient for exploiting the benefits of Lean and BIM; without integration, the scope for improvement is limited. Integration may be formal, as in IPD, but for the most part, integration in construction is achieved through informal collaborations among companies who learn to search out win-win opportunities.

Vision, leadership and focus

As we saw in Chapter 1, Tidhar has a clear declaration of values and a vision statement. In laying out the "Growing the Margin" initiative, Gil Geva connected the operational plan to the vision and the values when he declared that BIM and Lean were to be the instruments through which the vision of excellence and the value of customer service were to be achieved. As co-owner and CEO, he stated his commitment to pursue a future state in which Tidhar fully embraced Lean and BIM. In doing so, he made it clear to Tidhar's people that he had sufficient faith in them, in the technology and in the positive effect that Lean and BIM would have, to take the risks and to invest the resources required. For the most part, his leadership overcame the natural scepticism and resistance felt by many people.

Leadership of this kind is "top-down". The other kind, "bottom-up", is no less important. Identifying and empowering people within the ranks of the company who have the ability to lead their peers is also essential. The chapters of the book are full of examples. Comprehensive off site fabrication of MEP modules for Skanska Finland's HQ building, the early introduction of LPS at Tidhar's Yavne project, multiple improvements to construction methods and many more, were all the result of respect for people and providing the conditions in which to bring their own ideas to the fore.

Unlike the vast majority of construction companies, Tidhar displayed leadership by actively funding and supporting research. Most construction companies, if they fund research at all, pursue applied R&D to solve short-term technical problems. Tidhar took the longer-term view that supporting theoretical, blue-sky research in construction management could yield deeper and more long-lasting benefits, both for Tidhar and for the construction industry at large. In addition to the direct benefits of exposing its own people to new ideas and knowledge, the relationship with the Virtual Construction Lab at the Technion provided a steady stream of excellent engineering research students who see the company as a natural fit for employment after graduation.

Finally, maintaining focus – keeping one's eye on the ball – is also a factor. Given the endothermic nature of Lean Construction adoption, it requires constant attention. In Tidhar's case, as the company grew quickly, Tidhar's "C-Suite" turned their attention to other matters. At times this left the "bottom-up" change agents without the support and positive reinforcement they needed to persist. Some improvement measures stalled and others were never started. Lack of focus was at least one of the reasons behind some of the failures listed earlier in this chapter.

Box 22.1 The benefit of hindsight

Gil Geva reflected on his role in the "Growing the Margin" initiative, expressing the frustration of hindsight and giving a clear idea of how he perceived his role:

> If I could turn back the clock, I would have continued the monthly "Growing the Margin" steering committee meetings that we held over the first three years. With the rapid growth of the company, I became distracted and let them slip from my schedule. I lost leadership focus; I wasn't supporting the BIM and Lean change agents. What made matters worse was that when senior managers perceived that I was not personally engaged, they also did not put their full weight behind the new initiatives. When department managers start explaining why a goal I set is not possible, instead of looking for ways to achieve it, I know that my leadership is lacking. Here's my advice to any construction company CEO who plans to adopt Lean and BIM:
>
> - Plan for a long, painful and expensive journey.
> - Don't begin if you don't intend to follow through, if you're not willing to make mistakes along the way, and if you're not willing to foot the bill.
> - You must lead the process yourself; organization change of this magnitude cannot be delegated. Without your personal commitment and involvement, you cannot change your organization's culture or the way your people work. Show that you believe in what you are doing by walking the walk.
> - Be decisive. If this is a path that you believe in, don't be afraid to use your prerogative and authority as the CEO to overcome initial scepticism and foot dragging.
> - Go to the Gemba. Take part in planning meetings, delve into the details and don't accept "we can't do that, it's not possible" as an excuse.
> - Let people understand that making mistakes is okay if they are made in good faith. What's not acceptable is failing to try.

Unfortunately, the project nature of construction – where everyone is used to focusing their efforts on missions with short, fixed lifespans – may itself work against people's ability to remain focused on long-term processes. Fighting daily fires gets in the way, making it difficult to create the space to pull back, take the long view, continue working not only on the pressing but also the important: improving processes to prevent future fires.

The product and the production system

Create the conditions for seeing

> **Box 22.2 Pooh Bear**
>
> "Here is Edward Bear, coming downstairs now, bump, bump, bump, on the back of his head, behind Christopher Robin. It is, as far as he knows, the only way of coming downstairs, but sometimes he feels there really is another way, if only he could stop bumping for a moment and think of it."
>
> (A. A. Milne, *Winnie-the-Pooh*)

While "learning to see" equips people with an essential skill for Lean improvement, production problems are very difficult to see when the processes are not stable. Construction projects commonly have a WIP of 100 per cent of the spaces that are their products over a significant part of their duration, and variability of production rates and of product types is high. The flow of information in a building's design process is often erratic, with much iteration and rework. The first step in attempting to manage and control flows is to try to achieve stability and visibility. The Last Planner® System and the use of BIM go a long way to making the flows of design and construction more stable and more visible. Even if they do not entirely cure the proverbial headaches, they help reveal their root causes.

Once the tools are applied and conditions allow thinking and reflection, an important next step is to consider the production system carefully. What are the products? How do they flow? Chapter 15, "What flows in construction?", provided some guidance on this issue, concluding that there are three axes of flow – the portfolio of projects, the processes and the operations. This broader view underlines the need to extend production management up the value stream to improve downstream processes.

In Chapter 20, "Putting it all together", for example, we discussed the case of the failure to identify a space time conflict in the MEP work in the parking basement of one of the Rosh Ha'ayin buildings. The presence of a tower crane (see Figure 20.5) that unexpectedly invalidated the system coordination efforts of the project's co-located BIM team, showed that deep integration of construction planning and detailed design is needed. In the same chapter, we recounted Tidhar's efforts to work with its upstream suppliers to stabilize delivery and installation of prefabricated components to site, taking a leaf from Toyota's book by actively working to improve its suppliers.

Tidhar's apartment product

As discussed in Chapter 16, "Production planning and control in construction", part of Tidhar's Lean journey has required a shift from project-centric thinking to production-centric thinking. Rather than seeing Tidhar as a company that builds buildings, its people have slowly begun to see individual apartments as their primary product. This is a key distinction, since the two conceptions lead to very different ways of behaving.

If each project is indeed one-off, then the imperative is to "get the project done" as quickly and cheaply as possible, with wide latitude over the means to pursue the goal. The ability to learn and improve is limited in this mindset, since each project is distinct from the others. Contrast that with a production-focused mindset, in which the similarities rather than the differences between individual projects are emphasized. In this conception, there is a lot in common in the underlying products, the apartments. Each one will have roughly the same systems, a kitchen, a small number of bathrooms and bedrooms, a living room, laundry, flooring, windows, doors. When the shift is made to see apartments as the product, production flow becomes apparent and a potential target for improvement, as do the benefits of standardization of products and of construction methods.

Fira made the same conceptual shift by seeing its pipe refitting projects as a flow of apartment bathroom and kitchen retrofits. This was the key to redesigning its business around the idea that they could achieve a win win arrangement by promising their subcontractors a reliable and continuous flow of work, rather than contracting with them (i.e. hiring and firing them) on a project-by-project basis.

The project-to-apartment product shift also reflects the Lean view of customer-defined value. It finds expression in the effort to see the product and the processes through the eyes of the customer, not only through the eyes of the company. Whereas the company does indeed have a portfolio of projects to manage, each with its project manager, site engineers, superintendents, tower cranes, subcontractors, etc., customers don't care about all that. The customer wants his or her apartment, in a reasonable time, with good quality and at an affordable price. So the challenge for Tidhar, like any construction company in a similar situation, is to balance the production system of apartments (the customer-focused viewpoint) with the demands of the discrete projects (which have collective needs, like permitting, design, site logistics, etc.). Given the strongly project-biased starting point of most people in the industry, a focus on the production view, with the efficiencies and improvements it can bring, should be encouraged.

The apartment-as-product view is also supported by the use of BIM. Libraries of standard product families within BIM tools begin with discrete building components – columns, beams, doors, windows, etc. At the next level, a company can standardize parametric models of rooms – bathrooms, bedrooms, etc. Once an apartment is viewed as a product, a BIM model of each apartment becomes central to the process.

Carried further, viewing the goal of construction as production of spaces or rooms makes the leap to off site manufacture of building modules, supported by BIM modelling and managed as a flow that transcends individual projects, a much more natural idea.

Why do some people find BIM and Lean Construction so difficult to adopt?

Many companies have embarked on programmes to adopt BIM and Lean Construction practices. Many have failed, and among those that have succeeded, few if any have managed to retain all of the process improvements that they adopted or developed. Some practices stick and become established ways of doing business, others fall away as people settle back into "business as usual". This is dependent to some extent on the rate of employee turnover a company experiences and on its capacity for organizational learning. In the final analysis, perhaps the one key attribute of a Lean company is its ability to continuously improve. This attribute is at the heart of Lean production, but is notoriously difficult to achieve and sustain.

Tidhar is no different than most companies that struggle to make process change permanent. Many of the process improvement initiatives developed through Gil Geva's "Growing the Margin" effort have degraded over time. Throughout the book we have told stories of both successes and failures. Some practices, like the WWP meetings of the Last Planner® System, have endured and are established practice at all of Tidhar's projects. Yet the WWP meetings are the only aspect of LPS that has remained. Collaborative design with BIM in big rooms has become standard practice for significant projects, and is considered essential for the more complex ones. Likewise, BIM is now used for all client apartment customization. However, the discipline required to sustain product and process standardization through BIM object libraries has dissipated.

There are numerous reasons for the difficulties that many people experience when adopting Lean Construction and BIM, but there are perhaps two fundamental, underlying reasons why Lean in particular is more difficult to implement than most people realize at the outset. First, traditional construction management is results oriented, whereas Lean Construction is process oriented. Second, Lean Construction does not conform to either one of the two ways in which many construction managers and engineers perceive their profession; it rejects the technical rationality that lies behind management models such as CPM and the "thermostat" model of planning and control, yet it also rejects the idea that construction management is artful and intuitive because production in construction is inherently uncertain and unpredictable.

Results versus process orientation

Lean is heavily process oriented. The expression "the Toyota way" reflects a process, a way of doing things, a way of thinking about production. Seeing the waste requires looking at how things are done and identifying the opportunities for improvement. If a company's production process costs less than those of all of its competitors, managers in a result-oriented culture are happy and unconcerned with waste. But in a process-oriented culture, managers trained to see the waste will be highly unsatisfied if they perceive unnecessary movement of workers or materials, waiting for information or for equipment, or large batch sizes, irrespective of how competitive they already are in the market. The sources of motivation for improvement are quite different.

Lean Construction is no less process oriented. Like Lean thinking in general, Lean Construction is concerned with reducing waste. Construction methods are scrutinized for the seven wastes identified by Taiichi Ohno.[4] Lean Construction is also concerned with improving processes: how to reduce cycle times for locations in a building. VSM helps identify potential improvements that reduce the waste and the number of operations. As Koskela revealed, learning to see the flow of production is an essential skill for Lean practitioners. In the same vein, LPS is all about facilitating and coordinating processes.

Yet construction project managers still tend to be highly result oriented. Work is subcontracted, risks are transferred and people are measured according to the results they produce. The ultimate conditions for satisfaction in a construction contract are the cost, the schedule and the quality. Measurement of the process – whether through quality assurance programmes or certification – tend to be minor aspects in a culture that views subcontracting as purchase of a product rather than procurement of a service. Result-oriented behaviour is inherent in the traditional culture of construction, as witnessed by the fact that clients call a company that builds buildings "construction contractor", "general contractor" or "subcontractor". They are rarely called simply "the builders", and then only informally.

Result-oriented behaviour is common in all the levels of management of construction companies. It is still deeply rooted in the psyche of Tidhar, Skanska, Lease Crutcher Lewis and Fira, despite the conscious efforts made in these companies to add a process perspective. Consider Tidhar – in the early days of the company's history, Arye Bachar and Tidhar's senior engineers learned about production flow in the Aviv Gardens project, but the solution they found to manage production was the results-oriented "work backlog" method. "Work backlog" had the big advantage over the competition at the time in that it recognized that subcontractor trades must progress at similar production rates. To achieve this, they calculated the total amount of work for each trade in a building in the measurement units of that trade's work (e.g. m^2 of flooring), divided that by the standard time that was to be allocated to each crew and derived a production rate for each trade. In other words, they established the *result* that would be required for each trade at the end of each time period. The overriding message that managers and supervisors throughout the company gave to the trade crews was this mantra: "I don't care how you do it or where you do it, all I care about is that you meet your target work quantity each week."

Aviv Gardens had a very large volume of apartments with no variation – no design variation and no customer-initiated changes. Most projects now are quite different, with extensive variation. The "mass construction" approach, with the work backlog method, is no longer appropriate and is being replaced by Lean Construction and LPS. Yet the "work backlog" method remains a part of Tidhar's control department procedures, co-existing alongside LPS, even though many project managers have come to consider it superfluous. Old habits die hard.

In Chapter 16, "Production planning and control in construction", we explained how a result-oriented management style leads to local optimization that ignores the flow of work that should be at the heart of process-oriented production control. The problem is that not all work is made equal and trade crews can achieve short-term local efficiencies by doing the easy work first, which in most cases is the work with big, continuous batches. This behaviour helps them meet their cumulative target work quantities at first, but because work locations are not controlled, the flow of work becomes haphazard and productivity steadily declines.

Technical rationality versus art and intuition?

In Chapter 17 (Box 17.1), we introduced an intriguing paper published in *Anthropological Quarterly* in 1982, in which Herbert Applebaum succinctly described the sharp dichotomy between two complementary ways of practising the special mix of engineering and management that is construction management.[5] Applebaum described the sharp distinction between two groups of people: on the one hand, the chief engineer and his engineering staff in a site office who were responsible for the planning and control of the work of building a new university laboratory building; and on the other hand, the site superintendent and his team of trade foremen and supervisors who were responsible for the day-to-day coordination and execution of the work on the site itself. The former worked with plans, schedules, drawings, bills of quantities, documents and calculations. The latter worked with construction equipment, materials and negotiated with people to get things done.

The sharp distinction between these two groups, as Applebaum observed them, is a metaphor for two ways of thinking about construction management. The first views construction management primarily as a profession of "problem solving using methods based on rational models of the way engineers perceive the world to function", a world in which projects can be managed perfectly if you operate in accordance with the "project management body of knowledge".[6] The second views construction management as a profession in which intuitive behaviours are learned over time by individuals who, to begin with, have the necessary aptitude to cope with the uncertainty and irrational behaviour that are the norm in construction projects, a world in which the ability to improvise and solve problems, to fight fires, together with the ability to lead people, are the most important skills.

The real world of construction is neither the one nor the other. It also does not lie at some specific point along an imagined continuum between the two. It is both. It is clearly messy and unpredictable to a degree, but it is also amenable to understanding and manipulation according to theoretical models, provided that the model in use is sufficiently robust and applicable. The development of Lean Construction is in large part a response to the inapplicability of the prevailing traditional models, including CPM, in an effort to provide better ways of conceiving of production in construction that are indeed applicable and robust. Koskela's "transformation, flow and value" model is just such an attempt, and LPS is a practical tool rooted in the notions of flow.

Yet herein lies the challenge of adopting Lean Construction methods: they conflict with the first, technically rational way of thinking because they cast doubt on the singular truth of its foundational theories; and they conflict with the second, artful and intuitive way of thinking, because they suggest that variability and uncertainty can be managed and controlled *according* to models of production. It is not surprising that it meets resistance.

Getting started

There is no one best way to adopt Lean Construction and BIM. Every company functions within its own environment, with its own constraints and opportunities. Yet having said that, here are some key points to guide your efforts:

- Establish commitment from a leader with power, motivation and belief in the ideas. Without their commitment and ongoing support, BIM and Lean adoption attempts are likely to prove highly frustrating.
- Empower change agents within the organization and provide the authority and resources they need. These are people who have the knowledge, the motivation and the leadership skills to take personal responsibility for BIM and/or Lean implementation. Without Jan Elfving, Ilkka Romo, Ronen Barak, Lana Gochenauer and Otto Alhava, none of their respective companies would have achieved what they have.

- Work with BIM and Lean consultants. They cannot lead the change, but their knowledge and experience is very valuable. A Lean consultant need not necessarily have extensive experience in construction – a sound understanding of Lean thinking in general can help broaden one's view. Their role is to direct your attention and focus, teach you how to see and think but not to make your organization Lean for you.
- Go for low-hanging fruit at first to establish visible benefits, without losing sight of the next steps, and leverage successes to maintain momentum. In BIM, this means early implementation of design coordination and clash detection (witness Skanska's beginning, Tidhar's actions in the Yavne and Dawn Tower projects and Lease Crutcher Lewis' first steps with BIM). In Lean Construction, examples of low-hanging fruit are recycling of construction waste materials, the WWP meetings of LPS and 5S's efforts on site. Tidhar's big gain using LPS to halve the duration of the first wave of interior finishing works at Yavne's buildings 10 and 11 and Skanska's win with off site fabrication of MEP modules at Manskun Rasti are good examples.
- Learn and teach not only tools and techniques but also the underlying philosophy. BIM is a change of process, introducing the notions of VDC and thorough collaboration; Lean Construction requires people to develop a new way of thinking about production in construction. Books, rather than popular articles online, are a good place to begin.[7] Formal education, such as a BIM course or a Lean boot camp, is also essential.
- Get hands-on practical experience and demand that all relevant people in the company: learn to use BIM software, prepare a value-stream map for your own process, etc. Very few people need to become experts, but all people need basic skills.
- Appreciate the depth of the change you are pursuing and manage it professionally. Articulate and communicate the reasons for change to all. The organization will need to adapt or change its culture in order to succeed with BIM and Lean: identify what will cause change, what might threaten or hinder it, and how people can be helped to cope.
- Identify early adopters within your organization, support them, protect them from more conservative people and tout their successes.

And finally, once you have started, persevere. After the hurrah of initial wins has passed, the real challenge with Lean and BIM is in creating new best practice standards and rolling them out to all corners of the organization.

Postscript

Up to now, we have not answered a key question that may remain in readers' minds. In Chapter 1, we described Tidhar's "Growing the Margin" initiative. So, did Tidhar's profit margin ultimately increase? The company grew, in terms of turnover, staff and overall profit. But did the operating margin improve?

Gil Geva, Tidhar Group CEO, addresses this question and discusses Tidhar's plan going forward:

> We have made real progress in the six years since the "Growing the Margin" initiative was launched at the end of 2011. The good news is that in absolute

numbers, turnover has grown fivefold and profits have grown tenfold. We broke through the NIS1 billion turnover target in 2015, reaching NIS1.5 billion in 2017. From 180 employees in 2010, we have grown to 550. The quality of our product has improved, with buildings and apartments that meet our customers' needs better than they did in earlier projects. The company's reputation as a quality builder has strengthened and we attract the best people to work for us.

We have achieved strong design capabilities using BIM, and BIM is now the standard throughout all of our real estate development projects. We have proven our capability to handle very large residential projects, making that into a clear competitive advantage. In fact, Tidhar is now the largest single builder of apartments in Israel, and we are the biggest privately held construction company. Thanks in very large part to extensive use of BIM, the small division we began for fitting out our own office developments has become a very competitive business unit in its own right, taking on a large volume of contract work.

At the same time, there are also areas where we missed the mark. We still do not have a fully automated model-based quantity take-off and estimation process. Our purchasing department still uses an out-dated IT system. To my mind, we are still some way from achieving the Lean culture change we aimed for. And most significantly, we have not grown the profit margin of the construction division in percentage-point terms. Given the reduction in profitability across the market during this period, you could say that maintaining what we had is in itself a big achievement (as some of our competitors have gone out of business), but that's just an excuse. I believe we could have – and still can – improve production far more than we have. Accelerated growth has hampered our ability to educate and train our people. Too many project managers, site supervisors and subcontractors feel free to do things "their way" instead of the standard way we have invested time and energy to develop.

With the benefit of hindsight and a mountain of experience, we're now preparing a new strategic plan. Some of the highlights include:

- Renewed focus on Lean implementation to achieve the performance gains we know are possible.
- Establish standard ways of working and experimenting – instil the "Tidhar way" in the hearts and minds of our people, and more importantly, in the way they work.
- Personal and organizational development, with significant investment in education and training.
- Build the information infrastructure needed to support our culture – deploy a company-wide IT system for resource planning, customer management, education and training and standardization.
- Continue to improve and strengthen our BIM capabilities.
- Innovation – develop, adopt and implement innovative technologies to improve production, management, quality and customer service processes. We plan to work with start-up companies, allowing them to draw on our resources as they define the needs for new products and processes. We think of this as an "innovation ecosystem".

We've succeeded up to now by improving our product, by making the company's products and services more attractive to customers (both apartment buyers and construction contracting clients), and so we are more profitable in absolute terms. But going forward, we're going back to the original, basic goal: grow the profit margin by fundamentally improving our operating margin with Lean and BIM.

Notes

1. If you are already the tallest tree in the forest, why grow taller? In his book *The Greatest Show on Earth*, Richard Dawkins describes the evolutionary pressure of natural selection that drives trees in a forest to evolve to be taller and taller. By considering a hypothetical "friendship forest" in which all the trees agree to cooperate and all remain at the same height, he explains why this state is unsustainable. A single tree that mutates and grows taller will enjoy more sunlight than all the others, leading over time to all the trees evolving to grow taller, as long as the benefit of more light outweighs the penalty of the extra energy needed for growth. The mutation is random, but inevitable, and so all the trees evolve to be as tall as they can sustain (Dawkins, 2009).
2. When Tidhar began the "Growing the Margin" initiative in late 2011, it had 250 direct employees. By 2017, it had grown to 550 – more than double in the space of six years.
3. A process owner is a senior manager responsible for development, facilitation, standardization and implementation of a business process. For a thorough description, see Becker et al. (2013).
4. Defects, overproduction, waiting, transporting, movement, unnecessary processing and inventory. See Ohno (1988).
5. Applebaum (1982).
6. PMI (2008).
7. *The Toyota Way* (Liker, 2003), *Lean Thinking* (Womack and Jones, 2003) and the *BIM Handbook* (Eastman et al., 2011) provide a sound basis for Lean and for BIM.

Glossary

4D modelling Four-dimensional modelling, in which in addition to the three spatial dimensions (X, Y and Z) of the model, there is also a time dimension. This allows the construction sequence of the project to be visualized for planning and for process control.

5S Sort, set in order, shine, standardize, sustain (from the Japanese: seiri, seiton, seiso, seiketsu and shitsuke). A workplace organization method designed to create an environment conducive to creating value for customers.

Baranovich method A concrete casting method where the stone cladding elements for the exterior façade of the building are attached to the inside face of the exterior formwork. When the formwork is stripped, the building already has stone cladding, with no need to erect scaffolding to attach it later. In most applications, the stone is attached in a ground floor workshop, although this can also be done on the floor whose walls are under construction.

Batch size The number of independent products transferred together to a workstation for processing. In construction, batches comprise work spaces. Allocating a whole floor that contains a number of work spaces, or even an entire apartment building, to a single trade crew to work in exclusively is the equivalent of manufacturing with very large batch sizes.

Big room A product development technique where designers of different sub-systems are brought together to promote communication, collaboration and short-cycle problem solving.

Building Information Modelling (BIM) Computerizing/digitizing the process of designing and/or managing the product and process information of building construction projects. BIM is a fundamental advancement over older computer-aided design and drafting (CAD) technology, which it essentially replaces.

Clash detection During the BIM design phase, checking to see that different systems don't overlap in 3D space (clash), in order to prevent errors in design that can't actually be built.

Flow A Lean ideal state in which value is constantly being added to the product, with no wastes.

Flowline chart See line-of-balance (LOB) chart

Gemba In Japanese, the place in the company where value is actually being created for the customer. In construction, this is the site itself where the workers are building the project.

Hansei In Japanese, self-reflection performed after something happens (for example, project completion, safety near miss, quality problem discovered) to think of ways to improve the next time.

Integrated project delivery (IPD) A contracting method designed to align the interests of the owner, designer and contractor, so that win win outcomes and solutions are actively pursued together.

Kaizen In Japanese, continuous improvement. Making small, incremental changes to processes, so that the company can be a little bit better today than it was yesterday.

Kanban In Japanese, a signal that is used to create pull within a production system. Often used to replenish only what has been consumed, to prevent bull-whip effects, and limit the amount of WIP.

Last Planner™ System (LPS) A method for planning construction projects that focuses on subcontractor engagement and commitment to a "just-in-time" plan, to prevent over-production or overprocessing of the scheduling task. Detailed production planning is done at the lowest possible level, by those closest to the work.

Level of development (LOD) The levels of development for a project designed using BIM define the levels of design decision-making and definition that are to be achieved at each project milestone by each design discipline. As the project progresses, the levels of development increase, although the steps may be different for different design disciplines.

Line-of-balance (LOB) chart A chart showing a construction schedule as a series of sloped lines. The horizontal axis is the time axis and the vertical axis lists the project's location hierarchy. The sloped lines represent the production rate of each trade as the crew works through the locations.

Look-ahead planning/make-ready From LPS, a method designed to prepare the pre-requisites for work packages, so that work can actually be performed by the workers when scheduled without having to wait.

Model integrity Where a BIM model is prepared by multiple modellers, or where changes are made, discrepancies or inconsistencies can arise between different building systems. Model integrity is a measure of the degree to which the different components in a model are consistent with one another.

Nemawashi In Japanese, "preparing the roots for planting". A method of building support for a new idea in a company and consensus around the plan of action, based on multiple off-line conversations with interested parties.

Obeya Japanese for "big room". See above.

PDCA Plan-do-check-act, a cyclical method for systematic problem solving.

Pull A Lean ideal, in which work is performed only when needed by the next step in the process, and in the proper quantity. This prevents overproduction, inventory build-up and other wastes.

Re-entrant flow A situation in which the product must revisit a workstation for additional processing, rather than progressing cleanly down the manufacturing line. In construction, this is equivalent to subcontractors being called back to a location they have already worked in to complete work they could not perform previously. For example, electricians coming back to install fixtures after walls have been finished, this in turn after the electrician finished roughing in the conduits.

Revit families Revit is Autodesk Inc.'s BIM software. A Revit family is a 3D definition of a building object stored in a library of objects which a modeller can draw from instead of creating objects from scratch. A family is commonly parametric, so that one definition can be used to instantiate a wide range of possible objects. For example, a family of wooden double doors may be used to rapidly generate many instances of double doors with different dimensions.

Self-performed work Work that the general contractor does with their own workers, as opposed to work that a subcontractor is paid to do.

Shop drawings Details-oriented blueprints that are given to subcontractors in order to instruct them as to how to complete the work. Focused around only the information relevant for completing one task/trade.

Sequence of events (SOE) Sequence of events is a process improvement method that maps out the steps of a process in high resolution, so that the process can be more fully understood and thence improved.

Takt From the German word meaning "clock interval". A pull method in which production is organized into regular, repeatable intervals that can serve to make sure production is progressing on schedule.

Value In Lean, value is what the customer is willing to pay money for. "Value-adding" activities are those that change the form, fit or function of the product or its components in a way that adds value and progresses from raw materials to finished goods.

Value-stream mapping (VSM) A Lean tool for building understanding about how a multi-person process actually works (the "current state") and consensus about how to improve towards a shared vision of how the process could work (the "future state").

Virtual design and construction (VDC) The process of compiling a BIM model for a project, testing the expected behaviour of the designed building through computer simulation and planning and controlling construction using the model information.

Waste/non-value-adding In Lean, as opposed to value-adding, all other activities and parts of the process that do not add value are defined as waste. Waste is an opportunity for improvement if it can be minimized or eliminated.

Work in progress (WIP) Products or materials upon which work has begun but that have not been completed. In apartment construction, any apartment that has begun to take shape but is not yet ready for occupancy.

Bibliography

Akers, P. A. (2011). *2 Second Lean: How to Grow People and Build a Fun Lean Culture*. FastCap LLC, Bellingham, WA.

Akinci, B., Fischer, M. and Kunz, J. (2002a). "Automated Generation of Work Spaces Required by Construction Activities". *Journal of Construction Engineering and Management*, 128, 306–315.

Akinci, B., Fischer, M., Levitt, R. and Carlson, R. (2002b). "Formalization and Automation of Time–Space Conflict Analysis". *Journal of Computing in Civil Engineering*, 16, 124–134.

Alarcon, L. F., Diethelm, S., Rojo, O. and Calderon, R. (2005). "Assessing the Impacts of Implementing Lean Construction". *13th Annual Conference of the International Group for Lean Construction*. R. Kenley, ed., UNSW, Sydney, Australia, 19–21 July 2005, 387–393.

Alhava, O., Kiviniemi, A., and Laine, E. (2015). "Emotional Intelligence: Improving the Performance of Big Room". eWork and eBusiness in Architecture, Engineering and Construction, ECPPM 2016.

Alhava, O., Laine, E. and Kiviniemi, A. (2015). "Intensive Big Room Process for Co-creating Value in Legacy Construction Projects". *Journal of Information Technology in Construction (ITcon)*, 20(11), 146–158.

Applebaum, H. A. (1982). "Construction Management: Traditional versus Bureaucratic Methods". *Anthropological Quarterly*, 55, 224–234.

Ballard, G. (2000a). "Positive vs Negative Iteration in Design". *Proceedings Eighth Annual Conference of the International Group for Lean Construction*, IGLC, Brighton, UK, 17–19.

Ballard, G. (2000b). "The Last Planner™ System of Production Control". PhD Dissertation, The University of Birmingham, Birmingham, UK.

Ballard, G., Harper, N. and Zabelle, T. (2002). "An Application of Lean Concepts and Techniques to Precast Concrete Fabrication". In Formoso, C.T. and Ballard, G. (eds), 10th Annual Conference of the International Group for Lean Construction. Gramado, Brazil, 6–8 August.

Ballard, G., Harper, N. and Zabelle, T. (2003). "Learning to See Work Flow: An Application of Lean Concepts to Precast Concrete Fabrication". *Engineering, Construction and Architectural Management*, 10, 6–14.

Becker, J., Kugeler, M. and Rosemann, M. (2013). *Process Management: A Guide for the Design of Business Processes*. Springer, Heidelberg.

Bertelsen, S. and Sacks, R. (2007). "Towards a New Understanding of the Construction Industry and the Nature of its Production". *15th Conference of the International Group for Lean Construction*, C. Pasquire and P. Tzortzopoulous, eds, Michigan State University, East Lansing, Michigan, 46–56.

Brodetskaia, I., Sacks, R. and Shapira, A. (2013). "Stabilizing Production Flow of Finishing Works in Building Construction with Re-entrant Flow". *Journal of Construction Engineering and Management*, 139, 665–674.

BSI (2013). *PAS 1192-2:2013*. The British Standards Institution, London.

Christodoulou, S. and Scherer, R. (2017). *eWork and eBusiness in Architecture, Engineering and Construction: ECPPM 2016: Proceedings of the 11th European Conference on Product and Process Modelling (ECPPM 2016), Limassol, Cyprus, 7–9 September 2016*. CRC Press, Boca Raton, FL.

Compusoft GB Ltd (2017). "Innoplus web". www.compusoftgroup.com/en-gb/bathroom-design/innoplus-web (25 April 2017).

Construction Dimensions (1982). "Up it Goes – in Record Time". www.awci.org/cd/pdfs/8203_f.pdf (25 April 2017).

ConTex (2017). "ConTex – Global Technology Experts". *About Us*, http://contex-products.com/about-us/ (25 April 2017).

Dawkins, R. (2009). *The Greatest Show on Earth: The Evidence for Evolution*. Free Press, New York.

Dayan A. and Sacks R. (2017). "Cognition Enhancement Using Virtual Reality in Apartment Customization". In Proc. Lean & Computing in Construction Congress (LC3), Vol. 1 (CONVR), Heraklion, Greece.

Deming, W. E. (1982). *Out of the Crisis*. Massachusetts Institute of Technology, Cambridge, MA.

Dillon, A. P. and Shingo, S. (1985). *A Revolution in Manufacturing: The SMED System*. Taylor & Francis, London.

Do, D., Ballard, G. and Tillmann, P. (2015). "Part 1 of 5: The Application of Target Value Design in the Design and Construction of the UHS Temecula Valley Hospital". Project Production Systems Laboratory, University of California, Berkeley.

dos Santos, A. (1999). "Application of Production Management Flow Principles in Construction Sites". PhD, University of Salford, Salford, UK.

Drummond, E. and Alabama Power Company (2002). "The House that Love Built, Really FAST – and Just in Time for Christmas Kicker: Habitat for Humanity Breaks World Record Set by New Zealand". *PR Newswire*, www.prnewswire.com/news-releases/the-house-that-love-built-really-fast----and-just-in-time-for-christmas-kicker--habitat-for-humanity-breaks-world-record-set-by-new-zealand-77236282.html (25 April 2017).

Eastman, C. M., Teicholz, P., Sacks, R. and Liston, K. (2011). *BIM Handbook: A Guide to Building Information Modeling for Owners, Managers, Architects, Engineers, Contractors, and Fabricators*. John Wiley and Sons, Hoboken, NJ.

El-Khouly, I. A., El-Kilany, K. S. and El-Sayed, A. E. (2011). "Effective Scheduling of Semiconductor Manufacturing Using Simulation". *World Academy of Science, Engineering and Technology*, 5(7), 225–230.

Emiliani, M. L. (2008). "The Equally Important 'Respect for People' Principle". *Real Lean: The Keys to Sustaining Lean Management (Volume Three)*, M. L. Emiliani, ed., The CLBM, LLC, Wethersfield, CT, 121–138.

Factory Physics (2017). "End the (Tug of) War between Accounting and Operations". *FactoryPhysics – Strategy.Execution.Profit.*, https://factoryphysics.com/blog/end-tug-war-between-accounting-and-operations (27 April 2017).

Fischer, M., Khanzode, A., Reed, D. and Ashcraft, H. W. (2017). *Integrating Project Delivery*. John Wiley, Hoboken, NJ.

Fondahl, J. W. (1962). "A Non-Computer Approach to the Critical Path Method for the Construction Industry". Dept of Civil Engineering, Stanford University Press, Stanford, CA.

Ford, H. and Crowther, S. (2003). *Today and Tomorrow* (updated 1988 reprint of the 1926 original). Productivity Press, Portland, OR.

Goldin, M. (2007). "Lean Construction Management of High-Rise Apartment Buildings with Late Change Orders". MSc Thesis, Technion – Israel Institute of Technology, Haifa, Israel.

Goldratt, E. M. (1997). *Critical Chain*. North River Press, Great Barrington, MA.

Goldratt, E. M., Cox, J. and Whitford, D. (2004). *The Goal: A Process of Ongoing Improvement*. Gower, London.

Gurevich, U. and Sacks, R. (2013). "Examination of the Effects of a KanBIM Production Control System on Subcontractors' Task Selections in Interior Works". *Automation in Construction*, 37, 81–87.

Gurevich, U., Sacks, R. and Shrestha, P. (2017). "BIM Adoption by Public Facility Agencies: Impacts on Occupant Value". *Building Research & Information*, 1–21. http://dx.doi.org/10.1080/09613218.2017.1289029.

Hamzeh, F. (2009). *Improving Construction Workflow – The Role of Production Planning and Control*. UC Berkeley: Civil Engineering, Berkeley, CA.

Hamzeh, F., Tommelein, I. and Ballard, G. (2007). *The Last Planner® Production System Workbook: Improving Reliability in Planning and Work Flow*. Lean Construction Institute and the Project Production Systems Laboratory, University of California, Berkeley, CA.

Hendrickson, C. and Au, T. (1998). *Project Management for Construction: Fundamental Concepts for Owners, Engineers, Architects, and Builders*. Chris Hendrickson, Pittsburgh, PA.

Hoeft, S. E. and Pryor, R. W. (2016). *The Power of Ideas to Transform Healthcare: Engaging Staff by Building Daily Lean Management Systems*. CRC Press, Taylor & Francis Group, Boca Raton, FL.

Hopp, W. J. and Spearman, M. L. (1996). *Factory Physics*. IRWIN, Chicago, IL.

Hopp, W. J. and Spearman, M. L. (2011). Factory Physics, 3rd Edition.. Waveland Press, Long Grove, IL.

Imai, M. (1986). *Kaizen: The Key to Japan's Competitive Success*. McGraw Hill, New York.

Imai, M. (2012). *Gemba Kaizen: A Commonsense Approach to a Continuous Improvement Strategy*. McGraw-Hill, New York.

Josephson, P. E. and Hammarlund, Y. (1999). "The Causes and Costs of Defects in Construction: A Study of Seven Building Projects". *Automation in Construction*, 8, 681–687.

Kenley, R. and Seppänen, O. (2006). *Location-Based Management for Construction: Planning, Scheduling and Control*. Routledge, London.

Kenley, R. and Seppänen, O. (2010). *Location-Based Management for Construction: Planning, Scheduling and Control*. Spon Press, London; New York.

Kerem, C., Barak, R., Sacks, R. and Priven, V. (2013). "Learning to See – Managers Working in the Gemba as Part of the Tidhar Way Training Program". *Proceedings of the 21st Annual Conference of the International Group for Lean Construction*, C. T. Formoso and P. Tzortzopoulos, eds, Universidade Federal do Ceara, Fortaleza, Brazil, 957–966.

Kingman, J. F. C. and Atiyah, M. F. (1961). "The Single Server Queue in Heavy Traffic". *Mathematical Proceedings of the Cambridge Philosophical Society*, 57(4), 902.

Koch, C. (2007). "Collaboration on Industrial Change in Construction". In *People and Culture in Construction: A Reader*, Spon research, A. Dainty, S. Green and B. Bagilhole, eds, Taylor & Francis, London; New York, 106–124.

Koerckel, A. and Ballard, G. (2004). "Return on Investment in Construction Innovation – A Lean Construction Case Study". *13th Annual Conference of the International Group for Lean Construction*. R. Kenley, ed., UNSW. Sydney, Australia, 19–21 July 2005, 91–98.

Korb, S. (2016). "'Respect for People and Lean Construction: Has the Boat Been Missed?" *24th Annual Conference of the International Group for Lean Construction*. Pasquire, C., Alves, T. and Reginato, J., eds, IGLC, Boston, USA, 20–22 July 2016.

Korb, S. and Sacks, R. (2016). "One Size Does Not Fit All: Rethinking Approaches to Managing the Construction of Multi-Story Apartment Buildings". *24th Annual Conference of the International Group for Lean Construction*, Pasquire, C., Alves, T. and Reginato, J., eds, IGLC, Boston, USA, 20–22 July 2016.

Koskela, L. (1992). *Application of the New Production Philosophy to Construction*. Center for Integrated Facility Engineering, Department of Civil Engineering, Stanford University, Stanford, CA.

Koskela, L. (2000). "An Exploration towards a Production Theory and its Application to Construction". Dissertation for the degree of Doctor of Technology, Helsinki University of Technology, Helsinki.

Koskela, L. (2004). "Making Do – The Eighth Category of Waste". *Lean Construction – DK*. Bertelsen, S. and Formoso, C.T., eds, Helsingør, Denmark, 3–5 August, 1–10.
Koskela, L. (2017). "Why Is Management Research Irrelevant?" *Construction Management and Economics*, 35(1–2), 4–23.
Kotter, J. P. (1996). *Leading Change*. Harvard Business School Press, Boston, MA.
Lane, R. and Woodman, G. (2000). "'Wicked Problems, Righteous Solutions': Back to the Future on Large Complex Projects". *8th Annual Conference of the International Group for Lean Construction*. IGLC, Brighton, UK, 17–19 July 2000, 1–12.
Laufer, A. (1996). *Simultaneous Management*. American Management Association, New York.
Laufer, A. and Shohet, I. (1991). "Span of Control of Construction Foreman: Situational Analysis". *Journal of Construction Engineering and Management*, 117(1), 90–105.
LEAPCON™ (2015). "Technion LEAPCON™ Management Simulation Game". *Technion – IIT*, http://sacks.net.technion.ac.il/research/lean-construction/technion-leapcon-management-simulation-game/ (23 April 2017).
LePatner, B. B. (2007). "Broken Buildings, Busted Budget". *Broken Buildings, Busted Budgets – Book Reviews, Press & Endorsements*, www.brokenbuildings.com/press.html (25 April 2017).
LePatner, B. B. (2008). *Broken Buildings, Busted Budgets: How to Fix America's Trillion-Dollar Construction Industry*. University of Chicago Press, Chicago, IL.
Lewin, K. (1951). *Field Theory in Social Science: Selected Theoretical Papers*. Harper & Row, New York, NY.
Liker, J. K. (2003). *The Toyota Way*. McGraw-Hill, New York.
Liker, J. K. (2004). *The Toyota Way: 14 Management Principles from the World's Greatest Manufacturer*. McGraw-Hill, New York.
Little, J. C. and Graves, S. (2008). "Little's Law". In *Building Intuition*, International Series in Operations Research & Management Science, D. Chhajed and T. Lowe, eds, Springer US, 81–100.
Love, P. E. D. and Edwards, D. J. (2005). "Calculating Total Rework Costs in Australian Construction Projects". *Civil Engineering and Environmental Systems*, 22, 11–27.
MacLeamy, P. (2004). *Collaboration, Integrated Information and the Project Lifecycle in Building Design, Construction and Operation*. Construction Users Round Table (CURT), Cincinnati, OH. See www.lcis.com.tw/paper_store/paper_store/CurtCollaboration-20154614516312.pdf.
Macomber, H., Howell, G. A. and Reed, D. (2005). "Managing Promises with the Last Planner™ System: Closing in on Uninterrupted Flow". *13th Annual Conference of the International Group for Lean Construction*, Sydney, R. Kenley, ed., UNSW, Sydney, Australia, 19–21 July 2005, 13–18.
McDonald, D. F. and Zack, J. G. (2004). *Estimating Lost Labor Productivity in Construction Claims; Recommended Practice No. 25R-03*. American Association of Cost Engineers International, Morgantown, WV.
Morgan, J. and Liker, J. K. (2006). *The Toyota Product Development System: Integrating People, Process and Technology*. Productivity Press, New York.
NCC (2015). "New strategy for profitable growth at NCC". *NCC Media*, www.ncc.group/media/pressrelease/a148296cac91f7a2/new-strategy-for-profitable-growth-at-ncc/ (23 April 2017).
Ohno, T. (1988). *Toyota Production System: Beyond Large-Scale Production*. Productivity Press, Cambridge, MA.
Partouche, R., Sacks, R. and Bertelsen, S. (2008). "Craft Construction, Mass Construction, Lean Construction: Lessons from the Empire State Building", *Proceedings of the 16th Annual Conference of the International Group for Lean Construction IGLC16*, P. Tzortzopoulos and M. Kagioglou, eds, University of Salford, Manchester, UK, 183–194.
Pe'er, S. (1974). "Network Analysis and Construction Planning". *Journal of the Construction Division*, 100, 203–210.
Pound, E. S., Bell, J. H. and Spearman, M. L. (2014). *Factory Physics for Managers: How Leaders Improve Performance in a Post-Lean Six Sigma World*. McGraw-Hill Education, New York.

Priven, V. (2016). "The Impacts of 'Social Subcontract' and Last Planner™ System on the Workflows of Construction Projects". PhD Thesis, Technion – Israel Institute of Technology, Haifa, Israel.
Priven, V. and Sacks, R. (2016a). "Impacts of the Social Subcontract and Last Planner™ System Interventions on the Trade Crew Workflows of Multistory Residential Construction Projects". *Journal of Construction Engineering and Management*, 142(7), 1–14.
Priven, V. and Sacks, R. (2016b). "Effects of the Last Planner™ System on Social Networks among Construction Trade Crews". *Journal of Construction Engineering and Management*, 141(6).
Priven, V., Sacks, R., Seppänen, O. and Savosnick, J. (2014). "A Lean Workflow Index for Construction Projects". *22nd Annual Conference of the International Group for Lean Construction*, B. T. Kalsaas, L. Koskela and T. A. Saurin, eds, Lean Construction – NO, Oslo, Norway, 25–27 June 2014, 715–726.
Rogers, E. M. (2003). *Diffusion of Innovations*, 5th Edition. Free Press, New York.
Rother, M. (2010). *Toyota Kata: Managing People for Improvement, Adaptiveness, and Superior Results*. McGraw-Hill, New York.
Rother, M. and Shook, J. (2009). *Learning to See: Value-Stream Mapping to Create Value and Eliminate Muda*. A lean tool kit method and workbook, Lean Enterprise Institute, Cambridge, MA.
Rother, M., Shook, J. and Womack, J. P. (2003). *Learning to See: Value-Stream Mapping to Add Value and Eliminate Muda*. A Lean tool kit method and workbook, Taylor & Francis, London.
Rousseau, J. J. (2010). *Social Contract*. Free Press, New York.
Saari, S. (2011). *Production and Productivity as Sources of Well-Being*. MIDO OY, Espoo, Finland.
Sacks, R. (2008). "Production System Instability and Subcontracted Labor". In *Construction Supply Chain Management Handbook*, W. O'Brien, C. Formoso, K. London and R. Vrijhoef, eds, CRC Press/Taylor and Francis, Boca Raton; London, 8-1–8-19.
Sacks, R. (2016). "What Constitutes Good Production Flow in Construction?" *Construction Management & Economics*, 34(9), 641–656.
Sacks, R. and Goldin, M. (2007). "Lean Management Model for Construction of High-Rise Apartment Buildings". *Journal of Construction Engineering and Management*, 133(5), 374–384.
Sacks, R., Gurevich, U., and Shrestha, P., (2015). "Review of National Standards and Organizational Guides for BIM Adoption", Journal of Information Technology in Cosntruction, Vol 21, pp 479–503.
Sacks, R. and Harel, M. (2006). "An Economic Game Theory Model of Subcontractor Resource Allocation Behaviour". *Construction Management and Economics*, 24(8), 869–881.
Sacks, R., Esquenazi, A. and Goldin, M. (2007). " LEAPCON™: Simulation of Lean Construction of High-Rise Apartment Buildings". *Journal of Construction Engineering and Management*, 133, 529–539.
Sacks, R., Treckmann, M. and Rozenfeld, O. (2009). "Visualization of Work Flow to Support Lean Construction". *Journal of Construction Engineering and Management*, 135, 1307–1315.
Sacks, R., Koskela, L., Dave, B. A. and Owen, R. (2010). "Interaction of Lean and Building Information Modeling in Construction". *Journal of Construction Engineering and Management*, 136(9), 968–980.
Sacks, R., Barak, R., Belaciano, B., Gurevich, U. and Pikas, E. (2013). "KanBIM Lean Construction Workflow Management System: Prototype Implementation and Field Testing". *Lean Construction Journal*, 9, 19–34.
Sacks, R., Seppänen, O., Priven, V. and Savosnick, J. (2017). "Construction Flow Index: A Metric of Production Flow Quality in Construction". *Construction Management and Economics*, 35(1–2), 45–63.
Sheldrake, J. (2003). *Management Theory*, Cengage/Thomson Learning, London.
Shingo, S. and Dillon, A. P. (1989). *A Study of the Toyota Production System: From an Industrial Engineering Viewpoint. Produce What Is Needed, When It's Needed*. Taylor & Francis, London.
Teicholz, P. (2001). "Discussion: US Construction Labor Productivity Trends, 1970–1998". *Journal of Construction Engineering and Management*, 127, 427–429.

Tidhar (2008). "Vision and Values". www.tidhar.co.il/en/vision.aspx (23 April 2017).

Toyota Motor Corporation (2012). "Corporate Philosopy: Toyota Way 2001". www.toyota-global.com/company/history_of_toyota/75years/data/conditions/philosophy/toyotaway2001.html (25 April 2016).

Tribelsky, E. and Sacks, R. (2011). "An Empirical Study of Information Flows in Multi-disciplinary Civil Engineering Design Teams Using Lean Measures". *Architectural Engineering and Design Management*, 7, 85–101.

Viana, D. D., Mota, B., Formoso, C. T., Echeveste, M., Peixoto, M. and Rodrigues, C. L. (2010). "A Survey on the Last Planner™ System: Impacts and Difficulties for Implementation in Brazilian Companies". *18th Annual Conference of the International Group for Lean Construction*, K. Walsh and T. Alves, eds, Technion – IIT, Haifa, 497–507.

VICO (2016). "Virtual Construction 2007". *Vico Trimble*, www.vicosoftware.com/ (23 April 2017).

VisiLean (2017). "VisiLean". http://visilean.com/ (25 May 2017).

Womack, J. P. and Jones, D. T. (2003). *Lean Thinking: Banish Waste and Create Wealth in Your Corporation*. Simon & Schuster, New York.

Index of characters

The stories of Lean and BIM adoption in Tidhar, Skanska, Lease Crutcher Lewis, Fira and many other companies are the stories of the people who have worked to translate an abstract and idealized concept of how construction could be done into an often-messy reality.

As you read the book, you will come across many of them, by name. To the best of our ability, we have endeavoured to describe their roles and responsibilities where we first introduce each of them. However, if you should choose to read individual chapters, you may find yourself wondering who the person is and what their role was when the events took place. The table below is intended to help you recognize the person, identify their role and their affiliation, and to find the chapter in the book in which they are introduced. The list is sorted alphabetically according to first names, as many people are referred to by their first name alone in some places.

Character	Role	Affiliation	Appears in chapters
Adi Brayer	Project manager	Tidhar	17
Ahmed Hassin	Site superintendent	Tidhar	17
Alexander Laufer	Professor	Technion	17
Amir Putievsky	Project manager	Tidhar	3
Amit Dayan	MSc student	Technion	20
Anneli Holmberg	Management consultant	Wiltrain Oy	19
Antti Kauppila	Verstas process owner	Fira Oy	19
Arye Bachar	Chair and co-founder	Tidhar	1, 16, 22
Brad Schmidt	Vice president	Lean Centre of Expertise Japan at MetLife	3
Chaim Kirshon	Vice president engineering	Tidhar	3, 18
Cheni Kerem	Lean consultant	Lean Israel	3, 9, 18, 21
Dima Valdman	Site superintendent	Tidhar	18
Dudi Nasi	Site superintendent	Tidhar	11, 18

Character	Role	Affiliation	Appears in chapters
Eli Renzer	BIM modeller	Tidhar	14, 20
Enni Laine	Design manager	Fira Oy	19
Gal Salomon	BIM modeller	Tidhar	20
Gil Geva	Chief executive officer and co-founder	Tidhar	1, 3, 11, 14, 18, 20, 21, 22
Glenn Ballard	Adjunct professor	UC Berkeley	5, 17
Greg Howell	Lean Construction consultant	LCI	17
Guy Frumer	Estimating manager	Tidhar	3
Hadas Zaig	Customer change coordinator	Tidhar	20
Ilan Nachman	Vice president purchasing	Tidhar	3
Ilan Pivko	Architect	Pivko Architects	7
Ilkka Romo	Vice president for research and development	Skanska Finland	5, 22
Jan Elfving	Vice president for Lean Construction	Skanska Finland	5, 22
Jeff Cleator	Chief executive officer	Lease Crutcher Lewis	12
Juha Hetemaki	Chief executive officer	Skanska Finland	5
Lana Gochenauer	VDC manager	Lease Crutcher Lewis	12, 22
Lauri Koskela	Professor	University of Huddersfield	3, 17, 21
Lawrie Pugh	BIM modeller	Lease Crutcher Lewis	12
Maia Davidson	Design manager	Tidhar	11, 14, 20
Mark Starobinsky	Senior site engineer	Tidhar	14
Nadav Galai	Vice president business development	Tidhar	3
Olli Seppänen	Professor	Aalto University	21
Ortal Yona	Production engineer	Tidhar	18, 20, 22
Otto Alhava	Chief technology officer	Fira Oy	19, 22
Rafael Sacks	Professor	Technion	1, 2, 3, 14, 17, 21
Ronen Barak	Research and development – Lean and BIM	Tidhar	1, 3, 11, 16, 17, 20, 21, 22
Ronen Ben Dor	Regional manager	Tidhar	17
Sakari Pesonen	Lean Construction process owner	Fira Oy	19
Sara Troberg	BIM facilitator	Skanska Finland	5

Character	Role	Affiliation	Appears in chapters
Sven Bertelsen	Lean Construction author and consultant	Self-employed	3
Tal Hershkovitz	Chief executive officer construction	Tidhar	1, 3, 6, 9, 11, 18, 20, 22
Tanya Pankratov	BIM modeller	Tidhar	12
Vitaliy Priven	PhD student	Technion	3, 17, 18, 20, 21, 22
Zhu Hongxi	Construction crew leader	Tidhar	18
Zohar Benor	Chief executive officer	Lean Israel	1
Zohar Raz	VP design	Tidhar	3, 11, 20

Index

Page numbers in **bold** indicate a Box or Table, *italics* an illustration and n, an endnote